ALLE ZEIT WACH
SJ
1842

I. K. Egorov and C. S. David (Eds.)

Transgenic Mice and Mutants in MHC Research

With 91 Figures and 80 Tables

Springer-Verlag Berlin Heidelberg New York
London Paris Tokyo Hong Kong

IGOR K. EGOROV, Ph.D., D.Sci.
The Jackson Laboratory
Bar Harbor, ME 04609, USA

CHELLA S. DAVID, Ph.D.
Department of Immunology
Mayo Medical School
Mayo Clinic
Rochester, MN 55905, USA

ISBN-13:978-3-642-75444-9 e-ISBN-13:978-3-642-75442-5
DOI: 10.1007/978-3-642-75442-5

Library of Congress Cataloging-in-Publication Data. Transgenic mice and mutants in MHC research: proceedings of a conference held at the Jackson Laboratory, Bar Harbor, June 24–28, 1989 / edited by Igor K. Egorov and Chella S. David. ISBN-13:978-3-642-75444-9 (U. S.: alk. paper) 1. Major histocompatibility complex – Congresses. 2. Transgenic mice – Immunology – Congresses. I. Egorov, Igor K. II. David, Chella S. QR184.315.T73 1990 599'.029 – dc20.

Softcover reprint of the hardcover 1st edition 1990

2123/3130 (3011)-543210 – Printed on acid-free paper

Preface

This volume contains papers presented at the international conference on "Transgenic Mice and Mutants in MHC Research", held in Bar Harbor, Maine, in June, 1989. The meeting brought together eighty researchers working in the field. While mouse H-2 mutants have been known for many years their studies continue to contribute a great deal to our understanding of structure/function relationships and evolution of MHC molecules. Recently a new direction of research has emerged on regulation, expression and function of MHC genes using transgenic animals carrying exogenous MHC genes from the same or other species or engineered MHC genes. With the introduction of transgenic mice more specific questions about the various functions of MHC genes can be answered as, for example, the role of soluble and membrane-bound MHC molecules in self tolerance and elimination of specific T cell clones, recognition of human MHC antigens by the mouse immune system, the role of individual human MHC genes in disease susceptibility. New approaches for evaluation of the role of MHC linked and unlinked genes in susceptibility of mice to malignant tumors and their metastases have been also reported. Our purpose has been to provide a forum for discussion of these new developments. Many questions remain to be answered but the necessary tools have become available.

We thank the contributors and hope that the readers will benefit from this Proceedings. We wish to acknowledge the financial sponsorship for the meeting by The Pew Charitable Trust and Howard Hughes Medical Institute. Financial support for the meeting was also provided by Biogen, Inc., Boehringer Ingelheim Pharmaceuticals, Inc., Imperial Cancer Research Fund, T Cell Sciences, Inc., Bristol Myers Company, Cetus Corporation, Immunex Corporation, Abbott Laboratories, Lilly Research Laboratories, Mallinckrodt, Inc., and New Brunswick Scientific Co., Inc.

We are indebted to Mr. George Vose and Ms. Sharon Tate and their staff for excellent administrative assistance in providing a meeting in Bar Harbor that was so pleasant and smoothly organized. Our special thanks goes to Ms. Linda (Founier) Neleski who shared with us all burdens of the organization of this conference.

I.K. Egorov and C.S. David

Contents

I. Mutant Models of MHC Antigen Expression and Function (H-2)

Spontaneous Frequency of H-2 Mutations
R.W. MELVOLD and H.I. KOHN (With 8 Tables) 3

Microgene Conversion in the Evolution of the MHC Class I Multigene Family
C. WHEELER, D. MALONEY, S. WATTS, J. VOGEL, J. TOWNER, D. BALDWIN, J. RUFER, M. ARBEITMAN, H. CHAN, S. FOGEL, and R.S. GOODENOW (With 1 Figure and 2 Tables) 14

Random Mutagenesis by Oligonucleotides: A Probe for MHC Structure and Function
K. KATOH, R. MURRAY, D. MULLER, and J.A. FRELINGER (With 1 Figure and 2 Tables) 26

Altered Recognition of LCMV Antigen Presented on Mutant Class I Molecules
D. MULLER, M.M. HOBBS, R. MURRAY, K. PEDERSON, R. AHMED, and J.A. FRELINGER 33

Relationship Between Class II Structure and B Cell Signaling
G.A. BISHOP and J.A. FRELINGER (With 1 Figure and 1 Table) 36

Multivalent Requirement for the Stimulation of Alloreactive T Cells: Studies with Engineered Soluble MHC Class I Proteins In Vitro and In Vivo
D.H. MARGULIES, L.F. BOYD, S. KOZLOWSKI, L. KJER-NIELSEN, R. LOPEZ, J. MCCLUSKEY, J. SCHNECK, and R. HUNZIKER (With 4 Figures) 39

Novel MHC Variants Spliced by Overlap Extension
H.D. HUNT, J.K. PULLEN, Z. CAI, R.M. HORTON, S.N. HO, and L.R. PEASE (With 4 Figures and 3 Tables) 47

The Functional Significance of Amino Acid Polymorphisms in Class I MHC Molecules
J.K. PULLEN, H.D. HUNT, R.M. HORTON, and L.R. PEASE (With 1 Figure and 1 Table) 56

Products of Separate Genes Encoded Within Minor Histocompatibility Loci Stimulate Class I and Class II MHC-Restricted T Cells
D.C. ROOPENIAN and A.P. DAVIS (With 1 Figure and 1 Table) 59

Determination of Distinction Odortypes by Mutation of Class I and Class II MHC Genes
K. YAMAZAKI, G.K. BEAUCHAMP, L. THOMAS, J. BARD, and E.A. BOYSE (With 1 Table) 62

II. Mutants of MHC Genes as Probes of Immunological Function (HLA)

Analysis of HLA-B27-Specific T-Cell Epitopes with Site-Directed Mutants Mimicking HLA-B27 Polymorphism
S. ROJO, V. CALVO, D. LÓPEZ, B. GALOCHA, and J.A. LÓPEZ DE CASTRO (With 3 Figures and 3 Tables) 67

The Role of Amino Acid Position and Side Chain Structure in Serological and CTL-Defined Epitopes on the HLA-A2.1 Molecule
K.T. HOGAN, C. CLAYBERGER, N. SHIMOJO, W.E. BIDDISON, A.M. KRENSKY, and V.H. ENGELHARD (With 1 Figure and 4 Tables) 77

Two Amino Acids on the α Helical Region of the α1 Domain Form Multiple Epitopes Recognized by HLA-Bw52 Specific Human Cytotoxic T Cell
M. TAKIGUCHI, J. YAMAMOTO, H. HAYASHI, and K. KANO (With 2 Figures) 89

III. Regulation of MHC Gene Expression

Mouse Mammary Tumor Virus-Directed Gene Expression in Transgenic Mice
S.R. ROSS, Y. CHOI, C.-L. L. HSU, and J. DUDLEY (With 5 Figures and 3 Tables) 93

Regulation of Class I MHC Expression: In Vivo Function of Regulatory DNA Sequence Elements in Transgenic Mice
J.E. MAGUIRE, R. EHRLICH, W.I. FRELS, and D.S. SINGER (With 4 Figures) 102

Expression of a Non-Classical Class I Gene in Transgenic Mice
C. BORDALLO, W.I. FRELS, J. LUNNEY, A. ROSENBERG, S. RUDIKOFF, and D.S. SINGER (With 2 Figures and 2 Tables) 109

Regulation of Expression of Human MHC Class I Heavy (HLA-B7) and Light (hβ_2-m) Chain Genes in Transgenic Mice
J.W. CHAMBERLAIN, P.J. CONRAD, and S.M. WEISSMANN (With 8 Figures and 1 Table) 114

Cloned Trans-Acting Factors that Bind to the Regulatory Elements of the Major Histocompatibility Complex Class I Gene
K. OZATO, S.L. GLEASON, B.-Z. LEVI, S. HIRSCHFELD, P.H. DRIGGERS, J.R. FLANAGAN, D.L. ENNIST, K. HAMADA, and E. APPELLA (With 5 Figures) 125

Transcriptional Regulation of MHC Class II Gene Expression: Are MHC Class II Genes Coordinately Regulated?
H.-C. LIOU, L.B. IVASHKIV, C.J. KARA, and L.H. GLIMCHER
(With 1 Figure and 2 Tables) 133

The Effect of Copy Number on mRNA and Cell Surface Expression of an $A\beta^k$ Transgene
S. GILFILLAN, S. AISO, S.A. MICHIE, and H.O. MCDEVITT
(With 1 Figure and 3 Tables) 143

Construction of Retroviral Vectors for the Study of Tissue-Specific Regulation of Human MHC Class II Genes
G. DEL POZZO, A. MAFFEI, C. PERFETTO, N. OMBRA, and J. GUARDIOLA
(With 1 Figure and 1 Table) 155

IV. Expression and Function of Class I Genes in Transgenic Mice

Male Sterility in HLA-B27-Transgenic Mice
P. IVANYI and W. LOKHORST (With 1 Figure and 1 Table) 161

The Lyt-2 Accessory Molecule is Responsible for the Weak Mouse Anti-HLA Xeno-Response
G.J. HÄMMERLING, U. KALINKE, and B. ARNOLD (With 4 Figures) 165

Human HLA-B27 Antigen in Transgenic Mice can Function as a Major Histocompatibility Antigen
M. PLA, A. SAMAAN, D. GILLET, M. REBOUL, B. FRANGOULIS, A. OPOLSKI, M. CHOPIN, and L. DEGOS (With 4 Figures) 173

Cytotoxic T Cell Responses Against Human Class I Molecules in Normal and HLA-A2.1 Transgenic Mice
V.H. ENGELHARD, E.J. BERNHARD, M.J. HOLTERMAN, A.-X.T. LE, R. HENDERSON, J.P. RIDGE, S. STRUB, J. BARBOSA, and E. LACY
(With 9 Tables) 179

Analysis of the HLA-Cw3 Specific CTL Response of HLA-B7x Human β_2 Microglobulin Transgenic Mice
C. BARRA, B. PERARNAU, P. GERLINGER, M. LEMEUR, A. GILLET, P. GIBIER, and F.A. LEMONNIER (With 2 Figures and 2 Tables) 191

H-2-Restricted Recognition of Xeno-MHC Antigens by Primary Mouse Cytotoxic T Cells is the Exception Rather than the Rule
F. KIEVITS, J. WIJFFELS, W. LOKHORST, W.J. BOERENKAMP, and P. IVANYI
(With 2 Figures) 198

Copy Number and the Presence of Human β_2-Microglobulin Control Cell Surface Expression of HLA-B27 Antigen in Transgenic Mice with a 25 kb B27 Gene Fragment
E.H. WEISS, G. SCHLIESSER, W. KUON, M. LANG, G. RIETHMÜLLER, F. KIEVITS, P. IVANYI, and G. BREM (With 5 Figures) 205

Expression and Function in a Transgenic Mouse of an H-2K^{b} Variant Gene Obtained by Site-Directed Mutagenesis in a K^{bm1} Mutant Mouse
M. GENDRON, M. NAKAGAWA, K. HASENKRUG, S. NATHENSON, and A.W. SEITZ (With 3 Tables) ... 214

V. Expression and Function of Class II Genes in Transgenic Mice

Clonal Anergy in Transgenic Mice with Pancreatic Expression of MHC Class II I-E
L.C. BURKLY, D. LO, O. KANAGAWA, R.L. BRINSTER, and R.A. FLAVELL (With 2 Figures and 3 Tables) ... 219

Role of Ia Antigens in Clonal Deletion of T Cells: Studies with Recombinant and Transgenic Mice
C.S. DAVID and G. ANDERSON (With 4 Figures and 8 Tables) ... 229

Transgenic Mice with MHC Class II Genes: The Use in the Study of Allelic α/β Chain Pairing and the Production of Engineered Mice with Mutant I-A Genes
B.-Y. WEI, J. MARTIN, S. SAVARIRAYAN, R. LITTLE, and C.S. DAVID (With 2 Tables) ... 237

T Cell Recognition of Major Histocompatibility Complex Antigens in HLA Class II Transgenic Mice
S.K. LAWRANCE, L. KARLSSON, J. PRICE, V. QUARANTA, Y. RON, J. SPRENT, and P.A. PETERSON (With 7 Figures) ... 247

VI. Transgenic and Mutant Models of Disease

HLA-B27 and Arthritis
H.S. LUTHRA, C.L. NICKERSON, S. SAVARIRAYAN, and C.S. DAVID (With 2 Figures and 2 Tables) ... 259

HLA-B27 Transgenic Mice as Potential Models of Human Disease
J.D. TAUROG, R.E. HAMMER, S.D. MAIKA, K.L. SAMS, F.A.K. EL-ZAATARI, S.A. STIMPSON, and J.H. SCHWAB (With 4 Figures) .. 268

Mouse Mutants Affecting Growth and Metastasis of a Syngeneic Tumor
O.S. EGOROV, E. SARGENT, G.A. CARLSON, and I.K. EGOROV (With 5 Tables) ... 276

Transgene Induced MHC Mismatch Results in Rapid Natural Killer Cell Dependent Elimination of Lymphoma Cells
P. HÖGLUND, H.G. LJUNGGREN, C. ÖHLÉN, and K. KÄRRE (With 1 Figure and 2 Tables) ... 282

Highly Immunogenic Transformed Tumor Clones Expressing Allogeneic Class I Histocompatibility Gene Demonstrate a Specific Immunotherapeutic Affect Against the Parental Tumor
K.M. HUI, T. SIM, T.T. FOO, and A.A. OEI (With 4 Figures and 2 Tables) 285

Graft Specific MHC Class II Gene Expression During Allograft Response
D.S. KITTUR, R. XU, J.F. BURDICK, and W. ADLER (With 1 Figure) 296

Influence of MHC and Non-MHC Genes on Tumorigenesis and the Use of the Recombinant Congenic Strains as a Novel Tool for the Genetic Analysis of Tumor Susceptibility
P. DEMANT, L.C.J.M. OOMEN, C.J.A. MOEN, M.A. VAN DER VALK, A.A.M. HART, and L.F.M. VAN ZUTPHEN (With 2 Figures and 3 Tables) .. 299

Subject Index 309

Contributors

Adresses are given at the beginning of the respective contribution

Adler, W. 296
Ahmed, R. 33
Aiso, S. 143
Anderson, G. 229
Appella, E. 125
Arbeitman, M. 14
Arnold, B. 165
Baldwin, D. 14
Barbosa, J. 179
Bard, J. 62
Barra, C. 191
Beauchamp, G.K. 62
Bernhard, E.J. 179
Biddison, W.E. 77
Bishop, G.A. 36
Boerenkamp, W.J. 198
Bordallo, C. 109
Boyd, L.F. 39
Boyse, E.A. 62
Brem, G. 205
Brinster, R.L. 219
Burdick, J.F. 296
Burkly, L.C. 219
Cai, Z. 47
Calvo, V. 67
Carlson, G.A. 276
Chamberlain, J.W. 114
Chan, H. 14
Choi, Y. 93
Chopin, M. 173
Clayberger, C. 77
Conrad, P.J. 114
David, C.S. 229, 237, 259
Davis, A.P. 59
Degos, L. 173
Del Pozzo, G. 155
Demant, P. 299
Driggers, P.H. 125
Dudley, J. 93
Egorov, I.K. 276
Egorov, O.S. 276
Ehrlich, R. 102
El-Zaatari, F.A.K. 268
Engelhard, V.H. 77, 179
Ennist, D.L. 125
Flanagan, J.R. 125
Flavell, R.A. 219
Fogel, S. 14
Foo, T.T. 285
Frangoulis, B. 173
Frelinger, J.A. 26, 33, 36
Frels, W.I. 102, 109
Galocha, B. 67
Gendron, M. 214
Gerlinger, P. 191
Gibier, P. 191
Gilfillan, S. 143
Gillet, A. 191
Gillet, D. 173
Gleason, S.L. 125
Glimcher, L.H. 133
Goodenow, R.S. 14
Guardiola, J. 155
Hamada, K. 125
Hammer, R.E. 268
Hämmerling, G.J. 165
Hart, A.A.M. 299
Hasenkrug, K. 214
Hayashi, H. 89
Henderson, R. 179
Hirschfeld, S. 125
Ho, S.N. 47
Hobbs, M.M. 33
Hogan, K.T. 77

Höglund, P. 282
Holterman, M.J. 179
Horton, R.M. 47, 56
Hsu, C.-L.L. 93
Hui, K.M. 285
Hunt, H.D. 47, 56
Hunziker, R. 39
Ivanyi, P. 161, 198, 205
Ivashkiv, L.B. 133
Kalinke, U. 165
Kanagawa, O. 219
Kano, K. 89
Kara, C.J. 133
Karlsson, L. 247
Kärre, K. 282
Katoh, K. 26
Kievits, F. 198, 205
Kittur, D.S. 296
Kjer-Nielsen, L. 39
Kohn, H.I. 3
Kozlowski, S. 39
Krensky, A.M. 77
Kuon, W. 205
Lacy, E. 179
Lang, M. 205
Lawrance, S.K. 247
Le, A.-X.T. 179
Lemeur, M. 191
Lemonnier, F.A. 191
Levi, B.-Z. 125
Liou, H.-C. 133
Little, R. 237
Ljunggren, H.G. 282
Lo, D. 219
Lokhorst, W. 161, 198
López de Castro, J.A. 67
López, D. 67
Lopez, R. 39
Lunney, J. 109
Luthra, H.S. 259
Maffei, A. 155
Maguire, J.E. 102
Maika, S.D. 268
Maloney, D. 14
Margulies, D.H. 39
Martin, J. 237
McCluskey, J. 39
McDevitt, H.O. 143
Melvold, R.W. 3
Michie, S.A. 143
Moen, C.J.A. 299
Muller, D. 26, 33
Murray, R. 26, 33
Nakagawa, M. 214
Nathenson, S. 214
Nickerson, C.L. 259
Oei, A.A. 285
Öhlén, C. 282
Ombra, N. 155
Oomen, L.C.J.M. 299
Opolski, A. 173
Ozato, K. 125
Pease, L.R. 47, 56
Pederson, K. 33
Perarnau, B. 191
Perfetto, C. 155
Peterson, P.A. 247
Pla, M. 173
Price, J. 247
Pullen, J.K. 47, 56
Quaranta, V. 247
Reboul, M. 173
Ridge, J.P. 179
Riethmüller, G. 205
Rojo, S. 67
Ron, Y. 247
Roopenian, D.C. 59
Rosenberg, A. 109
Ross, S.R. 93
Rudikoff, S. 109
Rufer, J. 14
Samaan, A. 173
Sams, K.L. 268
Sargent, E. 276
Savarirayan, S. 237, 259
Schließer, G. 205
Schneck, J. 39
Schwab, J.H. 268
Seitz, A.W. 214
Shimojo, N. 77
Sim, T. 285
Singer, D.S. 102, 109
Sprent, J. 247
Stimpson, S.A. 268

Strub, S. 179
Takiguchi, M. 89
Taurog, J. D. 268
Thomas, L. 62
Towner, J. 14
Van der Valk, M. A. 299
Van Zutphen, L. F. M. 299
Vogel, J. 14
Watts, S. 14
Wei, B.-Y. 237
Weiss, E. H. 205
Weissmann, S. M. 114
Wheeler, C. 14
Wijffels, J. 198
Xu, R. 296
Yamamoto, J. 89
Yamazaki, K. 62

I. Mutant Models of MHC Antigen Expression and Function (H-2)

Spontaneous Frequency of H-2 Mutations

R. W. MELVOLD[1] and H. I. KOHN[2]

[1] Department of Microbiology/Immunology, Northwestern University Medical School, 303 East Chicago Avenue, Chicago, IL 60611, USA
[2] Department of Radiation Therapy, Shields Warren Radiation Laboratory, Harvard Medical School, 44 Binney Street, Boston, MA 02115, USA. *Current address:* 1203 Shattuck Ave., Berkeley, CA 92709, USA

INTRODUCTION

The availability of single gene mutations involving the H-2 complex has been of great value in analyzing the fine structure of the murine MHC, as well as in defining particular loci, or epitopes associated with the products of those loci, in a wide variety of immunological functions. The $H\text{-}2^{dm2}$ mutation, for example, was instrumental in the identification and subsequent analysis of the H-2L locus (Hansen *et al* 1977; Hansen *et al* 1983; Levy and Hansen 1980; McKenzie *et al* 1977). Mutations of $H\text{-}2K^b$ and $H\text{-}2D^b$ have been utilized to identify of epitopes involved in MHC restriction of cytotoxic T lymphocytes against a variety of cellular antigens and viruses, and the lone I region mutant $H\text{-}2^{bm12}$, involving the A_β gene of the $H\text{-}2^b$ haplotype, has been widely used in the fine dissection of genetic control of immune responsiveness by MHC class II loci (reviewed in Hansen *et al* 1983; Klein 1976; Klein 1978; Kohn *et al* 1978; LaFuse and David 1984; Levy and Hansen 1980; McKenzie *et al* 1977; Hansen *et al* 1984). The precise structural alterations in most of the mutations have been defined at the amino acid and/or nucleic acid sequence levels (Burnside *et al* 1984; Hemmi *et al* 1988; Martinko *et al* 1987; Nairn and Nathenson 1978; Nairn *et al* 1980; Nathenson *et al* 1986; Rubocki *et al* 1986; Sun *et al* 1985).

In vivo mutations of histocompatibility genes (as detected by skin grafting) have been observed at about 1 - 5 per thousand tested animals (H-2 and non-H-2 combined) from untreated or irradiated parents, but an overall specific mutation rate has been difficult to assign because the actual number of loci involved is not known (Bailey and Kohn 1965; Kohn 1973; Kohn and Melvold 1974; Melvold and Kohn 1975; Kohn *et al* 1976; Harnasch and Stumpf 1982; Egorov and Egorov 1988). H-2 mutation rates (untreated or irradiated) have been estimated as about 5×10^{-4} per gene per generation for H-2 genes, and between 10^{-6} and 10^{-7} for non-H-2 histocompatibility genes, based on an assumption of 30 non-H-2 loci (Melvold and Kohn 1975). Egorov and Blandova (1972) recovered two H-2 mutants among 5240 $H\text{-}2^a$ -, $H\text{-}2^f$ -, $H\text{-}2^d$ - or $H\text{-}2^b$ - bearing gametes (some mutagenized by DES) for a composite rate of 3.8×10^{-4} per gamete. Egorov and Egorov (1988) reported 2 H-2 mutations in 2600 unmutagenized gametes for a rate of 7.7×10^{-4} per gamete. While X-irradiation seems to have little influence on the observed rate of histocompatibility gene mutations (Kohn and Melvold 1976; Kohn *et al* 1976), chemical mutagens such as TEM, DES, ENU and procarbazine may increase these rates (Kohn 1973; Egorov and Blandova 1972; Harnasch and Stumpf 1982; Egorov and Egorov 1988).

Since 1966, our laboratories, in Boston and in Chicago, have screened over 100,000 mice, by reciprocal tail skin grafting, for mutations of histocompatibility genes (Kohn 1973; Kohn and Melvold 1975; Kohn *et al* 1976; Kohn and Melvold 1976; Melvold and Kohn 1975; Melvold *et al* 1982). Roughly a quarter of the recovered mutants have involved the H-2 complex. Aside from their effects on various immune responses and functions, the screening procedure itself has provided intriguing results, foremost among them the high frequency of spontaneous mutants and the recurrence of particular mutant phenotypes generated by independent mutational events (Geleibter *et al* 1986; Yamaga *et al* 1977*a,b*), both somatic and germ line. Previous reports on induced mutation rates (Bailey and Kohn 1965; Dunn and Kohn 1981; Egorov and Blandova 1972; Egorov and Egorov 1988; Kohn 1973; Kohn *et al* 1976; Kohn and Melvold 1976) are available and a comprehensive survey of radiation-induced and spontaneous mutation rates, in H-2 and non-H-2 histocompatibility genes, in our studies over the past 24

I.K. Egorov/C.S. David (Eds.)
Transgenic Mice and Mutants in MHC Research

years will be published elsewhere. In this report, we summarize our aggregate data, from 1966 to 1980 in Boston and from 1980 to the present in Chicago, which bears on the spontaneous mutation rates of class I and class II H-2 genes.

SCREENING FOR SPONTANEOUS MUTATIONS

A total of 113,779 mice have been screened by reciprocal tail skin grafting (Table 1) in Boston and in Chicago, which provide a basis for estimating spontaneous mutation rates involving the *H-2*^{b}, *H-2*^{d} and *H-2*^{k} haplotypes. The grafting of these 113,779 animals constituted a screening of 100,708 unmutagenized gametes bearing *H-2*^{b}, 106,952 bearing *H-2*^{d} and 8,184 bearing *H-2*^{k}, from which 20 H-2 mutations were identified (Table 2). Seventeen of the recovered mutations were of the "gain and loss" type, with the remaining three being of the "loss" type.

The screening utilized reciprocal tail skin grafts, a technique which has been described previously in detail (Bailey and Usama 1960; Kohn 1973). Briefly, split thickness, orthotopic grafts were exchanged between animals, examined for technical failures on the fifth day after grafting and monitored on a regular basis thereafter for 57-63 days in most cases, although some were monitored for 90-100 days. Skin and epidermis of the grafts were each graded on a scale of 1 (good condition) to 4 (absent) and the two scores totalled, with a sum of 6 or greater taken as a full rejection of the graft. The criteria for identifying and mapping these mutants has been described previously in detail (Kohn 1973; Kohn and Melvold 1975; Melvold and Kohn 1976).

Table 1. Animals Monitored for Spontaneous H-2 Mutations

Genotype	Haplotype	Animals tested
C57BL/6 (B6) [a]	b	16,168
C57BL/10 (B10) [b]	b	500
C3H.SW	b	110
A.BY	b	60
BALB/c (C) [c]	d	22,250
DBA/2 [d]	d	2,839
B6.C-*H-2*^{d}	d	268
B10.D2	d	111
(B6 x C)F1	b/d	54,904 [e]
(C x B6)F1	b/d	12,477
C3H/He [f]	k	2,895
CBA [g]	k	624
AKR [h]	k	448
B10.BR	k	125
Total		113,779

[a] primarily B6Kh, but also includes B6ByJ and B6/NCr.
[b] primarily B10ScNCr, but also includes B10Sn and B10NCr.
[c] primarily CKh, but also includes CByJ and C/NCr.
[d] includes both DBA/2J and DBA/2NCr.
[e] of these, 11,364 were derived from matings of normal B6 mothers and irradiated C fathers and are thus are included in the calculations for spontaneous mutation rates of *H-2*^{b}, but not of *H-2*^{d}.
[f] primarily C3H/HeNCr,but also includes C3H/HeJ and C3H/BiDa.
[g] includes CBA/J, CNA/NCr and CBA/CaHNCr.
[h] includes AKR/J and AKR/NCr.

MUTANTS RECOVERED

Nineteen animals bearing mutations of *H-2K*^{b} were recovered, all of the gain and loss type, resulting from 14 separate events. Two of these (*H-2*^{bm6} and *H-2*^{bm9}) were recovered as "clusters" of 3 and 5 animals, respectively, apparently resulting from somatic mutations involving parental gonads (Melvold and Kohn 1975; Geleibter *et al* 1986), and it is appropriate to count all 8 in calculating the mutation rates. The spontaneous mutation rate for K^b was approximately 2 x 10^{-4} (Table 3), in good accord with previously published estimates of 5.5 x 10^{-4} (Melvold and Kohn 1975). By contrast, the rates observed for the *I-A*^{b} and D^b alleles ranged from 1 - 2 x 10^{-5}, an order of magnitude lower. The 95% confidence limits for the mutations rates of K^b versus *I-A*^{b} and D^b did not overlap. *H-2*^{b} mutation rates range from one to two orders of magnitude above the spontaneous rates reported for several color genes in the mouse (Green *et al* 1965).

At least 10 other *H-2*^{b} mutants have been reported: *H-2K*^{bm1} and *H-2K*^{bm2} (Bailey and Kohn 1965), *H-2K*^{bm3} (Apt *et al* 1975), *H-2K*^{bm4} (Forman and Klein 1977), *H-2D*^{bm14} (Bailey and Cherry 1975), *H-2*^{bm15} (Egorov and Blandova 1972), *H-2K*^{bm23} (Egorov and Egorov 1984), *H-2K*^{bm26} and *H-2K*^{bm27} (Egorov and Egorov 1988), and *H-2D*^{bm28} (Melvold, unpublished data).

Table 2. Spontaneous H-2 Mutations Recovered

Mutation	Original Mutant	Type [a]	Locus & Allele
H-2^{bm5}	B6 female	GL	K^b
H-2^{bm6}	B6 (2 f, 1 m) [b]	GL	K^b
H-2^{bm7}	B6CF1 male	GL	K^b
H-2^{bm8}	B6 male	GL	K^b
H-2^{bm9}	B6CF1 (1 f, 4 m) [b]	GL	K^b
H-2^{bm10}	B6CF1 male	GL	K^b
H-2^{bm11}	B6CF1 male	GL	K^b
H-2^{bm12}	B6CF1 female	GL	*I-A*^{b} (A_β)
H-2^{bm13}	B6CF1 male	GL	D^b
H-2^{bm17}	B6 male	GL	K^b
H-2^{bm18}	B6CF1 female	GL	K^b
H-2^{bm19}	B6CF1 female	GL	K^b
H-2^{bm20}	B6CF1 female	GL	K^b
H-2^{bm22}	B6 male	GL	K^b
H-2^{bm24}	B10 [c]	GL	D^b
KB-98	B6 male	GL	K^b
H-2^{dm2}	C [c]	L	L^d
H-2^{dm3}	B6CF1 female	L	? [d]
H-2^{dm5}	CB6F1 female	GL	K^d
KH-166A	B6CF1 female	L	? [d]

[a] GL = gain & loss; L = loss only.
[b] occurred as "clusters", presumably due to somatic mutation in parental gonad.
[c] already segregating within stock when identified.
[d] mapped to H-2 complex, but not assigned to specific loci. For *H-2*^{dm3}, the K, I-A and L loci were excluded (Melvold, unpublished data). For KH-166A, no further mapping data was obtained.

MUTATION RATES

Spontaneous mutation rates for the K and L loci within the *H-2*^{d} haplotype (Table 3), based on 4 recovered mutant animals, were similar to those for *I-A*^{b} and D^b, about 1 - 2 x 10^{-5}. The K^d mutation was of the gain and loss type, while that involving L^d was a loss. Two additional loss mutants were assigned to *H-2*^{d}, but could not be assigned to specific loci before they were lost due to breeding difficulties. For *H-2*^{dm3}, the K, I-A and D loci were excluded (Melvold, unpublished data), and it may have been at the L locus. For KH-166A, no additional mapping information was obtained. In studies not included in this survey, five other *H-2*^{d} mutants have been recovered: *H-2*^{dm1}, a gain and loss involving both the D and L loci (Egorov 1967); *H-2*^{dm4}, a gain and loss mutation involving the K locus (Morgan *et al* 1981), *H-2*^{dm6}, a loss involving the D locus (Melvold *et al* 1983), KH-188, a loss involving the L locus (Melvold, unpublished data), and *H-2*^{dm7} (Egorov and Egorov 1988).

No mutations of the *H-2*^{k} haplotype were recovered in these experiments, although at least two *H-2K*^{k} mutants, *H-2*^{km1} and *H-2*^{kv1}, have been found elsewhere (Blandova *et al* 1975; Kohn *et al* 1978).

Table 3. Mutation Rates Within H-2 (observed mutations/gamete/per generation)

H-2^{b}: Locus	Gametes Tested	Mutations Observed	Mutation Rate (95% confidence limits)
K^b	100,708	19 [a]	1.9×10^{-4} $(0.9 - 2.2 \times 10^{-4})$
	100,708	13 [b]	1.3×10^{-4} $(0.8 - 2.2 \times 10^{-4})$
$I\text{-}A^b$	100,708	1	9.9×10^{-6} $(0.025 - 5.5 \times 10^{-5})$
D^b	100,708	2	2.0×10^{-5} $(0.2 - 7.2 \times 10^{-5})$
Total [a]	100,708	22	2.2×10^{-4}. $(0.9 - 2.3 \times 10^{-4})$
Total [b]	100,708	16	1.6×10^{-4} $(0.9 - 2.2 \times 10^{-4})$
***H-2*^{d}: Locus**	**Gametes Tested**	**Mutations Observed**	**Mutation Rate (95% confidence limits)**
K^d	106,952	1	9.3×10^{-6} $(0.24 - 52.1 \times 10^{-6})$
L^d	106,952	1	9.3×10^{-6} $(0.24 - 52.1 \times 10^{-6})$
Unassigned [c]	106,952	2	
Total	106,952	4	3.7×10^{-5} $(1.0 - 9.6 \times 10^{-5})$

[a] includes *H-2*^{bm6} (3 animals) and *H-2*^{bm9} (5 animals) clusters.
[b] counts *H-2*^{bm6} and *H-2*^{bm9} as only two mutant events.
[c] mapped to the H-2 complex, but not assigned to specific loci (*H-2*^{dm3} & KH-166A).

The composite mutation rates (counting *H-2*^{bm6} and *H-2*^{bm9} as clusters rather than as single mutations) for the *H-2*^{b} and *H-2*^{d} haplotypes were 2.2×10^{-4} and 3.7×10^{-5}, respectively, with the composite rate for *H-2* being 1.2×10^{-4} (26 mutants out of 215,844 tested gametes, including *H-2*^{k}). If *H-2*^{bm6} and *H-2*^{bm9} were counted as single mutations each, the rate for *H-2*^{b} was 1.6×10^{-4}, and the composite spontaneous mutation rate for the H-2 complex was 9.3×10^{-5}.

However, the preponderance (23 out of the 26 total recovered mutants) of the rate for *H-2*^{b} and for the entire *H-2* complex is attributable to a single allele, K^b (1.9×10^{-4}). The reason for K^b mutating spontaneously at a rate 10 times higher than other H-2 alleles is unknown. However, most (if not all) K^b mutants appear to have resulted from a type of micro-recombinational event whereby they have incorporated DNA sequences encoded in the D and/or Qa regions (Mellor *et al* 1983; Miyada *et al* 1985; Pease *et al* 1983; Schulze *et al* 1983; Weiss *et al* 1983), although this is not a universal view (Klein 1984), and it is may be that K^b is especially favorable for these events. Micro-recombination also appears to account for the $I\text{-}A^b$ (Mengle-Gaw *et al* 1984) and D^b mutations (Hemmi *et al* 1988), although at a much lower frequency. It has also been suggested that the frequency of micro-recombination may be differentially effected by sex, occurring primarily in female gonads (Loh and Baltimore 1984). However, it is unknown whether the exceptionally high mutation rate of K^b is intrinsic or the result of its genetic milieu, or perhaps both. Is the high mutation rate of K^b dependent upon being within the context of the rest of the $H\text{-}2^b$ haplotype? Thus it would be of interest to know whether K^b could continue to mutate at these high frequencies in isolation (that is, in the absence of the D/Qa regions of the $H\text{-}2^b$ haplotype), for example in a strain such as B10.A(5R) (K^b D^d), but such data are not available.

Not only is K^b the dominant factor in spontaneous H-2 mutation rates, but one mutant phenotype (the "*bg*" group) accounts for nearly half of the K^b mutants (Tables 4 and 5). Six of the K^b mutants recovered in these surveys (as well as a seventh, $H\text{-}2^{bm16}$, recovered by Melief) had identical substitutions of phenylalanine for tyrosine at position 116 and 4 of them also shared arginine for cystine substitutions at position 121 (Geleibter *et al* 1986; Yamaga *et al* 1977*a,b*). Regardless of the residue at position 121, all of the "*bg*" mutants have similar phenotypes, whether tested by skin grafting or by other *in vivo* or *in vitro* assays (reviewed in Nathenson *et al* 1986). All of these mutations arose from independent events (Melvold *et al* 1982). The mutation rate for the "*bg*" mutant phenotype alone (which includes the two clusters) is 13/100,708 or 1.3×10^{-4} in this survey, accounting for over half of the cumulative spontaneous mutation rate for the K^b allele (1.9×10^{-4}).

Table 4. Prevalence of the "*bg*" phenotype among K^b mutants

Allele	position 116	position 121
K^b	Tyr	Cys
"*bg*" mutants: all histocompatible with one another [a]		
K^{bm5} [b]	Phe	Cys
K^{bm6} [b]	Phe	Arg
K^{bm7} [b]	Phe	Arg
K^{bm9} [b]	Phe	Arg
K^{bm17} [c]	Phe	Arg
K^{bm20}	unknown	unknown
KB-98 [c]	Phe	Cys
non-"*bg*" mutants: all unique		
K^{bm8}, K^{bm10}, K^{bm11}, K^{bm18}, K^{bm19}, K^{bm22}		

[a] K^{bm16}, found by Melief, also has the Tyr to Phe and Cys to Arg changes at positions 116 and 121, respectively (Yamaga *et al* 1983*a*).
[b] Yamaga *et al* 1983*a*, 1983*b*.
[c] G. Pfaffenbach, personal communication.

Table 5. Mutation Rates of the "bg" mutant type (observed mutations/gamete/per generation)

Locus	Gametes Tested	Mutations Observed	Mutation Rate	(95% confidence limits)
K^b	100,708	19 [a]	1.9×10^{-4}	($0.9 - 2.2 \times 10^{-4}$)
	100,708	13 [b]	1.3×10^{-4}	($0.8 - 2.2 \times 10^{-4}$)
"*bg*" only	100,708	13 [a]	1.3×10^{-4}	($0.83 - 2.2 \times 10^{-4}$)
"*bg*" only	100,708	7 [b]	7.0×10^{-5}	($2.8 - 14.3 \times 10^{-5}$)
non-"*bg*"	100,708	6	6.0×10^{-5}	($2.2 - 13.0 \times 10^{-5}$)

[a] includes $H\text{-}2^{bm6}$ (3 animals) and $H\text{-}2^{bm9}$ (5 animals) as clusters.
[b] counts $H\text{-}2^{bm6}$ and $H\text{-}2^{bm9}$ as only two mutant events.

While the K region of the *H-2*^{b} haplotype displays an asymmetry in terms of its mutant phenotypes, a second interesting asymmetry involves the D region of the *H-2*^{d} haplotype (Table 6). If one surveys all of the H-2 mutants reported from various sources which have been assigned to specific loci, all of the deletion-type mutations reported, with the exception of *H-2K*^{bm26} (Egorov and Egorov 1988), involve *H-2D*^{d} and/or *H-2L*^{d} (Hansen *et al* 1977; McKenzie *et al* 1977; Melvold *et al* 1983; Egorov 1967; Nairn and Nathenson 1978; Rubocki *et al* 1986), while no substitution (including gene conversion) mutations have been found to involve these alleles. The "gain" associated with *H-2*^{dm1} actually results from a "hybrid' fusion molecule formed from a deletion extending from the middle of D^d to the middle of L^d (Burnside *et al* 1984; Sun *et al* 1985). K^b, K^d and D^b mutants, on the other hand, are entirely of the substitution type. Because substitution mutations can be detected whether homozygous or heterozygous, while deletion mutations (with the exception of *H-2*^{dm1}) can be detected only when homozygous, any systematic underestimation of mutation rates should be largely confined to D^d and L^d.

Table 6. Distribution of Mutations Due to Substitutions vs Deletion (includes all reported, mapped mutants)

Haplotype	K region	D region
H-2^b	20 substitutions [a] 1 loss [b]	4, all substitutions [c]
H-2^d	2, both substitutions [d]	5, all deletions [e]

[a] *bm1, 2, 3, 4* (all reviewed in Nathenson *et al* 1986), *5, 6 , 7, 8, 9,10, 11, 16, 17, 18, 19, 20, 21* (Kohn and Dunn 1981), *22, 23* (Egorov and Egorov 1984), *KB-98.*
[b] *bm*^{26},(Egorov and Egorov 1988).
[c] *bm13, 14* (Bailey and Cherry 1975), *24, 28* (Melvold, unpublished data).
[d] *dm4* (Morgan *et al* 1981), *dm5*.
[e] *dm1* (Burnside *et al* 1984; *Sun et al 1985), dm2, dm3* (found not to involve the K, I-A or L loci, Melvold, unpublished data), *dm6* (Melvold *et al* 1983), *KH-188* (Melvold, unpublished data).

Comparison of the spontaneous mutation rate for class I, as compared to class II, loci also show clear differences. In the *H-2*^{b} haplotype, the spontaneous rate for the two class II loci (A_α and A_β) is 5 x 10^{-6} locus/gamete/generation (1 mutation, for two loci, in 100,708 gametes) with 95% confidence limits of 0.13 - 2.8 x 10^{-6}. The comparable rates at the two class I loci in the same haplotype is 1.0 x 10^{-4} per locus/gamete/generation (counting *H-2*^{bm6} and *H-2*^{bm9} as clusters, for a total of 21 mutants for two loci in 100,708 gametes) with 95% confidence limits of 0.5 - 1.1 x 10^{-4}. If *H-2*^{bm6} and *H-2*^{bm9} are counted as single mutations rather than clusters, the equivalent rate for the class I loci is 7.4 x 10^{-5} per locus/gamete/generation with 95% confidence limits of 4.4 - 11.2 x 10^{-5}.

In the *H-2*^{d} haplotype, if we tentatively assign *H-2*^{dm3} to the D region, the spontaneous rate for class I is 9.3 x 10^{-6} per locus/gamete/generation (3 mutants at 3 loci in 106,952 gametes) with 95% confidence limits of 1.9 - 27.3 x 10^{-6}. No class II mutants were found at any of the four class II loci (A_α, A_β, E_α and E_β) of the *H-2*^{d} haplotype for an obvious observed rate of 0. The upper 95% confidence limit for this, given the sample size, is 2.9 x 10^{-6} locus/gamete/generation, which lies just within the estimated range for the class I loci. Thus, in the *H-2*^{d} haplotype, the difference between the spontaneous mutation rates of class I and class II loci is not as stark as it is in the *H-2*^{b} haplotype.

The differences of 1 - 2 orders of magnitude between the class I and II loci of *H-2*^{b} probably reflect at least two factors. The first of these is, again, the prevalence of the "*bg*" mutant phenotype. However, even if the "*bg*" phenotype is excluded altogether, the class I spontaneous mutation rate is still 4.0 x 10^{-5} per locus/gamete/generation (with 95% confidence limits of 1.7 - 7.8 x 10^{-5}), which is still an order of magnitude greater than that for class II. The second factor is that class I mutations are probably more readily detected by skin grafting than are class II differences.

Table 7. Class I vs. class II mutation rates (observed mutations/locus/gamete/per generation)

Haplo-type	Class	# of Loci	Gametes Tested	Mutations Observed	Rate (95% conf. lim.)
H-2^b	I	2 (K,D)	100,708	21 [a]	1.0 x 10^{-4} (0.5 - 1.1 x 10^{-4})
	I	2 (K,D)	100,708	15 [b]	7.4 x 10^{-5} (0.24 - 11.2 x 10^{-5})
	I	2 (K,D)	100,708	8 [c]	4.0 x 10^{-5} (1.7 - 7.8 x 10^{-5})
	II	2 (A_α,A_β)	100,708	1	5.0 x 10^{-6} (0.13 - 27.7 x 10^{-6})
H-2^d	I	3 (K,D,L)	106,952	3	9.3 x 10^{-6} (1.9 - 27.3 x 10^{-6})
	II	4 (A_α,A_β,E_α,E_β)	106,952	0	0 (0 - 2.9 x 10^{-6})

[a] *H-2*^{bm6} and *H-2*^{bm9} fully counted as clusters.
[b] *H-2*^{bm6} and *H-2*^{bm9} counted as single mutants.
[c] "*bg*" mutants excluded entirely.

Grafting of mutant strains has also provided the opportunity to find sequential or reverse mutations. Out of 6,722 *H-2*^{b} bearing and 2,496 *H-2*^{d} bearing gametes, no "mutations of mutations" were observed. The upper end of the 95% confidence limits for 0 mutants recovered (out of 9,218 gametes) would be 4 x 10^{-4}. This overlaps the forward mutation rate estimated for H-2 genes and so there is no reason to presume, *a priori*, that the reverse mutation rate is significantly different that the forward rate.

However, in the course of complementation testing to map the *H-2*^{bm18} mutation in 1978 (not included in the graft testing entered into this survey), one apparent reverse mutation of K^{bm1} to K^b was found. A female who should have been K^{bm1}/K^b had the phenotype of K^b/K^b, both by direct graft testing and by the fact that all 13 of her progeny inherited an apparently normal *H-2*^{b} haplotype from her (Table 8).

The husbandry and record keeping was carefully re-examined for any circumstances which might have led to the misidentification of the "revertant" animal, but none were found. Unfortunately, it was impossible to distinguish the presumed "reverted K^b" allele from the real K^b allele among her progeny, and so subsequent analysis was impossible.

In summary, the genes of the H-2 complex clearly generate spontaneous variants at a rate 1 or 2 orders of magnitude above the average rate observed for non-H-2 histocompatibility genes and for non-histocompatibility genes, in accord with previous reports (Egorov and Blandova 1972; Melvold and Kohn 1975). While the overall H-2 spontaneous rate (1.9 x 10^{-4} per gene per generation) is surprisingly high, it is skewed by the fact that the observed mutation rate of a single allele (K^b) to a single mutant phenotype ("*bg*") is 1.3 x 10^{-4} per gene per generation, accounting for nearly 70% of the entire observed spontaneous mutation rate of the H-2 complex. Excluding the "*bg*" mutations, the class I genes of the H-2 complex mutate spontaneously at a rate ranging roughly from 10^{-4} to 10^{-5} per gene per generation, about an order of magnitude greater than that observed for the class II genes.

Table 8. Origin and analysis of presumptive K^{bm1} to K^{b} revertant

A. Identification of Revertant

Parents: B6.C-*H-2*^{bm1} female x B6.C-*H-2*^{bm18}/*H-2*^{b} male (N3-B6)
Progeny: Reciprocally skin grafted with B6

# of Progeny	Response to B6	Response by B6	Presumptive H-2 Genotype
4	Reject	Rejected	*H-2*^{bm1}/*H-2*^{bm18}
10	Accept	Rejected	*H-2*^{bm1}/*H-2*^{b}
1 [a]	Accepted [b]	Accepted [b]	*H-2*$^{b?}$/*H-2*^{b}

[a] female # 138204.
[b] results confirmed by extensive regrafting.

B. Breeding Tests of Presumptive Revertant Female # 138204

Mating: B6.C-*H-2*$^{b?}$/*H-2*^{b} female 138204 x B6.C-*H-2*^{d} male
Progeny: Grafted onto B6CF1 hybrids and also received B6 grafts

Response of B6CF1 to progeny	Response of progeny to B6 [a]	Presumed Genotype
13/13 accepted	6/6 accepted	*H-2*^{b}/*H-2*^{d}
0/13 rejected	0/6 rejected	----

[a] these 6 animals were a subset of those accepted by the B6CF1 recipients.

REFERENCES

Bailey DW, Usama B. (1960) A rapid method of grafting skin on tails of mice. Transplant Bull 7:424-425

Bailey DW, Kohn HI (1965) Inherited histocompatibility changes in progeny of irradiated and unirradiated inbred mice. Genet Res 6:330-340

Bailey DW, Cherry M (1975) A new mutation at the D end of the H-2 complex. Jackson Laboratory Annual Report 47:53

Blandova Z, Mnatsakanyan YA, Egorov IK (1975) Study of H-2 mutations in mice. VI. M523, a new K end mutant. Immunogenetics 2:291-295

Burnside SS, Hunt P, Ozato K, Sears DW (1984) A molecular hybrid of the *H-2D*^{d} and *H-2L*^{d} genes expressed in the dm1 mutant. Proc Natl Acad Sci USA 81:5204-5208

Dunn GR and Kohn HI (1981) Some comparisons between induced and spontaneous mutation rates in mouse sperm and spermatozoa. Mutat Res 80:159-164

Egorov IK, Blandova ZK (1972) Histocompatibility mutations in mice: Chemical induction and linkage with the H-2 locus. Genet Res 19:133-143

Egorov OS, Egorov IK (1984) *H-2*^{bm23}, a new K^{b} mutant similar to, but not identical with *H-2*^{bm3}. Immunogenetics 20:83-87

Egorov IK, Egorov OS (1988) Detection of new MHC mutations in mice by skin grafting, tumor transplantation and monoclonal antibodies: a comparison. Genetics 118:287-298

Forman J, Klein J (1977) Immunogenetic analysis of H-2 mutations: VI. Cross-reactivity in cell-mediated lympholysis between TNP-modified cell from H-2 mutant strains. Immunogenetics 4:183-193

Geliebter J, Zeff RA, Melvold RW, Nathenson SG (1986) Mitotic recombination in germ cells generates two MHC genes determined to be identical by RNA sequence analysis: K^{bm9} and K^{bm6}. Proc Natl Acad Sci USA 83:3371-3375

Green EL, Schlager G, Dickie MM (1965) Natural mutation rates in the house mouse: Plan of study and preliminary estimates. Mutat Res 2:457-465

Hansen TH, Cullen SE, Melvold RW, Kohn HI, Flaherty L, Sachs DH (1977) Mutation in a new H-2 associated histocompatibility gene closely linked to H-2D. J Exp Med 145:1550-1558

Hansen TH, Ozato K, Sachs DH (1983) Heterogeneity of H-2D region associated genes and gene products. Adv Immunol 34:39-70

Hansen TH, Spinella DG, Lee DR, Shreffler DC (1984) The immunogenetics of the mouse major histocompatibility complex. Ann Rev Genet 18:99-129

Harnasch D, Stumpf R (1982) Histocompatibility gene mutation rates, spontaneous and induced by the chemical mutagen procarbazine. Mutat Res 92:251-262

Hemmi S, Geliebter J, Zeff RA, Melvold RW, Nathenson SG (1988) Three spontaneous *H-2D*^{b} mutants are generated by genetic micro-recombination (gene conversion) events: Impact on the H-2 restricted immune responsiveness. J Exp Med 168:2319 - 2335

Klein J (1976) An attempt at an interpretation of the mouse H-2 complex. Cont Top Immunol 5:297-336

Klein J (1978) H-2 mutations: Their genetics and effect on immune function. Adv Immunol 26:55-146

Klein J (1984) Gene conversion in MHC genes. Transplantation 38:327-329

Kohn HI (1973) H-gene (histocompatibility) mutations induced by triethylene-melamine in the mouse. Mutat Res 20:235-242

Kohn HI, Melvold RW (1975) Spontaneous histocompatibility mutations detected by dermal grafts: Significant changes in rate over a 10-year period in the mouse H-system Mutat Res 24:163-169

Kohn HI, Melvold RW, Dunn GR (1976) Failure of X-rays to mutate class II histocompatibility loci in BALB/c mouse spermatogonia. Mutat Res 37:237-244

Kohn HI, Melvold RW (1976) Divergent X-ray-induced mutation rates in the mouse for H and "7-locus" groups of loci. Nature 259:209-210

Kohn HI, Klein J, Melvold RW, Nathenson SG, Pious D, Shreffler DC (1978) The first H-2 mutant workshop. Immunogenetics 7:179-294

LaFuse WP, David CS (1984) Ia antigens. Genes, molecules and function. Transplantation 38:443-453

Loh DY, Baltimore D (1984) Sexual preferences of apparent gene conversion events in MHC genes of mice. Nature 309:639-640

Levy RB, Hansen TH (1980) Functional studies of the products of the H-2L locus. Immunogenetics 10:7-17

McKenzie IFC, Pang T, Blanden RV (1977) The use of H-2 mutants as models for the study of T cell activation. Immunological Rev 35:181-230

McKenzie IFC, Morgan GM, Melvold RW, Kohn HI (1977) BALB/c-*H-2*^{db}: a new H-2 mutant in BALB/cKh that identifies a locus associated with the D region. Immunogenetics 4:333-347

Mellor AL, Weiss EH, Ramachandran K, Flavell RA (1983) A potential donor gene for the bm1 gene conversion event in the C57BL mouse. Nature 306:792-795

Mengle-Gaw L, Conner S, McDevitt HO, Fathman CG (1984) Gene conversion between murine class II major histocompatibility complex loci. Functional and molecular evidence from the bm12 mutant. J Exp Med 160:1184-1194

Melvold RW, Kohn HI (1975) H-2 mutation rates: H-2 and non-H-2. Mutat Res 27:415-418

Melvold RW, Kohn HI (1976) Eight new histocompatibility mutations associated with the H-2 complex. Immunogenetics 3:185-191

Melvold RW, Kohn HI, Dunn GR (1982) History and genealogy of the *H-2K*^{b} mutants from the C57BL/6Kh colony. Immunogenetics 15:177-185

Melvold RW, Stuart PM, Busker AE, Beck-Maier B (1983) A new loss mutant in the H-2^{d} haplotype. Transplantation Proc 15:2045-2047

Miyada CG, Klofelt C, Reyes AA, McLaughlin-Taylor E, Wallace RB (1985) Evidence that polymorphism in the murine major histocompatibility complex may be generated by the assortment of subgene sequences. Proc Natl Acad Sci USA 82:2890-29894

Morgan G, Dellos H, McKenzie IFC, Melvold RW (1981) Study of an *H-2K*^{d} mutant strain: C.B6-*H-2*^{dm4}. Immunogenetics 12:555-560

Nairn R, Nathenson SG (1978) Structural studies of the H-2D products of the mouse mutant BALB/c-*H-2D*^{db} and the parental strain BALB/cKh-*H-2*^{d}. J Immunol 121:869-871

Nairn R, Yamaga K, Nathenson SG (1980) Biochemistry of the gene products from murine MHC mutants. Ann Rev Genet 14:241-277

Nathenson SG, Geliebter J, Pfaffenbach GM, Zeff RA (1986) Murine major histocompatibility complex class-I mutants: Molecular analysis and structure-function implications. Ann Rev Immunol 4:471-502, 1986.

Pease LR, Schulze DH, Pfaffenbach GM, Nathenson SG (1983) Spontaneous H-2 mutants provide evidence that a copy mechanism analogous to gene conversion generates polymorphism in the major histocompatibility complex. Proc Natl Acad Sci USA 80:242-246

Rubocki RJ, Hansen TH, Lee DR (1986) Molecular studies of murine mutant BALB/c-*H-2*^{dm2} define a deletion of several class I genes including the entire *H-2L*^{d} gene. Proc Natl Acad Sci USA 83:9606-9610

Schulze DH, Pease LR, Yokoyama K, Geier SS, Pfaffenbach GM, Geliebter J, Zeff RA, Rosenblatt BP, Nathenson SG (1983) Diversity and polymorphism in the MHC appear to be generated by a copy mechanism. Transplantation Proc 15:2007-2012

Sun YH, Goodenow RS, Hood L (1985) Molecular basis of the dm1 mutation in the major histocompatibility complex of the mouse: a D/L hybrid gene. J Exp Med 162:1588-1602

Weiss EH, Mellor A, Golden L, Fahrner K, Simpson E, Hurst J, Flavell RA (1983) The structure of a mutant H-2 gene suggests that the generation of polymorphism in H-2 genes may occur by gene conversion-like events. Nature 301:671-674

Yamaga KM, McGovern D, Nisizawa T, Melvold RW, Kohn HI, Nathenson SG (1983*a*) Biochemical studies of H-2K antigens from a group of related mutants. I. Identification of a shared mutation in B6-*H-2*^{bm5} and B6-*H-2*^{bm16}. Immunogenetics 17:19-29

Yamaga KM, McGovern D, Nisizawa T, Melvold RW, Kohn HI, Nathenson SG (1983*b*) Biochemical studies of H-2K antigens from a group of related mutants. II. Evidence that B6-*H-2*^{bm6}, B6.C-*H-2*^{bm7} and B6.C-*H-2*^{bm9} are identical. Immunogenetics 17:31-41

Microgene Conversion in the Evolution of the MHC Class I Multigene Family

C. Wheeler[1], D. Maloney[2], S. Watts[3], J. Vogel[4], J. Towner[1], D. Baldwin[1], J. Rufer[1], M. Arbeitman[1], H. Chan[1], S. Fogel[1], and R. S. Goodenow[5]

[1] Department of Molecular and Cellular Biology, University of California, Berkeley, CA 94720, USA
[2] Department of Plant Pathology, University of California, Berkeley, CA 94720, USA
[3] Cell Biology Department, Baylor College of Medicine, One Baylor Plaza, Houston, TX 77030, USA
[4] Department of Plant Biology, University of California, Berkeley, CA 94720, USA
[5] To whom correspondence should be addressed (415) 643-6804

INTRODUCTION

The class I gene family resides on mouse chromosome 17 and contains from 26 to 34 different loci (Hood et al., 1983). Most of these genes reside within the Qa and Tla subregions, have unknown functions, and are relatively non-polymorphic (Mellor et al., 1984; Watts et al., 1989a). However, genes residing within the H-2 region, whose products function in presenting antigenic peptides to cytotoxic T cells (CTL)(Townsend et al, 1986), exhibit extreme allelic polymorphism. Interestingly, most of the diversity displayed among class I alleles and loci is in exons 2 and 3, which encode the $\alpha1$ and $\alpha2$ regions of the protein (Jaulin et al., 1985).

CTL can recognize and respond to peptidic antigens only in conjunction with host-derived H-2 molecules (Zinkernagel and Doherty, 1979). Recent studies, including X-ray crystallographic analysis of human HLA-A2 (Bjorkman et al., 1987a), have revealed that H-2 molecules probably bind peptidic antigens within a cleft formed by the $\alpha1$ and $\alpha2$ regions, and these H-2/peptide complexes may then be recognized on target cells by the T cell receptor of appropriate CTL. Superimposition of the hyper-polymorphic residues onto the crystal structure of HLA-A2 (Bjorkman et al., 1987b), reveals that these residues generally lie within the presumed antigen binding cleft.

As a result, the means by which class I diversity, and H-2 polymorphism in particular, is generated and maintained poses several interesting questions at the level of the gene, as well as at the level of population dynamics. Evidence of reciprocal recombination (crossover) has been documented within the class I gene family. The number of class I genes in wild and inbred mouse strains is variable, indicating that expansion and contraction due to unequal crossover may have occured (see Eastman O'Neill et al., 1986). Hybrid class I genes apparently resulting from such events can be seen in some inbred mouse strains (Weiss et al., 1984; Mellor et al., 1985; Watts et al, 1989). Two H-2 mutations

I.K. Egorov/C.S. David (Eds.)
Transgenic Mice and Mutants in MHC Research

also appear to have arisen within inbred lines by crossover events. Whereas the dm1 mutation is likely to have resulted from intrachromosomal exchange between H-2L^d and H-2D^d, producing a hybrid H-2D/L gene (Sun et al., 1985), the dm2 mutation probably involved crossover between H-2 flanking sequences, resulting in the deletion of the H-2L^d locus (Rubocki et al., 1986). It would seem that crossovers within the class I gene family are relatively infrequent (or undetectable by the screening methods used), and may not account for H-2 polymorphism to as large a degree as other mutagenic events discussed below.

Sequence comparisons of H-2 genes have revealed that non-allelic genes share small patches of sequence similarity in positions which distinguish alleles (Klein, 1986). In addition, several H-2 mutants have been isolated and characterized, most of which derive from the H-2^b haplotype (See Table1). H-2 mutants appear to occur at a frequency on the order of 10^{-3} per generation, the highest rate of any mammalian genes recorded to date (Klein, 1986). The structure of most mutant H-2 genes reveal multiple, clustered base pair substitutions relative to the progenitor H-2 genes in the region of the antigen binding cleft (Nathenson et al, 1986). Moreover, these substituted sequences were found at homologous positions in other class I genes of both the mutant and parental haplotypes (Pease et al., 1983). These data suggest that "microgene conversion", or non-reciprocal transfer (similar to gene conversion in yeast) of very short regions of DNA into the H-2 genes might account for much of the observed polymorphism and general evolution of the class I gene family.

A recombinational bias within certain exons of class I genes, and possibly within the H-2^b haplotype, could be responsible for the localization of many H-2 mutants to the α1 and α2-encoding exons. Alternatively, selective forces might be responsible for the recovery of a limited set of mutants which affect these regions. Rejection of reciprocal skin grafts as the sole criterion by which H-2 mutants were originally identified itself may bias the types of mutants recovered (Nathenson et al, 1986). Thus, the collection of characterized mutants may not be truly representative of the mutagenic or recombinogenic potential of H-2 genes within the confines of preservation of essential H-2 function *in vivo*.

The possible role of selection biasing the types of H-2 mutants recovered is underscored by the fact that, although Qa genes seem to be preferential donors in H-2^b mutants (see Table 1), Qa-specific residues are never transferred in these mutants, implying a role for these residues in Qa-specific function (Watts et al., 1989b). Similarly, the new residues transferred in the bm series mutants can always be found in H-2 genes of other mouse strains, further suggesting that only certain residues can be tolerated within the antigen binding cleft region and still allow rapid graft rejection. This is also supported by the fact that the Qa-unique residues on the top of the α-helices available to the T cell receptor are not involved in any of the mutations characterized to date (Watts et al., 1989b).

In addition, the preponderence of H-2 genes as recipients with Qa genes as probable donors of transferred sequences could represent either a recombinational bias or an artifactual result of selective forces. Analysis of the true mutagenic and recombinogenic potential of class I genes free of selection for function, or even expression of viable molecules, is therefore essential to resolve these questions and determine the molecular parameters influencing H-2 polymorphism. Such analyses can only be achieved in *in vitro* or other systems where there is an absence of selective forces.

Table 1. Amino acid substitutions in the bm series of mutations and km2 are found in other H-2 genes (compiled from Watts et al., 1989c)

MUTANT	SUBSTITUTIONS	GENES
K^{km2}	$Ile^{95}Leu^{98}Ser^{99}$	$H2K^{b,q,w28,bm1,4,5,6,7,9,1016}$,$A^{216}$ $H\text{-}2K1^{k}$,$H\text{-}2D^{b,p,k}$,$Q1^{k}$,$Q4^{k,b}$,$Q10^{b,k}$
K^{bm1}	$Ala^{152}Tyr^{155}Tyr^{156}$	$H\text{-}2L^{d}$,$Q2^{k}$,$Q10^{b,k,q}$
K^{bm3}	$Ser^{77}Ala^{89}$	$H\text{-}2D^{q,b}$,$Q2^{k}$,$Q5^{k}$
K^{bm4}	$Glu^{173}Leu^{174}$	$H\text{-}2K^{d,(bm10)}$,$T5^{b}$,$H\text{-}2D^{k,p}$ $Q1^{k}$,$Q4^{b}$,$Q5^{k}$,$Q9^{b}$,$Q10^{b,k,q}$
$K^{bm5,16}$	Phe^{116}	$H\text{-}2K^{d}$, $H\text{-}2D^{d}$, $H\text{-}2L^{d}$, A^{216},$Q1^{k}$,$Q4^{k,b}$,$Q7^{b,d}$,$Q8^{b}$,$Q9^{b}$
$K^{bm6,7,9}$	$Phe^{116}Arg^{121}$	$H\text{-}2K^{d}$,$H\text{-}2D^{b}$,$Q4^{k,b}$,$Q7^{b,d}Q8^{b}$,$Q9^{b}$
K^{bm10}	$Ala^{163}Met^{165}$ $Glu^{173}Leu^{174}$	$K1^{b}$
K^{bm11}	$Ser^{77}Asn^{80}$	$H\text{-}2D^{b,p}$
K^{bm23}	$His^{75}Ser^{77}$	$H\text{-}2K^{q}$,$Q10^{q}$

Some important aspects of class I diversification have already been partially resolved relying primarily on DNA sequence analysis. Analysis of a spontaneous H-2 mutant from C3Hf mice ($H\text{-}2^{k}$) has revealed the same sort of multiple, clustered base pair substitutions in the mutant $H\text{-}2K^{km2}$ gene relative to the progenitor $H\text{-}2K^{k}$ gene as is seen in many of the $H\text{-}2K^{bm}$ mutants. Four base pair substitutions clustered within a 14 base pair region of exon 3 results in 3 amino acid substitutions within the region contributing to the bottom of the antigen binding cleft of the K^{km2} polypeptide (Vogel et al., 1988). Furthermore, $H\text{-}2D^{k}$ was shown to be the only gene in both the parental and mutant genomes that carries this same 14 nucleotide sequence, and in fact shares 45 base pairs of homology with $H\text{-}2K^{km2}$ in this general region of exon 3. Therefore, $H\text{-}2D^{k}$ was proposed to be the donor gene for the km2 conversion event. Since the substituted region in K^{km2} is at the bottom of the antigen binding cleft and is likely to be inaccessible to the T cell receptor molecules, identification of this mutant on the basis of graft rejection is likely the result of altered peptide antigen presentation rather than direct rejection of a foreign class I molecule. This demonstrated that microgene conversion was not restricted to a single locus within the b haplotype, and suggested that this phenomenon might be a general feature of H-2 gene evolution. Thus, microgene conversion together with point mutation and perhaps other classes of recombination, such as crossover and conversion of longer tracts of DNA, might shape the concerted evolution of the class I gene family as a whole.

STASTICAL ANALYSIS OF CLASS I EVOLUTION

The cloning and sequencing of all the H-2 and Qa genes from a single mouse strain has allowed a more global examination of variability within the class I gene family than had been previously possible (Watts et al., 1989a). Using the Stephens' algorithm designed to identify recombination events on the basis of the length spanned by patterns of nucleotides shared by groups of sequences, as well as the largest gaps within such shared sequences (Watts et al., 1989b), a preponderance of relatively small potential gene conversion events was seen among 23 class I sequences from H-$2^{b,d,k,p,q}$ haplotypes (Table 2). In contrast, relatively few examples of recombination involving long tracts of DNA were as readily apparent.

Table 2. Class I recombinants identified by computer algorithm

24 events characterized by small, clustered similarities.

12/24 "microgene conversion" events overlap α-helices or β-strands. 2 of these are silent substitutions/ 2 expected to have minimal affect on antigen binding.

6 events involve H-2 sequences as both donors and recipients.

10 events involve Qa sequences as both donors and recipients.

11 events involve H-2 and Qa sequences: 2 Qa to H-2, 3 H-2 to Qa.

Furthermore, all types of class I genes, H-2, Qa, and Tla, seemed to possess the capacity to serve as both donor and recipient based on this analysis, and the events were not significantly confined to the regions encoding the antigen binding cleft or 5' exons in H-2 genes (although frequencies of events occuring at different locations cannot be assessed in this type of analysis). Thus, microgene conversion seems to play a role in the evolution of all class I genes, and is not absolutely biased in terms of loci involved or location within affected genes. These data argue for the involvement of selective forces, whose mechanisms remain unclear, in fixing the existing allelic forms of class I genes.

Analysis of deletions in transfected class I genes integrated into the genome of Ltk- cells has revealed associations between dispersed palindromic DNA sequences and deletion breakpoints (Vogel et al., 1989). This highlights the possible role of specific DNA sequences within class I genes in enhancing strand breakage, ligation, and other events which may be important to recombinational processes.

It is perhaps ironic that in yeast, where gene conversion was originally demonstrated and has sincc been extensively characterized, there is a paucity of data on microgene conversion; only a single, rare putative event, remaining uncharacterized recombinationally, exists in yeast (Ernst et al., 1981). Given that the majority of recombination events seen in this organism involve stretches of DNA that are orders of magnitude longer than those proposed for microgene conversion (Judd and Petes, 1988) generates another question: Is the phenomenon of class I microgene conversion a unique property of mammalian genes, or a function of specialized recombination machinery in mammalian cells?

RECOMBINATION OF CLASS I GENES IN YEAST

In order to address the question posed above, as well as to dissect the parameters influencing recombination between class I genes, such as chromosomal position, extensive heterozygosity, and specific DNA sequences, and to shed more light on the mechanism of general meiotic recombination in eukaryotes, we have been examining recombination between class I genes inserted into yeast.

Our strategy has been to place two class I genes that are donor and recipient of a known microgene conversion event at identical positions on opposite homologues in a diploid yeast strain. For example, we inserted H-2K^k and H-2D^k (the gene pair that produced the H-2K^{km2} mutation) into separate Arg4 alleles on yeast chromosome VIII, and picked unselected random tetrads for analysis. We have followed recombination between the class I sequences within these tetrads using Southern blotting with general class I probes, RNase A cleavage protection of specific RNA probes, and DNA sequencing.

Employing Southern analysis of 150 random yeast tetrads, and 53 Arg4 convertants derived from 347 random tetrads, we have identified two recombinants involving class I sequences. These recombinant spores produce patterns on Southern blots which appear as hybrids between the parental patterns (See Fig. 1),and resemble typical yeast gene conversions, involving relatively long tracts of transferred DNA sequences with endpoints within the two regions of highest sequence similarity between H-2K^k and H-2D^k.

One tetrad contains a single spore harboring a hybrid class I gene created by a gene conversion endpoint within exon 4. The conversion tract was donated by H-2K^k into H-2D^k and extends 5' the entire length of the class I gene. This tetrad also harbors a conversion of Arg4; however, the Arg4 conversion occurs within a separate spore, is independent of the class I conversion, and is not derived from an associated crossover event. Thus, this tetrad has undergone a *bona fide* gene conversion within class I sequences involving at least 2.3 kb of transferred DNA (Fig. 1, lanes 1-8).

Another tetrad also contains a class I hybrid gene in a single spore, with a conversion endpoint within exon 3, between two polymorphic PstI sites (Fig 1, lanes 9-12). The conversion tract is donated by H-2D^k and extends 5' the entire length of the class I gene and probably into the flanking Arg4 sequences, which are also converted, thereby constituting a conversion of at least 1.5 kb of DNA.

Fig. 1. Configuration of class I genes inserted into yeast.

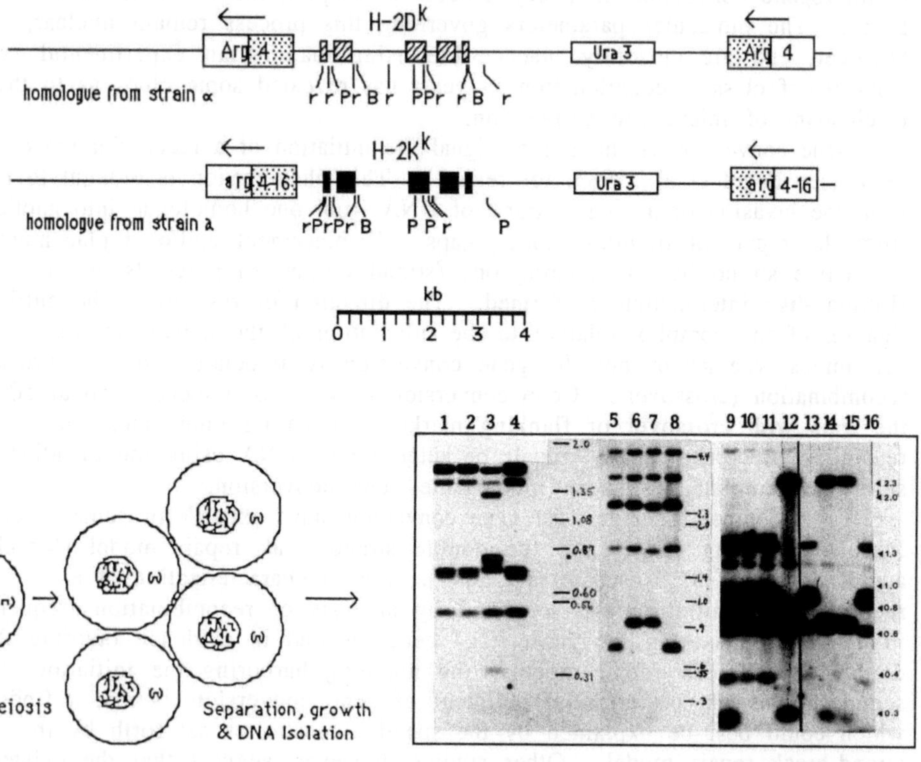

Lanes 1-8 show the four spores of one recombinant tetrad digested with RsaI(r) (1-4), and PstI (P)(5-8). Lanes 9-16 show the spores of another recombinant (9-12) and those of a normal tetrad (13-16) digested with PstI + BamHI (P+B).

RNase A cleavage protection assays were used in order to detect small sequence transfers within class I sequences which did not span polymorphic restriction enzyme sites. Analysis of 38 4-spored tetrads has revealed two recombinants which in some ways resemble microgene conversions as defined in mouse H-2 mutants. Two tetrads contain an H-2K^k spore which harbors a limited stretch of DNA characteristic of H-2D^k. These recombinants, whose precise structures will be reported elsewhere, appear to involve gene conversion tracts spanning from 6 to 49, and 25 to 215 nucleotides, respectively, falling within the range of distances spanned by microgene conversions reported from H-2 mutants.

DISCUSSION

Previous studies documenting the structure of *in vivo* H-2 mutants, as well as those analyzing the inherent variation in the class I gene family, point strongly to microgene conversion as a key process in shaping the evolution of these genes. The molecular parameters governing this process remain unclear, however, and are currently inaccessible using mammalian experimental systems. Analysis of class I recombination in yeast has revealed some clues as to the mechanism of microgene conversion.

Gene conversion is thought to signal the initiation of a recombination event in yeast (see Fogel et al., 1983, for review). This phenomenon is thought to result from the invasion of a single strand of DNA from one homologue into another, or from the repair of double-stranded gaps. Displacement and/or replacement of the native strand by an invading one (strand migration) proceeds and a Holliday-like intermediate is formed. The direction of resolution, the cutting and ligation of the complex relative to the orientation of the native strands, determines whether or not the gene conversion is associated with reciprocal recombination (crossover). Gene conversion in yeast is associated about 50% of the time with crossover of flanking markers, indicating randomness in the resolution process. Finally, repair of heteroduplex DNA using the invading strand as template results in identifiable gene conversion.

Two competing models for gene conversion have existed for some time, the Meselson-Radding model and the double strand-break repair model (Meselson and Radding, 1975, Szostak et al., 1983). Recent characterization of a recombination initiation site localized by analysis of recombination frequencies in deletion mutants in the yeast Arg4 gene resulted in evidence favoring the double strand-break repair model: the homolog harboring the initiation sequence was the preferrential recipient of gene conversion events, a finding which could best be explained by the specific constraints set forth by the double strand-break repair model. Other studies, however, suggest that the existence of double strand-breaks are a possible consequence, rather than a direct cause of recombination (Borts and Haber, 1987).

Using the heterology inherent to different class I loci, we have effectively marked the DNA around potential recombination initiation sites, facilitating precise mapping of these sequences. The possibility of microgene conversion occurring between class I genes makes them inviting candidates for analysis of not only recombination initiation, but also of termination sites. Thus, we have begun to study the effects of extensive DNA heterologies on the initiation and outcome of recombination events. Importantly, this approach should also allow us to assess the recombinational potential of class I genes in the absence of selective forces.

As mentioned above, other factors, particularly DNA sequences which serve to enhance recombination frequency (hotspots), may also influence recombination within class I genes. Examples of DNA sequences proposed to effect recombination include the chi sequence in *E. coli*, the EC-1 repeat [d(TG-AC)n] of β-globin, tetrameric repeats (CAGG) in the mouse I-E gene, and the G-C rich human minisatellite and yeast PMS (post-meiotic segregation) consensus sequences (Treco and Arnheim, 1986, Steinmetz et al., 1986, Nicholls et al., 1985, White et al, 1988). Class I genes contain many interesting sequence elements as well. In particular, the 5' regions of class I genes are quite rich in GC content, and contain sequences which resemble mouse I-E repeats, yeast PMS, and human minisatellite consensus sequences. Jaulin et al. (1985) have proposed that the CpG repeats in the 5' end of class I genes could serve as break sites for recombination initiation. In addition, H-2 genes and some Qa genes contain long tracts of alternating C and T residues, as well as Alu-like B1 and B2 repeats in their large intron and 3' flanking regions, which could be involved in some aspects of recombination (Singer et al., 1983; Nicholls et al., 1985).

Some of these sequences may actually represent initiation sites for recombination, and the studies of initiation at Arg4 discussed above suggest a mechanism for directional gene conversion (where one locus is a preferential recipient or donor of genetic information) which could be relevant to class I recombination and evolution. Directional conversion might occur if only one of the genes in a pair undergoing recombination harbored an initiation site. The gene containing the site(s) would be the recipient, provided that the rules dictated by the double strand-break repair model hold true. Thus, recombinants which initiated within well-marked regions of two interacting genes, along with analysis of the directionality of transfer, could serve to pinpoint the fine sequence constraints on theoretical recombination initiation sites, or other DNA sequences associated with recombination. Although no such recombinants have been recovered thus far, ours remains the only system in which such questions can be addressed.

Our initial results indicate that the class I/yeast recombination system may produce recombinants which resemble *in vivo* H-2 mutations at a higher frequency than those resembling typical yeast conversions (4% short tract conversion vs 0.3% long tract conversion), although the small sample size of tetrads screened for microconversions may well skew this frequency estimate. Never the less, this supports the notion that short non-reciprocal transfers of genetic information (such as H-2 microgene conversion) are dictated by the structure of gene pairs, and are not exclusively the properties of more global characteristics of mammalian genomes such as gross chromosomal configuration or specific recombination machinery. Moreover, most of the recombinants seen within this system occur within regions of highest sequence similarity between H-$2K^k$ and H-$2D^k$, in exons 3 and 4. This suggests that initiation and/or termination of at least some types of recombination events may occur preferentially within regions of relatively high sequence similarity. Conversely, initiation, termination, or strand migration during recombination might be inhibited by extensive DNA sequence heterology. Preliminary evidence suggests that removal of extensive heterozygosities increases the rate of recombination at the flanking Arg4 markers (D. Maloney, unpublished data). Since we see a low frequency of long tract gene conversions at class I (0.3%) as compared to conversion at Arg4 ($\geq$4%), some suppression may be induced by extensive heterologies affecting long-tract class I conversions as well.

It is tempting to suggest that microgene conversion may be a result of gene conversions that are initiated within pockets of high sequence similarity flanked by regions of extensive sequence heterology, such that strand migration is effectively blocked at these heterologies. Such blockage may only allow asymmetrical single strand exchange, which could preclude extension and resolution resulting in associated crossovers. Moreover, a high frequency of microgene conversion, as seems to occur between H-2 genes, might be facilitated by potential hotspots for initiation of recombination. An overview of the structures of the known bm mutants, km2, and their donors reveals that many of these genes exhibit extensive sequence heterology between proposed donor and host genes at one or both sides of the proposed transfer (Nathenson et al, 1986, Geliebter and Nathenson, 1988, Vogel et al., 1988).

Under the precepts of this model, the degree and dispersal of heterozygosities would dictate the length of the gene conversion tract - the more extensive the heterozygosities, the shorter the tract length. This could have the effect of producing still more diversity through recombination in cases where interacting genes exhibited extreme heterozygosity, since more complex patchwork genes might be formed. The stringency of selective forces, together with the recombination rate between two members within the family, would then determine the extent of class I diversity fixed at the population level. In light of their immunological role in antigen presentation and the diversity thought to be important to this function, H-2 genes might enjoy relatively relaxed negative selective forces constraining amino acid substitutions, in addition to potentially high rates of microgene conversion (Nagy et al., 1989). The structure of the two α-helices atop β-srands which contribute to this cleft appear to be especially tolerant of heterozygosity, since multiple amino acid residues are present at these locations within different H-2 molecules (see Watts et al., 1989c). Recombination events, such as microgene conversion, creating additional diversity within the region of H-2 genes encoding the antigen binding cleft might then be expected to be fixed in the population at a high rate, relative to recombination within different portions of the gene. Recombination events (such as microgene conversion) together with selective constraints could thereby allow an especially high incidence of diversification in and around the cleft. Thus, recombination has the potential to effectively maximize the antigen binding potential of H-2 molecules, thereby increasing the effective immune repertoire within a population. Future studies should clarify the mechanism by which this potential is realized, as well as the relative roles of selection and recombination in class I evolution.

C. Wheeler was supported by NIH predoctoral training grant AI26942A-01. This work was supported by NIH grants CA37099 and R1A127853, and March of Dimes grant 58698 to R. S. Goodenow.

REFERENCES

Bjorkman, P.J., M.A. Saper, B. Samraoui, W.S. Bennett, J.L. Strominger and D.C. Wiley (1987) Structure of the human class I histocompatibility antigen, HLA-A2. Nature 239:506-512.

Bjorkman, P.J., M.A. Saper, B. Samraoui, W.S. Bennett, J.L. Strominger and D.C. Wiley (1987) The foreign antigen binding site and T cell recognition regions of class I histocompatibility antigens. Nature 239:512-518.

Borts, R.H. and J.E. Haber (1987) Meiotic recombination in yeast: Alteration by multiple heterozygosity. Science 237:1459-1465.

Eastman O'Neill, A., K. Reid, J, C. Garberi, M. Karl, and L. Flaherty (1986) Extensive deletions in the Q region of the mouse major histocompatibility complex. Immunogenet. 24:368-373.

Ernst, J.F., J.W. Stewart and F. Sherman (1981) The cycl-11 mutation in yeast reverts by recombination with a non-allelic gene: Composite genes determining the iso-cytochromes c. Proc. Nat. Acad. Sci., USA 78:6334-6338.

Fogel, S., R.K. Mortimer and K. Lusnak (1983) Meiotic gene conversion in yeast: Molecular and experimental perspectives. pp. 65-107 in: Spencer, J.F.T., D.M. Spencer and A.R.W. Smith (eds.) Yeast Genetics: Fundamental and Applied Aspects, Springer Verlag, N.Y.

Geliebter. J. and S.G. Nathenson (1988) Microrecombinations generate sequence diversity in the murine major histocompatibility complex: Analysis of the K^{bm3}, K^{bm4} , K^{bm10} and K^{bm11} mutants. Mol. Cell. Biol.. 8:4342-4352. Hood, L., M. Steinmetz, and B. Malissen (1983)

Hood, L., M. Steinmetz, and B. Malissen (1983) Genes of the major histocompatibility complex of the mouse. Ann. Rev. Immunol. 1: 529-568.

Jaulin, C., A. Perrin, J.-P. Abastado, B. Dumas, J. Papamatheakes, and P. Kourilsky (1985) Polymorphism in mouse and human class I H-2 and HLA genes is not the result of random independent point mutations. Immunogenet. 22:453.

Judd, S.R. and T.D. Petes (1988) Physical lengths of meiotic and mitotic gene conversion tracts in Saccharomyces cerevisiae. Genetics 118:401-410.

Klein, J. (1986) Natural History of the Major Histocompatibility Complex. John Wiley and Sons, N.Y.

Mellor, A. L., E. H. Weiss, M. Kress, G. Jay, and R. A. Flavell (1984) A nonpolymorphic class I gene in the murine major histocompatibility complex. Cell 36:139.

Mellor, A. L., J. Antoniou, and P. J. Robinson (1985) Structure and expression of genes encoding murine Qa-2 class I antigens. Proc, Natl, Acad. Sci. USA 82:5920.

Meselson, M.S. and C.M. Radding (1975) A general model for genetic recombination. Proc. Nat. Acad. Sci., USA 72, 358-361.

Nagy, Z. A., P. V. Lehmann, F. Falcioni, S. Muller and L. Adorini (1989) Why peptides? Their possible role in the evolution of MHC-restricted T-cell recognition. Immunol. Today 10:132.

Nathenson, S. G., J. Geliebter, G. M. Pfaffenbach, and R. A. Zeff (1986) Murine major histocompatibility comples class I mutants: Molecular analysis and structure-function implications. Ann. Rev. Immunol. 4:471-502.

Nicholls, R. D., D. R. Higgs, J. B. Clegg, and D. J. Weatherall (1985) α^0-Thalassemia due to recombination between the α1-globin gene and an *Alu*I repeat. Blood, 65, 1434.

Pease, L. R., D. H. Schulze, G. M. Pfaffenbach, and S. G. Nathenson (1983) Spontaneous H-2 mutants provide evidence that a copy mechanism analogous to gene conversion generates polymorphism in the major histocompatibility complex. Proc. Natl. Acad. Sci. USA 80:242.

Rothstein, R., 1985 in DNA Cloning volume II, A Practical Approach, D. M. Glover, editor, IRL Press Limited, Oxford. pp. 52-58.

Rubocki, R. J., T. H. Hansen, and D. R. Lee (1986) Molecular studies of murine mutant BALB/c H-2^{dm2} define a deletion of several class I genes including the entire H-2Ld gene. Proc. Natl. Acad. Sci. U.S.A. 83:9606-9610.

Singer, D. S., R. Lifshitz, L. Abelson, Pl Nyirjesy, and S. Rudikoff (1983) Specific association of repetitive DNA sequences with major histocompatibility genes. Molec. Cell. Biol. 3:903.

Steinmetz, M., D. Stephan, and K. F. Lindahl (1986) Gene organization and recombination hotspots in the murine major histocompatibility complex. Cell 44:895-904.

Szostak. J., T. Orr-Weaver, R. Rothstein and F. Stahl (1983) The double-strand-breakmodel for recombination. Cell 23:25-35.

Sun, Y. H., R. S. Goodenow, and L. Hood (1985) Molecular basis of the dm1 mutation in the major histocompatibility complex of the mouse: a D/L hybrid gene. J. Exp. Med. 162:1588-1602.

Townsend, A. R. M., J. Rothbard, R. M. Gotch, G. Bahadur, D. Writh, and A. J. McMichael (1986) The epitopes of Influenza nucleoprotein recognized by cytotoxic T lymphocytes can be defined with short synthetic peptides. Cell 44:959-968.

Treco, D. and N. Arnheim (1986) The evolutionarily conserved repetitive sequence d(TG-AC)n promotes reciprocal exchange and generates unusual recombinant tetrads during yeast meiosis. Mol. Cell. Biol. 6:3934-3947.

Watts, S., A. Cranmer Davis, B. Gaut, C. Wheeler, L Hill and R. S. Goodenow (1989) Organization and structure of the Qa genes of the major histocompatibility complex of the C3H mouse: implications for Qa function and class I evolution. EMBO J. 8:1749-1759.

Watts, S., M. Kuhner, W. Klitz, and R. S. Goodenow (1989) Gene conversion in the evolution of the class I multigene family of the murine major histocompatibility complex. Genetics (submitted).

Watts, S., C. Wheeler, R. Morse, and R. S. Goodenow (1989) Amino acid comparison of the class I antigens of murine major histocompatibility complex. Immunogenetics (in press).

Weiss, E. H., L. Golden, K. Fahrner, A. L. Mellor, J. J. Devlin, H. Bullman, H. Tiddens, H. Bud, and R. A. Flavell (1984) Organization and evolution of the class I gene family in the major histocompatibility complex of the C57BL/10 mouse. Nature, (London) 310:650-655.

White, J. H., J. F. DiMartino, R. W. Anderson, K. Lusnak, D. Hilbert, and S. Fogel (1988) A DNA sequence conferring high postmeiotic segregation frequency to heterozygous deletions in *Saccharomyces cerevisiae* is related to sequences

associated with eucaryotic recombination hotspots. Molec. Cell. Biol. 8:1253.
Vogel, J.M., A.C. Davis, D.M. McKinney, M. McMillan, W.J. Martin and R.S. Goodenow (1988) Molecular characterization of the C3Hf/HeN H-2K^{km2} mutation. J. Exp. Med. 168:1781-1800.
Vogel, J.M., M. Nieto, A. Fischer, and R. S. Goodenow (1989) Overlapping palindromic sequences are associated with somatic deletion and meiotic recombination of MHC class I genes. Molec. Immunol. (in press).
Zinkernagel, R. M., and P. C. Doherty (1979) MHC restricted cytotoxic T cells: Studies on the biological role of polymorphic major transplantation antigens determining T cell restriction specificity, function, and responsiveness. Adv. Immunol. 27:51-75.

Random Mutagenesis by Oligonucleotides: A Probe for MHC Structure and Function

K. Katoh, R. Murray, D. Muller, and J. A. Frelinger[1]

[1] Department of Microbiology and Immunology and Lineberger Cancer Research Center, University of North Carolina, Chapel Hill, NC 27599-7290, USA

INTRODUCTION

The study of variants induced by mutagenesis has been extremely important in understanding many biological systems (Drake 1969). In the case of the mouse histocompatibility complex (MHC) H-2 mutants have been important in the definition of pleiotropic effects of single MHC genes. These mutants were laboriously isolated by screening for skin graft rejection (Melvold et al. 1982). The generation of such spontaneous mutants, however, is time consuming and difficult. Several laboratories have used *in vitro* selection techniques (by monoclonal antibodies) as an alternative, (Flores and Rajan 1977; Pious et al. 1982) others have used molecular genetics techniques (Shiroishi et al. 1985). These include using site directed mutagenesis and *in vitro* recombination. The *in vitro* recombinants have been crucial in establishing that the α1 and α2 domains of MHC class I molecules interact significantly to form the class I peptide binding structure recognized by T cells (Darsley et al. 1987). This was later confirmed using X-ray crystallography on the isolated HLA-A2 protein (Bjorkman et al. 1987a).

Our laboratory has used an alternative approach for mutant production, utilizing random mutagenesis by oligonucleotides. In our initial reports we used a mutant library covering the α1 region of the mouse MHC class I gene H-2D^{p}. Here we report an overview of our analyses of class I mutants.

PRODUCTION OF MUTANTS

Mutants were produced as previously described (Murray et al. 1988a, 1988b). Here, a synthetic restriction fragment consisting of the entire α1 protein coding domain and adjacent remaining sequence was synthesized under mutating conditions (Hutchinson et al. 1986). These twelve oligonucleotides were assembled and cloned into M13. Individual M13 clones were analyzed by DNA sequencing. Following sequencing, M13 RF were isolated, the α1 containing restriction fragments isolated, and used to replace the wild type fragment. Mutants are designated using the single letter amino acid code of the wild type amino acid, the amino acid number, and the new mutant amino acid. Thus, DP20(Y27N) is a replacement of tyrosine at position 27 by asparagine. We sequenced 120 clones and recovered 51 in the correct reading frame. Of these 51 clones 38 were mutants with substitutions in 67 positions (Murray et al. 1988b). The mutated genes were introduced into L cells by DNA

I.K. Egorov/C.S. David (Eds.)
Transgenic Mice and Mutants in MHC Research

mediated gene transfer using electroporation (Murray et al. 1988a). All cells were co-transfected with a tk gene regulated by the metallothionein promoter. Cells were routinely cultured in 0.1 mM Zn_2SO_4 in addition to HAT. Mutants DP14(A11V,E32Q), DP20(Y27N), DP164(R14L), DP174(P57Q), and DP181(D37H) have been previously described. The sequence of DP170(N30S) has also been previously described (Murray et al. 1988a). All of the mutants have been characterized by binding with three non-cross blocking D^p specific monoclonal antibodies. All transfected cells express essentially equivalent levels of the exogenous gene product and all react equivalently with all three monoclonals based both on radioimmunoassays and an analysis by flow cytometry. Flow cytometry using a monoclonal antibody specific for D^p is shown in Fig. 1. As seen in the right panels of Fig. 1 essentially all the cells express equivalent amounts of the transfected gene product, D^p. Monoclonal antibody reactive with K^k but not D^p is shown in the left panels of Fig. 1. Ltk and all other cells express K^k.

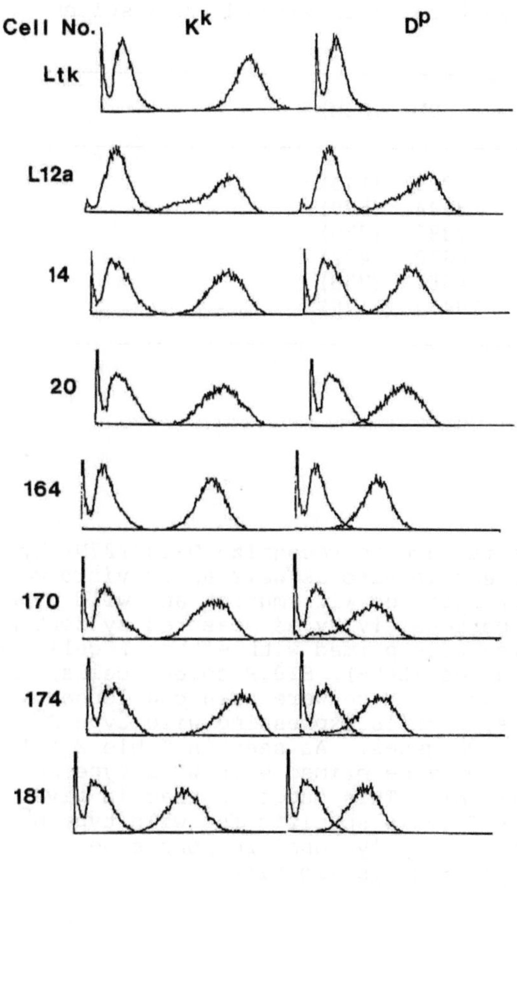

Fig. 1 Relative levels of endogenous K^k and transfected D^p class I molecules as assessed by flow cytometry. Primary monoclonal antibody for anti-K^k (16-1-2) and anti-D^p (7-16.0) was secondarily stained with goat anti-mouse IgG2a FITC (Fisher). The first peak in each panel shows background staining with the secondary antibody only. The untransfected cell line (Ltk) shows no staining with anti-D^p. L12a, expressing the wild type D^p molecule, stains with anti-D^p at approximately the same intensity as the endogenous class I molecule K^k. All cell lines transfected with mutant D^p genes show equivalent staining with anti-D^p

PRIMARY RESPONSE TO H-2 MUTANTS *IN VITRO*

We produced primary CTL effector cells against B10.P spleen cells by co-culturing C3H spleen cells with irradiated B10.P spleens. Cytotoxic cells were tested on ^{51}Cr labeled L cell targets. The recognition of five mutants covering six amino acid substitutions was tested. Only DP20(Y27N) was not killed by C3H anti-B10.P effector cells. Mutants DP14(A11V,E32Q), DP164(R14L), DP174(P57Q), DP181(D37H) were recognized as well as L cells expressing the wild type D^p gene by C3H anti-B10.P effectors (Murray et al. 1988a).
We wished to determine whether C3H cells could recognize the altered D^p molecules in a primary proliferative response. We co-cultured C3H spleen cells with untransfected L cells, B10.P spleen cells, L cells expressing the wild type molecule, DP14(A11V,E32Q), DP20(Y27N), and DP181(D37H). As seen in Table 1 all mutants, the wild type D^p molecule, and B10.P spleen cells stimulated a vigorous proliferative response by C3H spleen cells. Untransfected L cells (Ltk-) did not induce proliferation.

Table 1. Recognition of transfected L cells *in vitro* by C3H spleen cells

Stimulator Cell	CPM[1] (SEM)
Ltk-	389 (145)
B10.P	3044 (570)
WT Dp	7383 (830)
DP14	6850 (432)
DP20	4387 (759)
DP181	10434 (1178)

[1] [^3H]Thymidine incorporation.

RESPONSE TO H-2 MUTANTS *IN VIVO*

In order to determine whether the failure to recognize DP20(Y27N) by wild type specific clones was present *in vivo* as well as *in vitro* we examined the ability of L cells transfected with mutant and wild type D^p genes to cause delayed type hypersensitivity as measured by foot pad swelling (Minami et al. 1982). Mice were primed with either L cells not expressing any exogenous class I genes (Ltk-), B10.P spleen cells, or L cells expressing the wild type D^p gene. Mice were then challenged with L cells (Ltk-), B10.P spleen cells, L cells expressing wild type D^p gene, or L cells expressing mutant D^p genes. As seen in Table 2 all mutants were able to be recognized by mice primed with wild type D^p genes with the exception of DP20(Y27N). This suggests that *in vivo* DP20(Y27N) is poorly recognized by T cells specific for wild type D^p molecules. It is in agreement with the polyclonal responses we measured in cell mediated lympholysis assays (*in vitro*).

EFFECTS OF H-2 MUTATION ON RECOGNITION OF LCMV INFECTED TARGET CELLS

Our initial experiments had showed no alteration in the ability of LCMV specific CTL to recognize virus infected cells. All the mutants were apparently equally recognized by polyclonal CTL. To examine this further, we performed a series of cold target inhibition experiments. We used B10.P mice infected with LCMV as a source of effector cells and used L cells expressing the wild type D^p gene as labeled targets. Killing by effectors was tested for inhibition by unlabeled, LCMV infected, L cells expressing mutant D^p genes. Of the six sites examined in detail all were, on a per cell basis, as effective as wild type D^p in inhibiting LCMV specific, D^p restricted cytotoxicity (Murray et al., unpublished data). This was not due to differences in the expression of D^p genes on those cells since all express essentially equivalent levels of protein (See Fig. 1).

Table 2. Recognition of transfected L cells *in vivo* by C3H mice

Priming cell	Challenge cell	Footpad Swelling[1]
Ltk-	Ltk-	0.7 (0.6)
	B10.P	1.3 (0.2)
	Wt Dp	2.0 (0.4)
B10.P	Ltk-	1.7 (0.6)
	B10.P	11.5 (1.8)
	Wt Dp	9.7 (1.3)
	DP20	1.0 (0.4)
	DP181	8.7 (0.9)
Wt Dp	Ltk-	2.0 (0.8)
	B10.P	9.2 (0.9)
	Wt Dp	10.1 (1.2)
	DP14	6.2 (0.8)
	DP20	0.8 (0.4)
	DP174	5.7 (1.0)
	DP181	9.5 (0.9)

[1]millimeters^{-3} (SEM).

In an effort to examine whether different viral epitopes might be differentially recognized on mutant class I proteins we used two strains of LCMV, Armstrong and Pasteur and their segmental recombinants 17 and 41. Similar to the bulk response of B6 mice, we see essentially no difference between any of the viruses when tested using P/J or B10.F effector cells. CTL produced by infection of B10.P mice show a pattern similar to that produced in BALB/c mice (Ahmed et al. 1984). Perhaps the response is sufficiently polyclonal that multiple epitopes always stimulate multiple responding T cell clones so that we cannot determine the difference between Armstrong and Pasteur.

In order to investigate the possibility that there might be interactions between T cell receptor genes and MHC genes in the response to LCMV we utilized B10.P, B10.F, and P/J mice. All of these mice express what ought to be identical H-2^p MHC molecules. However, the P/J and B10.F and B10.P have different background genes including those for T cell receptors. When these mice were examined for their ability to recognize D^p mutants infected with LCMV, we noticed that P/J mice recognized DP170(N30S) in a heteroclitic fashion. That is, CTL's raised against LCMV presented by wild type D^p were much more efficient at lysing DP170(N30S) than they were at lysing wild type D^p. In contrast, T cells from B10.P mice recognized LCMV infected cells expressing wild type D^p better than DP170(N30S) (see Muller et al., this publication). Thus, depending on the particular combination of T cell receptor and MHC molecule, it is possible that mutations not only fail to disturb the interaction but may in fact cause better interactions between MHC molecules and T cell receptors or the immunogenic peptide and MHC.

DISCUSSION

The effects of the mutations in the α1 region of the D^p gene have been extremely interesting. Mutations at amino acid 27 which is located on the floor of the predicted peptide binding site causes significant alteration in allospecific recognition while essentially leaving class I restricted recognition for LCMV unchanged. This is particularly interesting in that it is difficult to explain in terms of simple alteration of peptide binding. Clearly, binding of the dominant LCMV peptides must be unaffected by this mutation while the binding of whatever peptides make up allo must be significantly altered. It is interesting to consider this since the side chain of the tyrosine residue 27 is reported not to point up into the binding site but to be a contact residue with β2-microglobulin (Bjorkman et al. 1987a). We had anticipated that perhaps this altered the association of β2-microglobulin (Allen et al. 1986). However, when we immunoprecipitated the DP20(Y27N) molecule we saw no alteration in association of β2-microglobulin and the DP20(Y27N) heavy chain (Murray et al., unpublished data).

The mutation in tyrosine 27 is worthy of particular comment since tyrosine 27 is completely conserved in every MHC molecule sequenced. Therefore, it would not have been a target for site directed mutagenesis. We would have predicted mutations in this position would result in a largely non-functional molecule. We hypothesize that the location of tyrosine 27 affects either binding of self peptides or else affects the contact residues for the T cell receptor in unbound molecules. The data measuring peptide binding to class II molecules suggests that the vast majority of MHC molecules are associated with peptide. However, the fact that exogenous peptides can be bound suggests that at least some peptides must be exchangeable and if they are exchangeable, empty class I molecules might exist at least transiently. These few empty class I molecules might be targets for allospecific T cells. The other possibility is that the spectrum of self peptides bound by D^p is altered by this mutation and it is that change in the spectrum of peptides which we detect. Low affinity interactions with self peptides might be easily subject to alteration while the higher affinity binding of LCMV specific peptides might be unaffected.

The heteroclitic response of P/J mice to DP170(N30S) is also interesting. Again, residue 30 is not predicted to be one of the contact residues for either the T cell receptor or for antigen (Bjorkman et al. 1987b) It too is located on the floor of this predicted binding site. In contrast to the deleterious change of tyrosine 27 to an asparagine, the alteration of asparagine 30 to a serine is not obviously a significant change (substituting a polar hydroxyl group for a polar amide group). It is difficult to interpret this in terms of binding directly by the T cell receptor. Thus, it must change the binding of the peptide. Perhaps, this mutation rearranges the bottom of the binding site so that one or several residues, which previously had not participated in binding, now form a site with increased affinity for LCMV specific peptides.

Mutations produced by random mutagenesis have shown surprising phenotypes. These would not have been predicted by examination of the structure and therefore would not have been chosen for site directed mutagenesis. Changes of both increasing and decreasing affinity are reported here and suggest the usefulness of this approach.

Acknowledgements

This work was supported by NIH grants AI 20288 to J.A.F., and an Individual Research Service Award F32 AI 107932 to D.M.. R.M. was a Linberger fellow.

REFERENCES

Ahmed R, Byrne JA, Oldstone MBA (1984) Virus specificity of cytotoxic T lymphocytes generated during acute lymphocytic choriomeningitis virus infection: Role of the H-2 region in determining cross-reactivity for different lymphocytic choriomeningitis virus strains. J Virol 51:34-41

Allen H, Fraser J, Flyer D, Calvin S, Flavell R (1986) β2-microglobulin is not required for cell surface expression of the murine class I histocompatibility antigen H-2D^b or of a truncated H-2D^b. Proc Natl Acad Sci USA 83:7447-7451

Bjorkman PJ, Saper MA, Samraoui B, Bennett WS, Strominger JL, Wiley DC (1987a) Structure of the human class I histocompatibility antigen, HLA-A2. Nature 329:506-512

Bjorkman PJ, Saper MA, Samraoui B, Bennett WS, Strominger JL, Wiley DC (1987b) The foreign antigen binding site and T cell recognition regions of class I histocompatibility antigens. Nature 329:512-518

Darsley MJ, Takahashi H, Macchi MJ, Frelinger JA, Ozato K, Appella E (1987) New family of exon-shuffled recombinant genes reveals extensive interdomain interactions in class I histocompatibility antigens and identifies residues involved. J Exp Med 165:211-222

Drake JW (1969) Mutagenic mechanisms. Ann Rev Gen 3:247-269

Flores C, Rajan TV (1977) H-2 antigen variants in a cultured mouse leukemia cell line. II Effect of selection against a K-end gene. Immunogenetics 5:295-308

Hutchison III CA, Nordeen SK, Vogt K, Edgell MH (1986) A complete library of point substitution mutations in the glucocorticoid response element of mouse mammary tumor virus. Proc Natl Acad Sci USA 83:710-714

Melvold RW, Khon HI, Dunn GR (1982) History and genealogy of the H-2K^b mutants from the C57BL/6Kh colony. Immunogenetics 15:177-185

Minami M, Okuda K, Sunday ME, Dorf ME (1982) H-2K-, H-2I- and H-2-D-restricted hybridoma contact sensitivity effector cells. Nature 297:231-233

Murray R, Hutchison III CA, Frelinger JA (1988a) Saturation mutagenesis of a major histocompatibility complex protein domain: Identification of a single conserved amino acid important for allorecognition. Proc Natl Acad Sci USA 85:3535-3539

Murray R, Pederson, Prosser H, Muller D, Hutchison III CA, Frelinger JA (1988b) Random oligonucleotide mutagenesis: application to a large protein coding sequence of a major histocompatibility complex class I gene, H-2D^p. Nucleic Acids Res 16:9761-9773

Pious D, Krangle MS, Dixon LL, Parham P, Strominger JL (1982) HLA structural gene mutants selected with an allospecific monoclonal antibody. Proc Natl Acad Sci USA 79:7832-7836

Shiroishi T,.Evan GA, Appella E, Ozato K (1985) In vitro mutagenesis of a mouse MHC class I gene for the examination of structure-function relationships. J Immunol 134:623-629

Altered Recognition of LCMV Antigen Presented on Mutant Class I Molecules

D. MULLER[1], M. M. HOBBS[1], R. MURRAY[1], K. PEDERSON[1], R. AHMED[2], and J. A. FRELINGER[1]

[1] Department of Microbiology and Immunology, University of North Carolina at Chapel Hill, Chapel Hill, NC 27599-7290, USA
[2] Department of Microbiology, University of California, Los Angeles, CA 90024, USA

INTRODUCTION

Mutations in the major histocompatibility complex (MHC) have been used to delineate the structure and function of these polymorphic molecules (Klein 1986). The original descriptions of these cell surface glycoproteins were concerned with their function as transplantation antigens (allorecognition). However, a more recently recognized function of the class I MHC protein is in the presentation of viral antigens to MHC compatible cytotoxic T lymphocytes (H-2 restriction). We have applied a method of generating random mutations in the class I MHC molecule H-2D^p to further determine regions of the molecule important for these immunologic functions.

METHODS

Mice: B10.P, B10.F, and B10.F(14R) strains were bred in our colony at UNC-Chapel Hill. P/J mice were obtained from The Jackson Laboratory (Bar Harbor).

Virus: Lymphocytic choriomeningitis virus (LCMV) strains Armstrong 3 (ARM, AA) and Pasteur 7B (PAST, PP) as well as recombinant strains 17 (PA) and 41 (AP) were used for the generation of CTL and infection of target cells (Riviere et al. 1986).

Target Cells: WT D^p is a transfected cell line expressing the wild-type D^p gene. L cells were co-transfected using electroporation with the mutant D^p gene and a thymidine kinase gene and maintained under selection with HAT. The following H-2D^p mutants were used DP14(A11V, E32Q), DP20(Y27N), DP164(R14L), DP170(N30S), DP174(P57Q), DP181(D37H). Details on the construction of these mutants can be found in the paper by Katoh and colleagues (this publication) and Murray et al. (1988a, 1988b).

Cytotoxicity Assay: Mice were infected i.p. with 5 X 10^5 PFU of virus and spleens were harvested 7 days later, spleen cells from two mice were pooled. Target cells were infected at a MOI of 1.0 and used 48 hours later. Target cells were labeled with ^{51}Cr; 2 X 10^4 cells were seeded per well with the indicated number of effector cells and incubated for 4.5h (Murray et al. 1988a).

I.K. Egorov/C.S. David (Eds.)
Transgenic Mice and Mutants in MHC Research

RESULTS

Cytotoxic T lymphocytes (CTL) recognize viral antigens presented on molecules encoded by the major histocompatibility complex. Mutations in class I molecules have been used to characterize areas within the molecule which are important in the presentation of antigenic peptides and interactions with accessory molecules. Prior work from our laboratory has shown that a single amino acid change (tyrosine 27 to asparagine) in a conserved region of a class I murine molecule, H-2D^p, resulted in decreased allogeneic recognition without effects on the presentation of viral antigen (Murray et al. 1988a). We have extended these observations using several additional strains of mice and LCMV.

These experiments show that a single amino acid change, asparagine 30 to serine (DP170), in the D^p class I molecule resulted in increased killing by LCMV specific D^p restricted CTL. Further, this heteroclitic response was seen in H-2D^p mouse strains P/J, B10.F, and B10.F(14R) but not B10.P. We tested the six different class I mutants using two parental (Armstrong and Pasteur) and two reassorted (17-P/A and 41-A/P) strains of LCMV. All four LCMV strains were recognized similarly by all effector populations from P/J and B10.F mouse strains. Effector cells from the B10.P strain showed more variability.

Overall, no single epitope could be assigned to the long or the short segment of the LCMV genome. The amino acid substitution in DP170(N30S) is particularly interesting since it is not directed toward the peptide binding site predicted by the HLA-A2 structure (Bjorkman et al. 1987a, 1987b). This suggests that subtle alterations in the folding of the binding site can have important effects on peptide binding and/or CTL recognition of the class I-peptide complex.

SUMMARY

Effector Cell Mouse Strain	Overall Response	Response to Homologous vs Heterologous Virus	*Heteroclitic Response
B10.P	Med-High	Variable	WT D^p>DP170
P/J	Low	No Difference	*DP170>WT D^p
B10.F	High	No Difference	*DP170>WT D^p
B10.F(14R)	Med-High	No Difference	*DP170>WT D^p

DISCUSSION

These preliminary findings raise intriguing questions regarding the nature of H-2 restricted recognition of viral antigens. The overall low response of P/J mice may point to a T cell receptor bias, although we cannot rule out a difference in the endogenous D^p gene in the P/J strain since the transfected target gene was derived from a B10.P mouse.

The heteroclitic response (increased over wild type) to the DP170(N30S) gene product may also be due to a T cell receptor bias. Alternatively, this response may be due to changes in peptide binding or processing, or altered interaction with accessory molecules. This altered response is surprising since the location of this mutation, amino acid 30, is at the bottom of the antigen binding site, with the side chain pointing away from the site and contacting the α3 domain, as predicted by the HLA-A2 model (Bjorkman et al. 1987a, 1987b). Further, the amino acid substitution, asparagine to serine, is conservative.

B10.P was the only strain to show an altered response to target cells infected with a different strain of LCMV. The response of P/J, B10.F, and B10.F(14R) strains resembled the CTL response seen for $H\text{-}2^b$ mice, in that the response did not differ for heterologous combinations of infection of effectors and targets. The B10.P response resembled the response of $H\text{-}2^d$ mice, in which Armstrong effectors were less able to kill Pasteur infected targets (Ahmed et al. 1984).

Acknowledgements

This work was supported by NIH grants AI 20288 to J.A.F., and an Individual Research Service Award F32 AI 107932 to D.M.. R.M. was a Lineberger fellow.

REFERENCES

Ahmed R, Byrne JA, Oldstone MBA (1984) Virus specificity of cytotoxic T lymphocytes generated during acute lymphocytic choriomeningitis virus infection: Role of the H-2 region in determining cross-reactivity for different lymphocytic choriomeningitis virus strains. J Virol 51:34-41

Bjorkman PJ, Saper MA, Samraoui B, Bennett WS, Strominger JL, Wiley DC (1987a) Structure of the human class I histocompatibility antigen, HLA-A2. Nature 329:506-512

Bjorkman PJ, Saper MA, Samraoui B, Bennett WS, Strominger JL, Wiley DC (1987b) The foreign antigen binding site and T cell recognition regions of class I histocompatibility antigens. Nature 329:512-518

Klein J (1986) Natural history of the major histocompatibility complex. Wiley, New York

Murray R, Hutchison III CA, Frelinger JA (1988a) Saturation mutagenesis of a major histocompatibility complex protein domain: Identification of a single conserved amino acid important for allorecognition. Proc Natl Acad Sci USA 85:3535-3539

Murray R, Pederson, Prosser H, Muller D, Hutchison III CA, Frelinger JA (1988b) Random oligonucleotide mutagenesis: application to a large protein coding sequence of a major histocompatibility complex class I gene, $H\text{-}2D^p$. Nucleic Acids Res 16:9761-9773

Riviere Y, Southern PJ, Ahmed R, Oldstone MBA (1986) Biology of cloned cytotoxic T lymphocytes specific for lymphocytic choriomeningitis virus V. Recognition is restricted to gene products encoded by the viral S RNA segment. J Immunol 136:304-307

Relationship Between Class II Structure and B Cell Signaling

G. A. BISHOP and J. A. FRELINGER[1]

[1] Department of Microbiology and Immunology, University of North Carolina, Chapel Hill, NC 27599-7290, USA

INTRODUCTION

The B cell clone CH12.LX can be induced to differentiate to Ig secretion in the presence of antigen (sheep red blood cells, SRBC) and T cells or monoclonal antibodies (Mabs) specific for the E^k molecules of CH12.LX (1). The A^k molecules of CH12.LX do not serve this function of B cell signal transmission, although CH12.LX A^k shows no mutations in protein sequence (2), and functions normally in antigen presentation and Mab binding (1,2). To determine if A molecules of other haplotypes could overcome this failure of A^k-mediated signaling, CH12.LX cells were transfected with genes encoding new A_α and/or A_β molecules. Results of these experiments suggest portions of the A_β molecule which are important to the function of B cell signaling.

MATERIALS AND METHODS

CH12.LX cells were transfected with plasmids containing the genes encoding $A_\alpha{}^b$ (3), $A_\beta{}^b$ (4), $A_\beta{}^d$ (5) or $A_\beta{}^u$ (6). All plamids either contained the $\underline{neo}^r$ gene or were cotransfected with the plasmid pSV2neo (7), to allow selection of transfectants by growth in the antibiotic G418. Transfection was by electroporation. Expression of transfected A molecules in G418-resistant cells was detected by immunofluorescence with Mabs.

Transfectants of CH12.LX are designated with a suffix listing the transfected Ia molecule and its haplotype, followed by the number of the cell line. For example, cells expressing both chains of the A^b molecule are called CH12.AB. If only a single Ia chain is expressed, the chain is designated as A (α) or B (ß), before the haplotype designation. Hence, cells expressing $A_\alpha{}^k$ paired with $A_\beta{}^b$, $A_\beta{}^d$, and $A_\beta{}^u$ are designated CH12.ABB, CH12.ABD, and CH12.ABU, respectively.

Differentiation to Ig secretion was measured in CH12.LX and transfected cells by direct plaque assay on a lawn of SRBC, as described previously (1,2).

RESULTS AND DISCUSSION

Stably transfected sublines of CH12.LX were established, expressing the products of $A_\beta{}^b$, $A_\alpha{}^b$, $A_\alpha{}^b$ + $A_\beta{}^b$, $A_\beta{}^d$, and $A_\beta{}^u$ genes. When DNA encoding only a single Ia chain was transfected, the newly expressed molecule paired with the endogenous $A_\alpha{}^k$ or $A_\beta{}^k$ chains of CH12.LX (assessed by 2-D gel electrophoresis; data not shown). For each cell type, several independently derived cell lines were tested for antigen-specific, MHC restricted induction of IgM secretion. Transfection of the genes encoding $A_\beta{}^b$, $A_\beta{}^d$, or $A_\beta{}^u$ resulted in cell lines expressing $A_\alpha{}^k A_\beta{}^b$ (CH12.ABB), $A_\alpha{}^k A_\beta{}^d$ (CH12.ABD), or $A_\alpha{}^k A_\beta{}^u$ (CH12.ABU) hybrid molecules at the cell surface. Mabs specific for $A_\alpha{}^k$ or $A_\beta{}^b$ induced antigen-specific IgM secretion of CH12.ABB cells (8; Fig.1), while Mabs specific for $A_\beta{}^k$ were ineffective. Hence, when the endogenous $A_\alpha{}^k$ molecules of CH12.LX were allowed to pair with $A_\beta{}^b$ rather than $A_\beta{}^k$, the resultant hybrid molecule overcame the A-specific signaling defect seen previously in CH12.LX (1,2). That $A_\beta{}^k$ is responsible for the failure of the endogenous A molecules of CH12.LX to transmit a differentiative signal was further confirmed in our recent study showing that CH12.LX cells transfected with the gene encoding $A_\alpha{}^b$ (CH12.AAB) do not receive signals through the hybrid Ia molecule $A_\alpha{}^b A_\beta{}^k$ (8). However, if both the $A_\alpha{}^b$ and $A_\beta{}^b$ genes are transfected together, differentiative signals are successfully transmitted through the resulting $A_\alpha{}^b A_b{}^b$ molecule (8). CH12.ABD and CH12.ABU cell lines were also tested for induced differentiation, to determine if A molecules of other haplotypes could transmit signals to CH12.LX. Results using a typical

I.K. Egorov/C.S. David (Eds.)
Transgenic Mice and Mutants in MHC Research

CH12.ABD cell line are shown in Fig. 1. In contrast to CH12.ABB cells, cells expressing A_β^d (Fig. 1) or A_β^u (data not shown) paired with A_α^k did not transmit differentiative signals through the hybrid A molecules. Data from all the cell lines tested are summarized in Table 1.

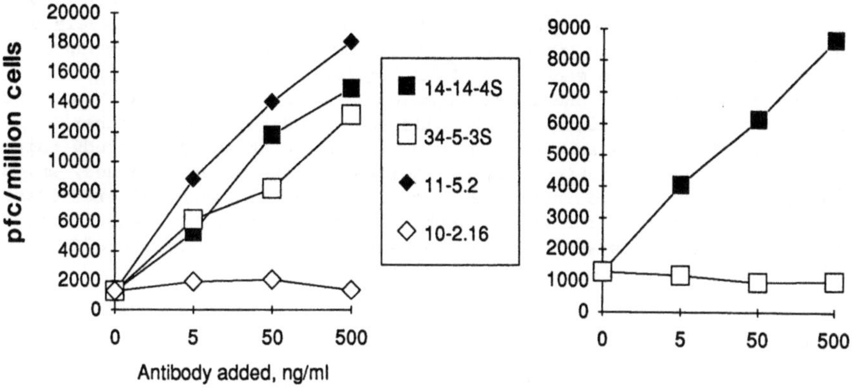

Figure 1. Differentiation induced in transfected cells by Mabs + antigen. Cells were cultured with SRBC and Mabs specific for E^k (14-4-4S), A_β^b/A_β^d (34-5-3S), A_α^k (11-5.2) or A_β^k (10-2.16). Pfc were determined at 48 hours of culture. Left panel, CH12.ABB1 cells. Right panel, CH12.ABD4 cells.

Table 1. Class II-Specific Signal Transmission in Transfectants of CH12.LX

Cell Type	n[1]	Ia-Mediated Differentiation[2] Positive[3]	Negative[4]
CH12.LX	4	E^k	A^k
CH12.ABB	2	E^k, $A_\alpha^k A_\beta^b$	A^k
CH12.AAB	2	E^k	A^k, $A_\alpha^b A_\beta^k$
CH12.AB	2	E^k, $A_\alpha^k A_\beta^b$, A^b	A^k, $A_\alpha^b A_\beta^k$
CH12.ABD	10	E^k	A^k, $A_\alpha^k A_\beta^d$
CH12.ABU	5	E^k	A^k, $A_\alpha^k A_\beta^u$

[1]n=number of experiments (CH12.LX) or number of cell lines tested (transfectants). [2]Pfc response measured following culture with SRBC and Ia-specific Mabs. [3]Positive=pfc induction using Mabs specific for either the α or β chains of these Ia molecules. [4]Negative=no pfc induction using Mabs specific for either the α or β chains of these molecules.

We compared the protein sequences of A_β molecules of the k, d, u, and b haplotypes to identify any residues for which A_β^b differed in sequence from the other three haplotypes (6, 9; numbered as in 10). Sequence at such residues might affect the ability of a particular A_β molecule to function in B cell signaling. A_β^b differs from the other haplotypes at four positions (residues 12, 47, 89, and 197). At positions 47 and 197, haplotypes k, d, and u share the same amino acids (tyr^{47} and arg^{197}), while b has a nonconservative substitution (his^{47} and trp^{197}). Thus, at position 47, an easily phosphorylated tyr residue, with an aromatic side chain, is substituted by a his residue, whic shares neither of these features. At position 197, the highly hydrophillic arg residue is substituted by a highly hydrophobic trp residue. Appropriate site-directed mutants are currently being prepared to test the effect of changes at these positions on B cell signaling by Ia.

In addition to the four positions shown in Fig. 2, other structural features of Ia molecules may influence their function as B cell signal transducers.

Particular α and β chain combinations may function differently, an effect which may operate separately from combinatorial effects on relative levels of surface expression. We are currently designing experiments to test this possibility. It should be noted that while we observed variable levels of surface expression of hybrid class II molecules between individual cell lines, these levels did not correlate with the ability of a particular transfected class II molecule to transmit a differentiative signal to the B cell (8). Indeed, although CH12.ABB cell lines expressed hybrid $A_\alpha{}^k A_\beta{}^b$ molecules at only 10-30% of the levels of the endogenous $A_\alpha{}^k A_\beta{}^k$ molecules, these $A_\alpha{}^k A_\beta{}^b$ molecules transmitted differentiative signals as well as did the endogenous $E_\alpha{}^k E_\beta{}^k$ molecules (Fig. 1).

Findings implicating the class II molecule as a B cell signal transmitter in normal mouse (11-13) and human (14) B cells have been reported by other laboratories. CH12.LX is an excellent clonal model with which to study this function of class II. Mutations can be engineered into class II genes, which can subsequently be transfected into CH12.LX cells. The resulting altered class II molecules can be examined for their function in transmitting differentiative signals to CH12.LX. This approach can be used to perform a structure/function analysis of the Ia molecule as a B cell signal receptor.

REFERENCES

1. Bishop, G.A. and Haughton, G. (1986) Proc. Natl. Acad. Sci. (USA) 83:7410.
2. Bishop, G.A., McMillan, M.S., Haughton, G. and Frelinger, J.A. (1988) Immunogenetics 28:184.
3. Landais, D., Matthes, H., Benoist, C. and Mathis, D. (1985) Proc. Natl. Acad. Sci. (USA) 82:2930.
4. Widera, G. and Flavell, R.A. (1985) Proc. Natl. Acad. Sci. (USA) 82:5500.
5. Malissen, B.N., Shastri, N. and Hood, L. (1986) Proc. Natl. Acad. Sci. (USA) 83:3958.
6. Estess, P., Begovich, A.B., Koo, M., Jones, P.P. and McDevitt, H.O. (1986) Proc. Natl. Acad. Sci. (USA) 83:3594.
7. Southern, P. and Berg, P. (1982) J. Mol. Appl. Gen. 1:327.
8. Bishop, G.A. and Frelinger, J.A. (1989) Proc. Natl. Acad. Sci. (USA), in press.
9. Choi, E., McIntyre, K., Germain, R. and Seidman, J.G. (1983) Science 221:283.
10. Klein, J. (1986) Natural History of the Major Histocompatibility Complex, John Wiley and Sons, N.Y., p.362.
11. Niederhuber, J.E., Frelinger, J.A., Dugan, E., Coutinho, A., and Shreffler, D.C. (1975) J. Immunol. 115:1672.
12. Cambier, J.C., Newell, M.K., Justement, L.B., McGuire, J.C., Leach, K.L. and Chen, Z.Z. (1987) Nature 327:629.
13. Baluyut, A.R. and Subbarao, B. (1988) J. Mol. Cell. Immunol. 4:45.
14. Palacios, R., Maza-Martinez, O. and Guy, K. (1983) Proc. Natl. Acad. Sci. (USA) 80:3456.

Multivalent Requirement for the Stimulation of Alloreactive T Cells: Studies with Engineered Soluble MHC Class I Proteins In Vitro and In Vivo

D. H. MARGULIES[1], L. F. BOYD[1], S. KOZLOWSKI[1], L. KJER-NIELSEN[2], R. LOPEZ[1,3], J. MCCLUSKEY[2], J. SCHNECK[1], and R. HUNZIKER[1,4]

[1] Molecular Biology Section, Laboratory of Immunology, National Institute of Allergy and Infectious Diseases, National Institutes of Health, Bethesda, MD 20892, USA
[2] Department of Immunology and Pathology, Monash Medical School, Melbourne, Australia
[3] Howard Hughes Medical Institute/NIH Research Scholar.
[4] A Fellow of the Leukemia Society of America.

INTRODUCTION

The murine class I major histocompatibility antigens play critical roles in the interaction of class I-restricted T cell receptors with their stimulatory or target antigen presenting cells. One approach taken by our laboratories has been to investigate the biological activity of genetically engineered, soluble, homogenous, purified class I molecules in the stimulation and inhibition of allospecific T cell hybridomas (Margulies et al, 1986; McCluskey et al, 1988; Schneck et al, 1989a). More recently we have established a system for comparing the role of the cell surface multivalent display of the MHC molecule with these monovalent soluble analogues expressed *in vivo* in transgenic mice. In this brief review, we hope to recount the logic by which the initial soluble class I MHC molecules were generated, summarize previously published as well as unpublished data on their expression and function, and present some preliminary evidence suggesting that transgenic mice expressing soluble analogues of the H-2D^d protein in a C57Bl/6 background will be useful for studies of the molecular and biochemical basis of the generation of immunological tolerance to class I molecules.

EXON SHUFFLING WITH Q10^b

The production of soluble analogues of membrane bound proteins by molecular biological techniques derives from the knowledge that hydrophobic transmembrane regions and "stop transfer" signals of type I membrane proteins are necessary for their stable membrane expression (for review see Margulies et al 1987). For a number of membrane molecules, introduction of a premature termination codon proximal to the hydrophobic amino acids that encode the transmembrane region has been sufficient for the secretory expression of the engineered protein (Sayre et al 1989; Berger et al , 1988). Initial studies by Zuniga and Hood (1986), studying the H-2L^d murine transplantation antigen, suggested that such premature termination mutants of a genomic H-2L^d clone, indeed were secreted from transfectant L cells, but not at a rate sufficient to be useful for protein purification.

The Q10^b gene has evolved from a 13 nucleotide deletion of a gene homologous to the classical class I genes, and directs the synthesis of a protein with a frameshift mutation and premature termination such that the Q10^b protein is effectively secreted. This molecule, whose natural function is unknown, is synthesized in the liver and accumulates

I.K. Egorov/C.S. David (Eds.)
Transgenic Mice and Mutants in MHC Research

at concentrations of 50 to 200 µg/ml in the serum of those mouse strains that possess the active gene (Lew et al 1986). Based on the accumulated experience that *in vitro* recombinant MHC class I molecules could be readily generated by exon shuffling techniques (Evans et al 1982; Margulies et al 1983), and that the α1 and α2 domains of MHC class I molecules are critical for interaction with allospecific and antigen specific MHC-restricted T cell receptors (Margulies and McCluskey, 1985), a general scheme for the production of genes encoding soluble analogues of class I MHC molecules was devised (see Fig. 1). This strategy entails production of chimeric genes encoding the leader (L), α1, and α2 domains of the polymorphic MHC class I gene of interest, and the α3 and carboxy-terminal regions of Q10^{b}. This method has proven effective for the production of soluble H-2D^{d}/Q10^{b}(Margulies et al 1986), H-2K^{b}/Q10^{b} (McCluskey et al 1988); HLA-A2/Q10^{b} (Schneck et al 1989a), H-2L^{d}/Q10^{b} (Lopez, McCluskey, and Margulies, unpublished), HLA-Cw3/Q10^{b} (Kahn-Perles et al 1989), HLA-B7/Q10^{b} (Kahn-Perles et al 1989), as well as for a number of the H-2K^{b} mutants (Schneck and Margulies, unpublished). In all the examples studied in our laboratory, with the exception of the H-2L^{d}/Q10^{b} molecule, transfection of the *in vitro* recombinant genes directed the secretion of the soluble molecules at rates of about 0.25 µg/10^{6} cells/24 hr. In the case of the H-2L^{d}/Q10^{b} molecule, the rate of secretion was significantly lower, and prompted us to carry out the kinetic experiments described below.

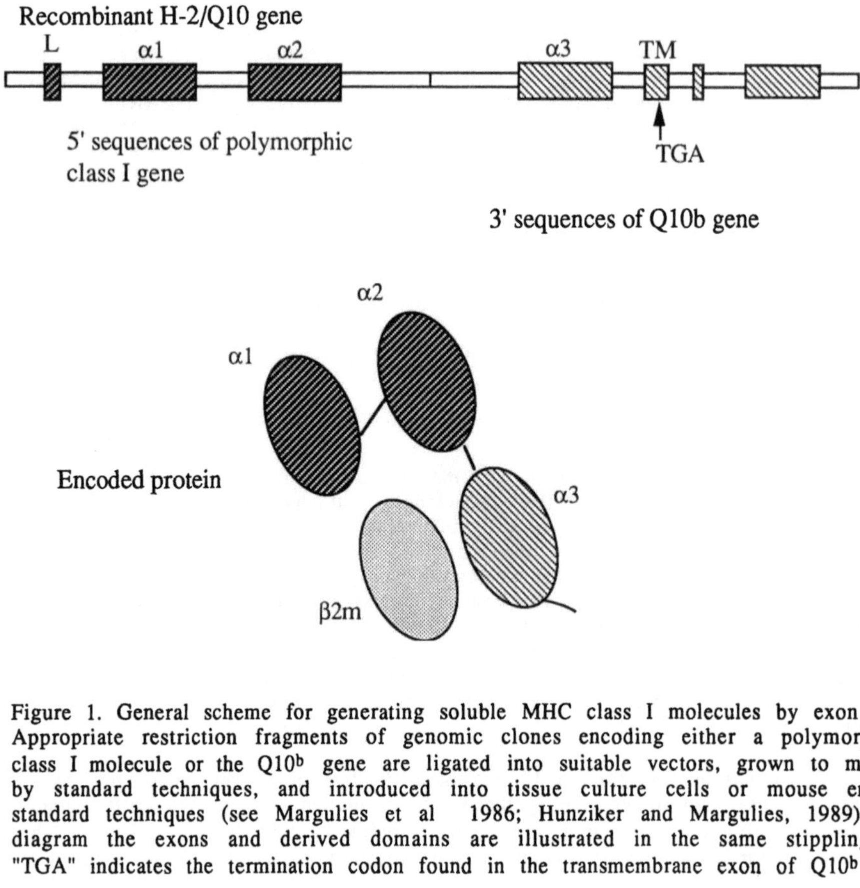

Figure 1. General scheme for generating soluble MHC class I molecules by exon shuffling. Appropriate restriction fragments of genomic clones encoding either a polymorphic MHC class I molecule or the Q10^{b} gene are ligated into suitable vectors, grown to mass culture by standard techniques, and introduced into tissue culture cells or mouse embryos by standard techniques (see Margulies et al 1986; Hunziker and Margulies, 1989). In this diagram the exons and derived domains are illustrated in the same stippling pattern. "TGA" indicates the termination codon found in the transmembrane exon of Q10^{b}

KINETICS OF SECRETION OF THE H-2L^d/Q10^b MOLECULE

Quantitation of the protein product secreted both by bulk populations of L cells transfected with the H-2L^d/Q10^b gene and by a panel of transfectant clones revealed that the rate of accumulation of the H-2L^d/Q10^b soluble product in the cell culture supernatant was approximately one fifth to one tenth that of either the H-2D^d/Q10^b or H-2K^b/Q10^b gene products. Pulse chase experiments of the intracellular soluble molecules indicated that the initial rate of synthesis of the H-2L^d/Q10^b molecules was the same as that of the other soluble analogues, but that only one fifth to one tenth of this material effectively chased into the cell supernatant (Fig. 2.; Lopez, McCluskey, and Margulies, unpublished observations). Thus, structural differences between H-2L^d and H-2D^d or H-2K^b in the $\alpha 1$ and $\alpha 2$ domains leads to selective degradation of the H-2L^d/Q10^b molecules at some critical site in intracellular transport. This result may explain the difficulty experienced by others in generating soluble analogues of the H-2L^d molecule that would be useful for protein purification. (Zuniga and Hood 1986).

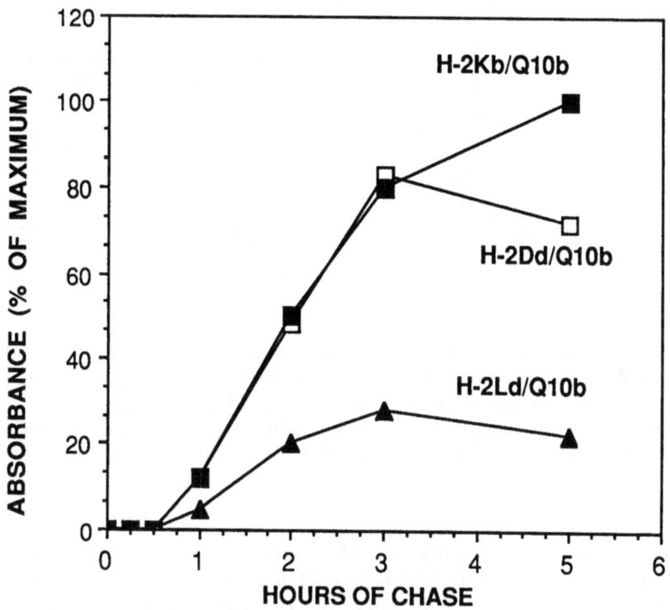

Figure 2. Kinetics of secretion of H-2L^d/Q10^b, H-2D^d/Q10^b, and H-2K^b/Q10^b. Clonal L cell transfectant cell lines CD2.10.1 (H-2L^d/Q10^b), DMT44.7.4(H-2D^d/Q10^b), and JT21.4.14 (H-2K^b/Q10^b) were pulse labelled with ^{35}S-methionine and chased with medium containing an excess of cold methionine. Aliquots of cell free supernatant were processed for immunoprecipitation and gel electrophoresis. Following fluorography and autoradiography, the H-2/Q10 heavy chain band was quantitated by densitometry. Details of these methods are described in Margulies et al, 1986 and McCluskey et al, 1988

EXPLORING THE FUNCTION OF THE MHC CLASS I MOLECULE WITH SOLUBLE ANALOGUES

The modest accumulation of the soluble H-2D^d/Q10^b molecule and of the H-2K^b/Q10^b molecule in transfectant cell supernatants was sufficient to allow purification, by immunoadsorbent chromotography, of milligram quantities of these two proteins. Material purified from serum-free tissue cultures consists of equimolar amounts of the respective recombinant H-2/Q10 heavy chain non-covalently assembled with the light chain, β2-microglobulin (Margulies et al, 1986; McCluskey et al, 1988). Gel filtration high pressure liquid chromotography (HPLC) in non-denaturing solvent (phosphate buffered saline, PBS) demonstrates that immunoadsorbent purified soluble class I protein is homogeneous and elutes from the column in a position consistent with the postulated monovalent, heterodimeric molecular weight of about 55 kDa (see Fig. 3).

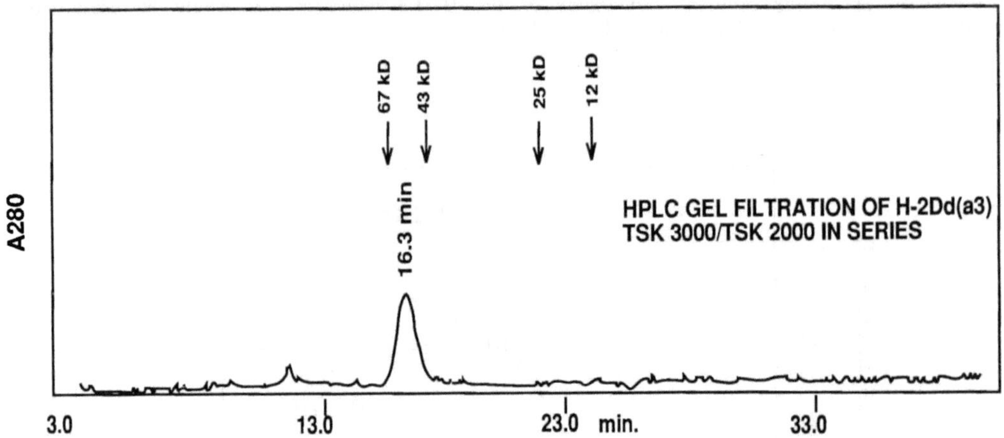

Figure 3. HPLC elution profile of purified H-2D^d(α3) protein analyzed on TSK 3000 column (4.6 x 250 mm) coupled in series with a TSK 2000 column (4.6 x 250 mm). The column was developed in phosphate buffered saline at a flow rate of 1.0 ml/ min. The positions of elution of molecular weight markers bovine serum albumin (67 kD), ovalbumin (43 kD), chymotrpsinogen (25 kD), and cytochrome C (12 kD) are indicated

Early studies investigated the ability of the purified H-2D^d/Q10^b to stimulate an alloreactive T cell hybridoma, 3DT52.5 (McCluskey et al, 1988). From these experiments it was clear that the soluble H-2D^d analogue, in aqueous solution, was incapable of stimulating the release of IL-2 from the T cell hybridoma. Nevertheless, when H-2D^d/Q10^b was made multivalent, either as the soluble, dextran conjugate, or as the insoluble, agarose-coupled or polystyrene-adsorbed form, these same protein preparations stimulated IL-2 secretion from the hybridoma. The degree of stimulation was routinely 1 to 10% of the stimulation by cell surface membrane H-2D^d-expressing cells, which raised several questions. First, do the amino acid differences between H-2D^d and Q10^b in the α3 domain account for this quantitative difference in the stimulation by the soluble molecule? H-2D^d and Q10^b differ at 9 positions in the α3 domain. Two positions in the α3 domain have been implicated as critically related to CD8 dependent interactions of the class I molecule. Potter and his colleagues (Potter et al, 1989) have suggested that a substitution of lysine for glutamic acid at position 227 of H-2D^d eliminates normal CD8 dependent interaction with T cells, and Salter et al (1989) have demonstrated that substitution of alanine for valine at position 245 in the α3 domain directly affects CD8 binding. The amino acid differences between H-2D^d and Q10^b do not include differences at position 245 and are conservative (glutamic acid in H-2D^d, aspartic acid in Q10^b) at position 227. To evaluate whether the α3 domain of H-2D^d would be more effective in the in vitro stimulation of the T cell hybridoma by a purified, engineered, class I molecule, we constructed H-2D^d(α3), a soluble analogue containing the α1, α2, and α3 domains of H-2D^d and the carboxyterminal 27 amino acids of Q10^b (Kozlowski, Kyer-Nielsen, McCluskey, and Margulies, unpublished observations). Preliminary studies suggest that this soluble analogue is at least as effective as the H-2D^d/Q10^b molecule when adsorbed to plastic plates for stimulating the T cell hybridoma, and may be significantly more resistant to the denaturation that occurs under these conditions (Kozlowski and Margulies, unpublished.).

INHIBITION OF ALLOSTIMULATION BY SOLUBLE H-2K^b/Q10^b

A full understanding of the nature of the interaction of MHC class I molecules with the T cell receptor, and insight into the nature of allostimulation can only be gained as we develop a quantitative sense of these interactions. The initial observation that the H-2K^b-reactive T cell hybridoma HTB 157.7 was stimulated to an intermediate level by H-2K^{bm10}-transfected mouse L cells suggested that soluble H-2K^b/Q10^b might be inhibitory (Schneck et al, 1989a). This indeed was the case, with half-maximal inhibition occuring at a concentration of 0.1μM. Since half-maximal inhibition by a monovalent ligand reflects the binding constant, K_d, of that ligand for its receptor, this allowed us to estimate that the affinity of the soluble H-2K^b/Q10^b molecule for the H-2K^b-reactive T cell receptor of this hybridoma, HTB157.7, was about 10^{-7} M. Consideration of the possibility that the purified H-2K^b/Q10^b molecule might still be complexed to self generated antigenic peptides or that only a small fraction of the H-2K^b/Q10^b molecules might be biologically active leads to the conclusion that this is a maximal estimate for the K_d (i.e. a *minimal* estimate for the strength of interaction), and that a true affinity might be as strong as 10^{-9} M or greater (Schneck et al, 1989a). Evaluation of various models for alloreactivity described in detail by several authors (Matzinger and Bevan, 1977; Bevan, 1984; Claverie and Kourilsky, 1986), would suggest that the model proposed by Matzinger and Bevan (1977), in which only a small proportion of MHC molecules are bound to the appropriate self peptide to allow recognition by the allospecific T cell receptor, is the most compatible with such a high affinity interaction.

INHIBITION OF ALLOSTIMULATION BY PEPTIDE K^{b}163-174

The observation that transfected L cells expressing the H-2K^{bm10} gene were less efficient in stimulating the H-2K^b-reactive T cell hybridoma than those expressing the wild type H-2K^b gene led to the hypothesis that the region of the H-2K^b molecule that was mutated in this mutant played a significant role in the interaction of H-2K^b with the hybridoma's T cell receptor. Thus, the mutation in H-2K^{bm10} was defined by nucleotide sequence determination to be a complex mutation in which five of the twelve amino acid residues from amino acids 163 to 174 had been substituted (see Fig. 4). We reasoned that if mutation of this region, a part of the molecule presumably exposed to solvent as suggested by the three-dimensional structure proposed for the HLA-A2 molecule (Bjorkman et al , 1987), significantly affected T cell interaction, then a synthetic peptide representing the wild type sequence in this region would be effective in inhibiting the H-2K^{bm10}-dependent stimulation of the hybridoma. Peptide K^{b}163-174 was an effective inhibitor of the hybridoma, with half-maximal inhibiton occurring at a concentration of 10^{-4} M. (Schneck et al 1989a).

```
Position no.      163                    174

       H-2Kb      T C V E W L R R Y L K N

       H-2Kbm10   A - M - S - - - - - E L
```

Figure 4. Sequence comparison of H-2K^b and H-2K^{bm10} in the region of the bm10 mutation. Sequence determination is described by Schneck et al(1989a)

These results prompted us to examine the activity of peptide K^b 163-174 in the inhibition of anti-H-2K^b alloreactive CTL lines, clones and bulk cultures. Surprisingly, peptide K^b 163-174, but not appropriate control peptides, effectively inhibited bulk, line, and clonal H-2K^b-mutant anti-H-2K^b CTLs (Schneck et al 1989b; Munitz et al 1989).

THE ROLE OF MONOVALENT MHC CLASS I MOLECULES IN THE GENERATION OF SELF TOLERANCE

The general conclusion from our studies in several allogeneic systems was that anti-H-2D^d and anti-H-2K^b T cell receptors could interact with soluble analogues of the respective target MHC class I molecules. Stimulation of the anti-H-2D^d hybridoma could only be achieved by multivalent preparations of the soluble H-2D^d analogues while inhibition of the anti-H-2K^b hybridoma resulted from incubation wtih the monovalent soluble H-2K^b. (Presumably because of more stringent requirements for stimulation, the H-2K^b-reactive hybridoma, unlike the H-2D^d-reactive one, cannot be stimulated by multivalent H-2K^b/Q10^b alone.)

To investigate further the interaction of potentially alloreactive T cells with soluble, monovalent MHC class I molecules, we asked whether expression of soluble, polymorphic MHC class I molecules, in adequate concentration, expressed at an appropriate time in ontogeny, would serve as the tolerogen or educating element that the same molecule expressed normally on cell surfaces would provide. Transgenic mouse technology provides the ideal system for a test of such a hypothesis. In particular, the description by Yoshioka et al (1987) of a C57Bl/6 mouse transgenic for the wild type cell surface expressed H-2D^d provided us with the necessary control strain for the direct comparison with mice expressing soluble analogues of H-2D^d. Thus, we have produced, by microinjection of homozygous C57Bl/6 fertilized eggs, two transgenic mouse strains (Hunziker and Margulies, 1989). One of these expresses the H-2D^d/Q10^b molecule described above which contains the $\alpha1$ and $\alpha2$ domains of H-2D^d and the $\alpha3$ and carboxyterminal 27 amino acid residues of Q10^b. The other expresses the H-2D^d($\alpha3$) molecule which contains the $\alpha1$, $\alpha2$, and $\alpha3$ domains of H-2D^d, and only the carboxyterminal 27 amino acids of Q10^b. Assay of the serum of founder animals and their heterozygous transgenic progeny by Western blotting, ELISA, or by immunoprecipitation indicates that the soluble analogues accumulate at levels of about 100 μg/ml. Since the gene constructs contain the same amount of 5' flanking region as used in the membrane H-2D^d transgenics (strain D8), we expect that the tissue specific expression as well as the expression during development will be approximately the same as that of D8 strain. Thus, direct assessment of the ability of the soluble H-2D^d transgenic lines to generate anti-H-2D^d specific CTL, and anti-H-2D^d antibody responses should allow a test of the hypothesis. Preliminary results suggest that these animals, transgenic for the soluble H-2D^d encoding genes, are capable of generating CTL against the membrane H-2D^d molecule, that is, they are not tolerant.

Further quantitation should permit us to draw more definitive conclusions about the mechanisms whereby soluble as compared with membrane bound polymorphic membrane associated class I molecules do or do not confer a tolerogenic signal to maturing T cells in the thymus.

Similar experiments reported here by Hämmerling and his colleagues (Arnold et al 1988) demonstrated the expression of a soluble analogue of H-2K^k in (C57Bl/6 x DBA/2)F2 transgenic mice. Their animals express the transgene product at about 300 ng/ml (approximately 1/300 the level of our transgenics), and their functional immunological analysis was complicated by multiple genetic background factors. Nevertheless, they too observed a lack of tolerance to the membrane form of H-2K^k in their transgenic animals. Further studies in both laboratories should clarify these issues.

CONCLUSION

In summary, then, studies of the function of MHC class I molecules, using soluble analogues, have led us to 1) an appreciation of the requirement for multivalent interaction in the stimulation of an H-2D^d-reactive T cell hybridoma; 2) the demonstration of inhibition of stimulation of an H-2K^b-reactive T cell hybridoma leading to an estimate of the binding constant, K_d, of that hybridoma's T cell receptor for the H-2K^b/Q10^b molecule of about 10^{-7} M; 3) the identification of peptide K^b 163-174 as inhibitory in the stimulation not only of the alloreactive hybridoma but also of CTL lines, clones, and bulk cultures; and 4) the generation of several strains of mice transgenic for the soluble analogues of H-2D^d on a homozygous C57Bl/6 background the immunological behavior of which will provide new insights into the molecular and cellular mechanisms by which immunological tolerance is achieved.

REFERENCES

Arnold B, Dill O, Küblbeck G, Jatsch L, Simon MM, Tucker J , and Hämmerling GJ (1988) Alloreactive immune responses of transgenic mice expressing a foreign transplantation antigen in a soluble form. Proc Nat Acad Sci USA 85: 2269-2273.

Berger EA, Fuerst, and Moss B (1988) A soluble recombinant polypeptide comprising the amino-terminal half of the extracellular region of the CD4 molecule contains an active binding site for human immunodeficiency virus. Proc Natl Acad Sci U S A 85: 2357-2361.

Bevan MJ (1984) High determinant density may explain the phenomenon of alloreactivity. Immunol Today 5: 128-130.

Bjorkman PJ, Saper MA, Samroui B, Bennett WS, Strominger JL, and Wiley DC (1987) Structure of the human class I histocompatibility antigen, HLA-A2. Nature 329: 506-512.

Claverie J-M and Kourilsky P (1986) The peptide self model: a reassessment of the role of the major histocompatibility complex molecule in the restriction of the T-cell response. Ann Inst Pasteur Immunol 137D: 425-442.

Evans GA, Margulies DH, Shykind B, Seidman JG, and Ozato K (1982) Exon shuffling: mapping polymorphic determinants on hybrid mouse transplantation antigens. Nature 300: 755-757.

Hunziker R and Margulies DH (1989) Mice transgenic for a gene encoding a soluble, polymorphic major histocompatibility complex antigen. In: First N (ed) Transgenic animals. Butterworths, Stoneham, Mass (in press).

Kahn-Perles B, Barra C, Jehan J, Fourel D, Barad M, Maryanski JL, and Lemonnier FA (1989) Cytotoxic T lymphocyte recognition of secreted HLA class I molecules. J Immunol 142: 3021-3025.

Lew AM, Maloy WL, and Coligan JE (1986) Characteristics of the expression of the murine soluble class I molecule (Q10). J Immunol 136: 254-258.

McCluskey J, Boyd LF, Highet PF, Inman J, and Margulies DH (1988) T cell activation by purified, soluble, class I MHC molecules, requirement for polyvalency. J Immunol 141: 1451-1455.

Margulies DH, Evans GA, Ozato K, Camerini-Otero RD, Tanaka K, Appella E, and Seidman JG (1983) Expression of H-$2D^d$ and H-$2L^d$ mouse major histocompatibility antigen genes in L cells after DNA-mediated gene transfer. J Immunol 130: 463-470.

Margulies DH, Lopez R, Boyd LF, and McCluskey J (1987) Engineering soluble major histocompatibility molecules: why and how. Immunol Res 6: 101-116.

Margulies DH and McCluskey J (1985) Exon shuffling: new genes from old. Surv Immunol Res 4: 146-159.

Margulies DH, Ramsey AL, Boyd LF and McCluskey J (1986) Genetic engineering of an H-$2D^d$/$Q10^b$ chimeric histocompatibility antigen: purification of soluble protein from transformant cell supernatants. Proc. Nat. Acad. Sci. U.S.A. 83: 5252-5256.

Matzinger P and Bevan MJ (1977) Why do so many lymphocytes respond to major histocompatibility antigens? Cell Immunol 29: 1-5.

Munitz TI, Schneck J, Coligan JE, Maloy WL, Henrich JP, Sharrow SO, Margulies DH, and Singer A (1989) A peptide derived from the alpha-helical region of class I MHC blocks CTL engagement of the class I MHC molecule. Cold Spring Harbor Symp Quant Biol 54: in press.

Potter T, Rajan TV, Dick RF, and Bluestone JA (1989) Substitution at residue 227 of H-2 class I molecules abrogates recognition by CD8-dependent, but not CD8-independent, cytotoxic T lymphocytes. Nature 337: 73-75.

Salter RD, Norment AM, Chen BP, Clayberger C, Krensky AM, Littman DR, and Parham P (1989) Polymorphism in the $\alpha 3$ domain of HLA-A molecules affects binding to CD8. Nature 338: 345-347.

Sayre PH, Hussey RE, Chang HC, Ciardelli TL, Reinherz EL (1989) Structural and binding analysis of a two domain extracellular CD2 molecule. J Exp Med 169: 995-1009.

Schneck J, Maloy WL, Coligan JE, and Margulies DH (1989a) Inhibition of an allospecific T cell hybridoma by soluble class I proteins and peptides: estimation of the affinity of a T cell receptor for MHC. Cell 56: 47-55.

Schneck J, Munitz T, Coligan JE, Maloy WL, Margulies DH, and Singer A (1989b) Inhibition of allorecognition by an H-$2K^b$-derived peptide is evidence for a T cell binding region on an MHC molecule. Proc Nat Acad Sci USA: in press.

Yoshioka T, Bieberich C, Scangos G, and Jay G (1987) A transgenic class I antigen is recognized as self and functions as a restriction element. J Immunol 139: 3861-3867.

Zuniga M and Hood L (1986) Clonal variation in cell surface display of an H-2 protein lacking a cytoplasmic tail. J Cell Biol 102: 1-10.

Novel MHC Variants Spliced by Overlap Extension

H. D. Hunt[1], J. K. Pullen[1], Z. Cai[1], R. M. Horton[2], S. N. Ho[1], and L. R. Pease[1,2]

[1] Department of Immunology and [2] Department of Biochemistry and Molecular Biology, Mayo Clinic, Rochester, MN 55905, USA

INTRODUCTION

We have developed a new technique, based on the polymerase chain reaction (Ho 1989), called "splicing by overlap extension" (SOE), which simplifies site-directed mutagenesis and allows the introduction of larger systematic changes in gene structure in a manner independent of the distribution of restriction endonuclease sites in the targeted area of the gene (Ho 1989). An added feature of SOE is that it provides a straight-forward strategy for the generation of chimeric genes from unrelated sources, such that our ability to make genetic constructs is limited by knowledge of how protein domains will interact and not by the structure of the genes that encode them (Horton 1989). We have used the SOE strategy to generate novel MHC class I variants designed to investigate the antigen presenting properties of the encoded glycoproteins. Here we present examples of how SOE has been applied in the generation of these constructs and the characterization of their gene products using mAbs.

SITE-DIRECTED MUTAGENESIS

The SOE strategy for introducing changes into a gene is shown in Fig. 1 and has been presented in detail previously (Ho 1989). The novel concept in this approach is that two DNA fragments can be combined into one fragment by primer extension, provided they contain a region of terminal complementarity. As shown in Fig. 1, two overlapping fragments (product AB and CD) of a gene are amplified by the polymerase chain reaction (PCR). The two fragments are isolated and combined in a subsequent reaction and are spliced together by primer extension. Inclusion of outside primers (primers "A" and "D" in the boxed area of Fig. 1) in the SOE reaction allows the spliced product to be amplified by PCR. The region of complimentarity which mediates the splicing of the two fragments is determined by the internal oligonucleotide primers ("C" and "B") used to generate the two primary PCR fragments. Incorporation of the desired nucleotide mismatches into these primers results in the planned site-directed substitutions. The position of the primers can be moved 5' or 3' to introduce substitutions anywhere within the targeted gene segment. Furthermore, the primers can be constructed such that they overlap only at their termini, lengthening the region where substitutions can be introduced in a single reaction. To date the largest mutation we have introduced is the replacement of six nucleotides

I.K. Egorov/C.S. David (Eds.)
Transgenic Mice and Mutants in MHC Research

distributed throughout a 36 nucleotide stretch (variant BM-8-wt, below). In principle, a substantially larger stretch could be targeted.

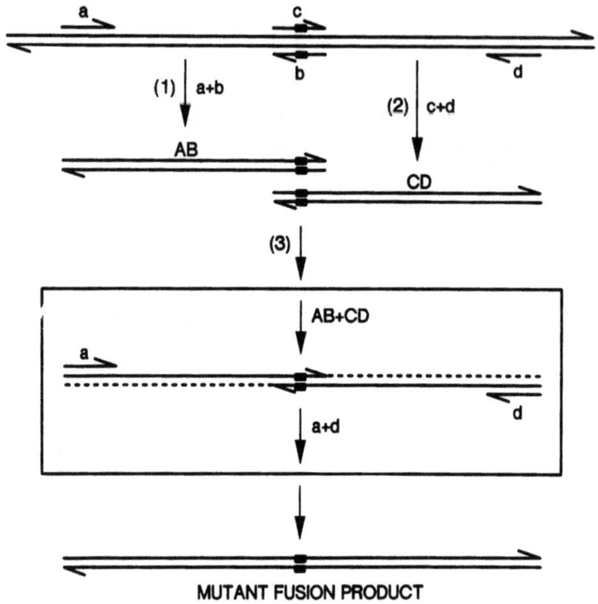

Fig. 1. General strategy for site directed mutagenesis using PCR mediated splicing by overlap extension (SOE)

One limitation of site-directed mutagenesis that has not been solved by the SOE strategy is the manipulation of large gene sequences. Segments in the range of 1 kb can be routinely mutagenized; however, genes are frequently significantly larger than 1 kb. Therefore, we have worked with a cassette model that enables us to target smaller DNA fragments, to introduce substitutions into the fragment using SOE, and to return the modified segment into the context of the remainder of the gene using flanking restriction endonuclease sites designed into the cassette structure of the targeted gene.

GENERATION OF K^{bm8} VARIANTS USING SOE

Variants of the K^{bm8} gene were generated by SOE to address the functional significance of the individual amino acids that comprise the clustered substitution distinguishing the spontaneous mutant glycoprotein from the parent gene molecule. The DNA sequence of each of these variant genes was determined after the gene segment was cloned. Sequence analysis of 11 clones (from the bm8 series, as well as from a similar project analyzing bm10 variants), representing eight different variant constructs, revealed only a single error in the approximately 4000 nucleotides analyzed.

Table 1. Variants of K^{bm8} Generated by SOE

	AMINO ACID POSITION			
K^{bm8} VARIANTS	22	23	24	30
K^b	TYR	MET	GLU	ASP
BM8-22	PHE			
BM8-23		ILE		
BM8-24			SER	
BM8-30				ASN
BM8-22,24	PHE		SER	
BM8-23,30		ILE		ASN
BM8-WT	PHE	ILE	SER	ASN

The flanking primers ("A" and "D" in Fig. 1) used to generate the bm8 variants contained restriction endonuclease sites (*Sal*I and *Hind*III, respectively) that permitted the PCR amplified fragments to be reintroduced into the context of the rest of the class I gene for expression. The expression system is comprised of a recombinant class I gene derived from 5' sequences of the K^b gene and 3' sequences from the L^d gene (Fig. 2). Each variant gene was transfected into L cells and clones expressing the variant construct were isolated using the α-3 specific mAb 28-14-8 as described previously (Pullen 1989).

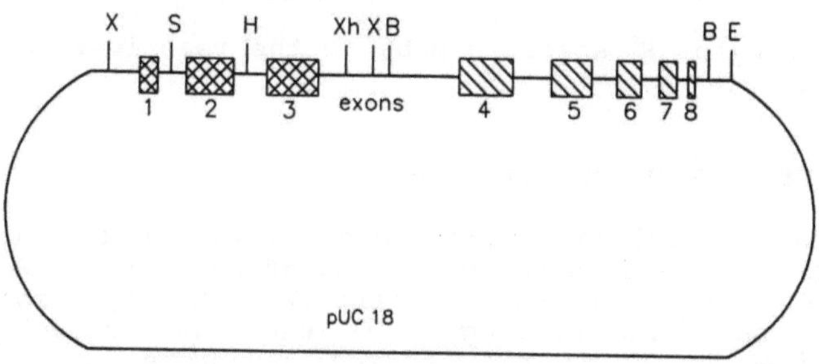

Fig. 2. MHC class I expression vector. Cross hatching (exons 1, 2, and 3) are composed of K^b and the variants made by SOE. The diagonal lines (exons 4-8) are from the L^d gene and contain the sequences necessary for the expression of the 28-14-8 mAb epitope

The serologic profile of the bm8 series of variants was analyzed by comparing a panel of K^b specific monoclonal antibodies to the common $L^d\alpha$-3 epitope. The mutations introduced along the floor of the beta-pleated sheet of the putative antigen recognition site of the class I glycoprotein had no effect on the serological profile as indicated by the similarities in the pattern of ratios shown in Table II. These ratios adjust for differences in surface expression of the various transfectants. Since several of the determinants recognized by these antibodies are conformational in nature, the observed pattern of reactivity indicates that no significant changes in overall conformation were introduced by changing individual amino acids or subsets of amino acids characteristic of the bm8 mutation. The results of CTL analysis using the L cell transfectants as targets are presented in the companion manuscript by Pullen in this volume.

Table 2. Relative Expression[a] of K^b Serologic Epitopes

L Cell Transfectant	28-14-8 $L^d_{\alpha 3}$	EH-144 $K^b_{\alpha 1}$	B8-24-3 $K^b_{\alpha 1}$	K10-56 $K^b_{\alpha 1,2}$	28-13-3 $K^b_{\alpha 1}$
K^b	1.0	0.7	1.0	0.8	1.2
bm8-Wild Type	1.0	0.7	1.0	0.9	1.0
bm8-22, 14	1.0	0.7	1.0	1.0	1.2
bm8-23	1.0	0.8	1.0	1.0	1.0
bm8-30	1.0	0.9	0.9	0.9	0.9
bm8-23, 30	1.0	0.8	1.0	1.3	1.3

[a]ratio of the reactivity of K^b specific mAbs to the reactivity of $L^d_{\alpha 3}$ specific mAb 28-14-8.

SHUFFLING GENE FRAGMENTS BETWEEN RELATED GENES

The SOE strategy for generating site-directed mutations can be extended to generating chimeric genes comprised of parts from two different homologues. The strength of this approach in comparison to traditional exon shuffling strategies is that the segments shuffled is determined based on the desired structure of the product rather than by the structure of the starting materials. In fact the term "exon shuffle" reflects the fact that usually enzyme sites located in introns are the most convenient sites available to generating this class of variant. In SOE, genes are spliced together using overlapping oligonucleotide primers which can be derived from any portion of the gene, exon or intron. In the example shown here, a construct was designed that would contain the α-helix encoding regions of the K^b gene in the context of the ß-pleats and loop structures of the L^d gene.

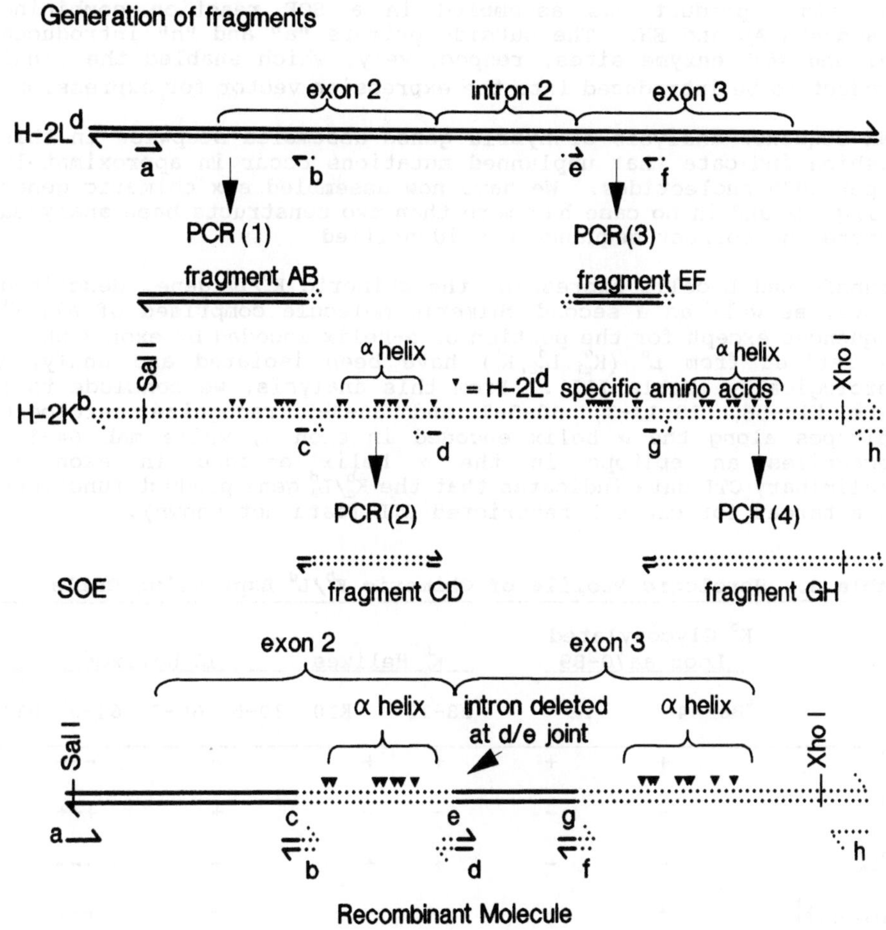

Fig. 3. Generation of the chimeric K^b_α/L^d_β gene using SOE strategy

The chimeric K^b_α/L^d_β gene was constructed, as described previously (Horton 1989), using internal oligonucleotides that overlap in regions of homology between the two class I genes, but where no convenient enzyme sites were located (Fig. 3). Because both exons

2 and 3 of the class I genes encode segments of the α-helix and ß-pleated regions, the desired construct was derived from the shuffle of four gene fragments, two from each gene. In the process, intron 2 sequences were deleted. The chimeric gene was assembled in a stepwise fashion: fragment AB was SOE'n to CD producing fragment AD, while in parallel fragments EF and GH were SOE'n to produce EH. The final product was assembled in a SOE reaction combining fragments AD and EH. The outside primers "a" and "h" introduced *Sal*I and *Xho*I enzyme sites, respectively, which enabled the final product to be introduced into the expression vector for expression.

DNA sequence analysis of hybrid genes assembled stepwise in this fashion indicate that unplanned mutations occur in approximately 1 per 2000 nucleotides. We have now assembled six chimeric genes using SOE and in no case has more than two constructs been analyzed before the correct sequence was identified.

Transfected L cells expressing the chimeric K^b_α/L^d_β gene, described above, as well as a second chimeric molecule comprised of all K^b sequences except for the portion of α-helix encoded by exon 3 which is derived from L^d ($K^b_{\alpha1}, L^d_{\alpha2}, K^b_\beta$) have been isolated and analyzed serologically (Table III). From this analysis, we conclude that mAbs K10-56, 28-13-3, 30-5-7 and 66-3-5 recognize serologic epitopes along the α helix encoded in exon 3, while mAb 64-3-7 recognizes an epitope in the α helix encoded in exon 2. Preliminary CTL data indicates that the K^b_α/L^d_β gene product functions as a target for class I restricted CTL (data not shown).

Table 3. Serologic Profile of Chimeric K^b/Ld Expressing Cells

	K^b Glycosylated Loop aa70-89		K^b_α Helixes		L^d_α Helixes			
	[a]B8-24	K9	28-13	K10	30-5	66-3	64-3	B11
K^b	+	+	+	+	-	-	-	-
L^d	-	-	-	-	+	+	+	+
$K^b_\alpha L^d_\beta$	-	-	+	+	-	-	-	-
$K^b_{\alpha1}, L^d_{\alpha2}, K^b_\beta$	+	+	-	-	+	+	-	+

[a]B8-24=B8-24-3, K9=K9-136, 28-13=28-13-3, K10=K10-56, 30-5=30-5-7, 66-3=66-3-5, 64-3=64-3-7.

GENERATING CHIMERIC GENES BY SOE

The ability to combine gene fragments by SOE can be extended to the generation of novel constructs that are comprised of segments of totally unrelated gene sequences. To accomplish this, the complementary internal oligonucleotide primers are designed such that at least one of the two primers bridges the region where the two unrelated gene segments are to be joined. The chimeric class I/class II construct, assembled as shown in Fig. 4, illustrates the power of this approach. In this case, we have SOE'n a class I leader peptide to the α_1 domain of a class II gene, which was in turn spliced to a class I linker region designed to join the α_1 region to the β_1 region from a second class II gene. The construct was completed by adding a second class I linker region that was designed to permit the hybrid segment to be cloned into the class I expression vector at the SalI and *Xho*I sites indicated in Fig. 1.

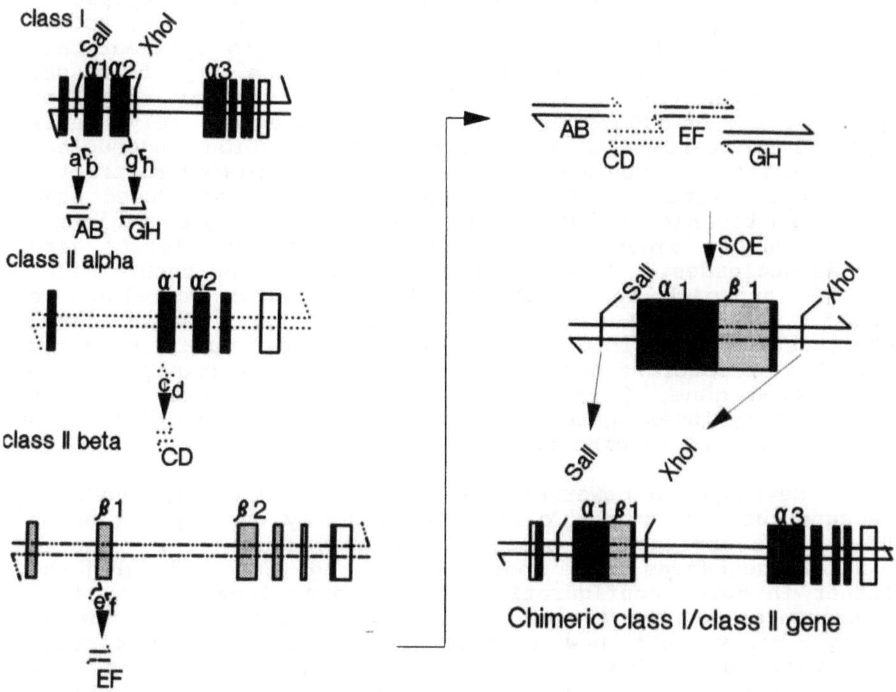

Fig. 4. Generation of chimeric class I/class II genes using SOE technology

The chimeric class I/class II gene was sequenced and transfected into L cells. Northern blot analysis as well as PCR characterization of cDNA derived from the isolated clones indicated that the hybrid gene is expressed and yields the expected mRNA species (data not shown). Although the class I serologic epitope determined by the α-3 region of the hybrid gene could be detected on the surface of transfected cells, no class II epitopes were found. Immunoprecipitation of ^{35}S-Met labeled product indicated that the hybrid protein was only about 27 kd, considerably smaller than the expected 45 kd product (data not shown). Together, these findings suggest that the hybrid glycoprotein is cleaved from the full length precursor to produce a smaller protein.

CONCLUSION

The structural basis of gene and protein function can be addressed by manipulating the structure of isolated genes using recombinant DNA technology. Until recently our ability to analyze structure/function correlates has been limited by the complexity of the techniques required to introduce meaningful structural changes into targeted genes. Strategies used to systematically localize specified functions to regions of a gene or protein, such as the deletion of protein binding motifs found in regulatory regions or the shuffling of exons among homologues, rely heavily on the serendipitous distribution of restriction endonuclease cleavage sites. In the absence of conveniently placed restriction sites, appropriate nucleotide sequences can be introduced using established techniques for site-directed mutagenesis or regions of the gene can be digested away from a distant restriction site, using exonucleases. Site-directed mutagenesis has been a highly effective approach for analyzing the structural correlates of function for highly characterized genes, but the available techniques are labor intensive. Since any single mutant construct probes the structure/function relationships of a narrow portion of the targeted gene, frequently, numerous mutants are needed to systematically investigate the structural requirements of function in less well defined circumstances.

We have developed a powerful new approach for generating novel gene constructs in vitro, called "splicing by overlap extension" (SOE). This approach is based on the polymerase chain reaction and allows sequences to be modified, deleted, inserted, or joined together in novel configurations in a manner independent of the DNA sequences at the target site. The versatility of the SOE strategy means that new genetic constructs can be designed exclusively, according to the properties desired in the final product, without consideration to practical limitations imposed by the DNA sequences of the building materials.

We have used SOE to construct a series of site-directed mutants, to shuffle intra-exon regions among homologous genes where no convenient restriction sites existed, and to generate class I/class II hybrid genes. In each case, quantitative yields of engineered

DNA molecules were isolated in a few simple steps. Because of the design of our expression system, the final products of the SOE reactions could be cloned directly into the expression system without intermediate cloning steps. Although, in many cases, existing technologies could have been used to generate these constructs, the necessity for introducing restriction sites at the junctional regions to be spliced (provided the DNA sequences present at the target region would permit such manipulations) would have been very demanding in terms of time and other resources. Using the SOE approach, these constructs were assembled, sequenced, and transfected in a small fraction of the time that ordinarily would have been required.

The most dramatic example of the power of SOE is illustrated by the assembly of the hybrid class I/class II gene outlined in Fig. 4. This complex chimera was constructed and cloned in two weeks. Although the hybrid gene was expressed, the protein behaved in an unpredictable manner in that a new proteolytic cleavage site appears to have been generated. It is clear from this example that our current ability to manipulate the structure of the gene is far ahead of our ability to predict how chimeric proteins will function.

REFERENCES

Ho SN, Hunt HD, Horton RM, Pullen JK, Pease LR (1989) Site-directed mutagenesis by overlap extension using the polymerase chain reaction. Gene 77:51-59

Horton RM, Hunt HD, Ho SN, Pullen JK, Pease LR (1989) Engineering hybrid genes without the use of restriction enzymes: gene splicing by overlap extension. Gene 77:61-68

Pullen JK, Hunt HD, Horton RM, Pease LR (1989) The functional significance of two amino acid polymorphisms in the antigen-presenting domain of class I MHC molecules: molecular dissection of Kbm3. J. Immunol. (in press)

The Functional Significance of Amino Acid Polymorphisms in Class I MHC Molecules

J. K. PULLEN[1], H. D. HUNT[1], R. M. HORTON[2], and L. R. PEASE[1,2]

[1] Department of Immunlogy, and [2] Department of Biochemistry and Molecular Biology, Mayo Clinic, Rochester, MN 55905, USA

INTRODUCTION

The x-ray crystal model of the HLA-A2 molecule depicts an antigen presenting domain consisting of two α-helices under which are found eight antiparallel ß-strands. Furthermore, the position, orientation, and functional contribution of many amino acid side chains has been proposed (Björkman 1987a,b). In general, amino acid residues that point inward/upward from the α-helices or point upward from the ß-strand floor are thought to be involved with either peptide binding or T-cell receptor (TcR) interaction. Amino acid side chains that point outward from the α-helices, project downward from the ß-strand floor, or are located on loops outside of the antigen recognition site (ARS) are thought to be silent with regard to TcR interaction or peptide binding (Björkman 1987b).

TESTING THE MODEL BY MOLECULAR DISSECTION OF K^{bm3} AND K^{bm8}

The goal of the study presented herein was to determine whether the human class I model was extrapolatable to the mouse class I MHC molecule. To test this, the naturally occurring K^{bm} mutant series was chosen for study. Each K^{bm} mutant expresses H-2K glycoproteins containing unique single sets of clustered amino acid substitutions that distinguish them from the parental K^b molecule (Nathenson 1986). These differences are capable of inducing skin graft rejection between parental and mutant mice, and generating MLR *in vitro*. As suggested by the model the K^{bm} mutants contain substitutions which, based on their homologous position in the HLA-A2 model, would contribute to either peptide binding or TcR interaction while other substitutions appear to be functionally silent (Björkman 1987b). For example as shown in Figure 1, the K^{bm3} molecule contains two amino acid substitutions (77;Asp→Ser,89;Lys→Ala) which distinguish it from the parental K^b molecule. The amino acid substitution at position 77 is predicted to be involved with ligand binding while the substitution at 89 is located in a position where it is predicted to have no direct effect on antigen presentation. Similarly, as depicted in Figure 1, the K^{bm8} molecule contains four amino acid substitutions (22,23,24,30) which distinguish it from the parent and these changes are located on the second ß-strand in the α_1 domain (Nathenson 1986). Amino acid 22 (Tyr→Phe) and 24 (Glu→Ser) are predicted to project into the ligand binding site and are predicted to effect peptide binding. Conversely, amino acid 23 (Met→Ile) and 30 (Asp→Asn) project downward and are predicted to be silent with regard to peptide binding.

I.K. Egorov/C.S: David (Eds.)
Transgenic Mice and Mutants in MHC Research

RESULTS

To determine the relative contribution of the individual amino acid substitutions involved in generating the K^{bm3} and K^{bm8} molecules, a panel of L-cell clones expressing mutagenized K^{b} molecules was generated and analyzed by using cytotoxic T-lymphocytes (Pullen 1989 and Table I). The results demonstrated that amino acid substitutions at positions 22 and 24 were responsible for conferring the complete bm8 alloreactive phenotype, while position 77 was responsible for conferring the entire bm3 alloreactive phenotype. Additionally, the results illustrated that amino acid substitutions at positions 23, 30 and 89 were functionally silent.

Table I. Summary of CTL Analysis

CLONED LINE[a]	AMINO ACID POSITION						LYSIS OF CLONES BY CTL POPULATIONS			
	22[b]	23[c]	24[b]	30[c]	77[b]	89[c]	B6αbm3	bm3αB6	B6αbm8	bm8αB6
K^{b}	Y	M	E	D	D	K	-	++++	-	++++
K^{bm3}	-	-	-	-	S	A	++++[d]	-		
K^{bm3-77}	-	-	-	-	S	-	+++	-		
K^{bm3-89}	-	-	-	-	-	A	-	+++		
K^{bm8}	F	I	S	N	-	-			++++	-
K^{bm8-30}	-	-	-	N	-	-			-	++++
K^{bm8-23}	-	I	-	-	-	-			-	++++
$K^{bm8-22,24}$	F	-	S	-	-	-			++++	-

a All cell lines expressed transfect K^{b} glycoproteins at comparable levels as determined by mAb binding and FACS analysis.

b Predicted to project into the antigen binding cleft.

c Predicted to be functionally silent.

d CTL assays were performed using standard ^{51}Cr assays over a wide range of effector to target ratios. (++++) 100% lysis equivalent to lysis of blasted spleen cells from parent or mutant mice; (+++) 75% of parental lysis; (++) 50% of parental lysis; (+) 25% of parental lysis; (-) lysis equivalent to L-cells expressing the TK gene only.

CONCLUSIONS

The results from the experiments described above have allowed us to conclude that in general the human HLA-A2 crystal model is extrapolatable to the mouse class I molecule concerning the amino acid positions studied here. Amino acids 22, 24, and 77, all of which project into the ligand binding site and are predicted to influence peptide binding, have strong effects on alloreactivity. Furthermore, amino acids 23, 30 and 89, which are located outside the ARS and predicted to be silent, appear to be silent with regard to alloreactivity.

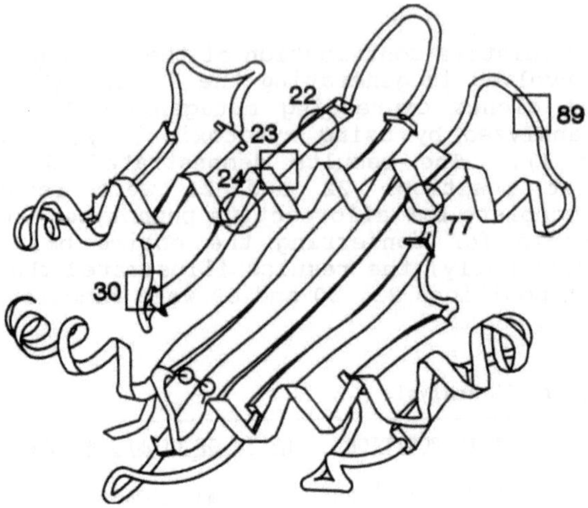

Figure 1. Schematic diagram of the HLA-A2 class I molecule as determined by Björkman. The corresponding amino acid positions in the murine class I molecule that distinguish K^b from K^{bm3} and K^{bm8} are shown. Functionally silent shown by squares and functionally productive shown by circles.

REFERENCES

Björkman PJ, Saper MA, Samraoui B, Bennett WS, Strominger JL, Wiley DC (1987a) Structure of the human class I histocompatibility antigen, HLA-A2. Nature 329:506.

Björkman PJ, Saper MA, Samraoui B, Bennett WS, Strominger JL, Wiley DC (1987b) The foreign antigen binding site and T cell recognition regions of class I histocompatibility antigens. Nature 329:512.

Nathenson SG, Geliebter J, Pfaffenbach GM, Zeff RA (1986) Murine major histocompatibility complex class-I mutants: Molecular analysis and structure-function implications. Annu. Rev. Immunol. 4:471.

Pullen JK, Hunt HD, Horton RM, Pease, LR (1989) The functional significance of two amino acid polymorphisms in the antigen-presenting domain of class I MHC molecules: Molecular dissection of K^{bm3}. J. Immunol. (In Press).

Products of Separate Genes Encoded Within Minor Histocompatibility Loci Stimulate Class I and Class II MHC-Restricted T Cells[1]

D. C. ROOPENIAN and A. P. DAVIS[2]

[1] Supported by N. I. H. grants # AI 24544 and 2 S07 RR05545 and A. C. S. grant # IN115.
Its contents do not necessarily represent the views of the N. I. H.
[2] The Jackson Laboratory, Bar Harbor, ME 04609, USA

Minor histocompatibility (H) loci encode polymorphic gene products that act as tissue transplantation barriers. More than 40 of these loci have been identified through the production of minor H congenic strains of mice; however, because of the inherent complexity of the cellular processes leading to tissue rejection, the nature of the polymorphic genes comprising minor H loci has remained elusive. We have been using in vitro approaches employing minor H antigen-specific T cell clones as genetic probes for analyzing minor H loci. The purpose of this report is to summarize our recent results and to discuss how the results provide a novel view of the genetic structure of minor H loci.

T cell responses were generated between sets of congenic strains differing at minor H loci *H-3* and *H-4*, located on Chromosomes 2 and 7, respectively. Cloned T cell lines were isolated from cultures of these cells and were analyzed for nominal antigen specificity, major histocompatibility complex (MHC)-restriction, cell-mediated lympholysis (CML) and CD4/8 phenotype. The results from these studies, summarized in Table 1, indicated that both CD4$^+$ class II MHC-restricted T helper (TH) and CD8$^+$ class I MHC-restricted cytotoxic T lymphocyte (CTL) subsets were generated during immune responses to *H-3* and *H-4*.

Table 1. Characteristics of Cloned T Cells Specific for *H-3* and *H-4* Antigens

Specificity	MHC Restriction	CML	Phenotype	Type
anti-*H-3*^{a}*	H-2D^b	+	CD4$^-$8$^+$	CTL
"	H-2K^b	+	CD4$^-$8$^+$	CTL
"	H-2IA^b	-	CD4$^+$8$^-$	TH
anti-*H-4*^{a}	H-2D^b	+	CD4$^-$8$^+$	CTL
"	H-2IA^b	-	CD4$^+$8$^-$	TH

**H-3*^{a} specific T cells were derived from in vivo primed and in vitro restimulated T cells from the responder-stimulator strain combinations: B10.LP-*H-3*^{b}/Sn anti-C57BL/10Sn (B10) or (B10.LP-*H-3*^{b}/Sn X BALB/cJ) F1 anti-B10 (Roopenian and Davis 1989); *H-4*^{a} specific T cells were derived from B10.129-*H-4*^{b}/Sn anti-B10 cells or (B10.129-*H-4*^{b}/Sn X P/J) F1 anti-B10 cells (Davis and Roopenian submitted).

I.K. Egorov/C.S. David (Eds.)
Transgenic Mice and Mutants in MHC Research

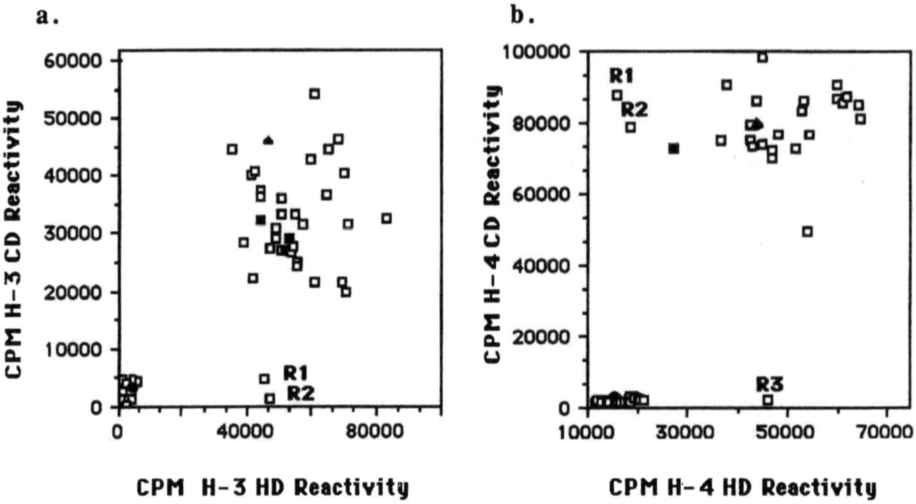

Fig. 1. Segregation analysis of HD and CD antigens as monitored by the ability of cells from F2 or backcross mice to stimulate proliferation of cloned TH and CTL. (Roopenian and Davis 1989, Davis and Roopenian, submitted). Each symbol represents mean ^{3}H-thymidine incorporation of an individual mouse. **a,** Analysis of *H-3* genes studying 40 (B10.UW-*H-3^b*/Sn x B10)F2 mice (open boxes); positive control B10 (closed squares) and F1 (closed triangles); negative control B10.UW-*H-3^b*/Sn (closed circles). **b**, Analysis of *H-4* genes studying 48 (129/J X B10) X 129/J backcross (open boxes); positive control B10 (closed squares) and F1 (closed triangles); negative control B10.129-*H-4^b*/Sn (closed circles). Putative recombinant mice are designated by **R**.

These cloned T cells were used to address the following question: for a given minor H locus, are the antigens recognized by CTL encoded by the same gene as those recognized by TH? The approach for addressing this question was to use the above-described minor H antigen-specific T cell lines as probes for studying the genetic segregation of TH-defined (HD) and CTL-defined (CD) antigens in backcross and F2 mice. To aid in positioning of HD and CD genes, strains chosen for these crosses differed by visually detected landmark loci. In the case of *H-3*, [B10.UW-*H-3b* X B10 (*H-3^a*)] F2 mice were studied, which made it possible to exploit polymorphisms at Chromosome 2 loci wellharrig (*we*) and undulating (*un*). In the case of *H-4*, [(129/J (*H-4^b*) X B10(*H-4^a*)] F1 X 129/J backcross mice were chosen, which made it possible to use pink-eyed dilution (*p*) and (*c*) albino as Chromosome 7 landmark loci. Figure 1 shows representative data from studies of *H-3* (**a**) and *H-4* (**b**) using proliferative assays to monitor the expression of minor H antigens in F2 and backcross mice. In most cases the F2 and backcross mice expressed CD and HD antigens concordantly, as would be predicted if the antigens were encoded by a single gene or by two or more closely-linked genes. There were, however, rare mice that expressed only one type of antigen suggesting that the genes stimulating TH and CTL are closely linked but separate. Segregation of landmark genes was consistent with the results from T cell typing and suggested that the genetic event leading to loss of expression of the T cell-defined genes was a classical recombinatorial event (not shown). Based on these studies, the following gene order is suggested for the *H-3* locus: (CD, *we*) - HD - *un*.(Roopenian and Davis 1989). (Order of genes enclosed within parenthesis is not known). For the *H-4* locus the following gene order is suggested: HD - *p* - CD - *c* (Davis and Roopenian submitted).

The novel finding from these studies is that, in the two cases that we have examined thus far, minor H antigen congenic partner strains differ by two types of genes. The products of one gene type stimulates class I MHC restricted CTL and products of the other stimulate class II MHC restricted "helper" T cells. A obvious suggestion from these data is that the phenomenology of minor H loci and the antigens produced by them is consistent with current conceptual frameworks of differential protein processing pathways for "self" antigens presented in the context of class I and class II MHC proteins (Morrison et al 1986, Germaine 1986). For this reason, the minor H congenic strains should be useful for studying the processing events leading to MHC protein presentation of polymorphic self proteins. An implication of our results that is also noteworthy is based on findings that TH-CTL collaboration seems particularly important for the generation of CTL against minor H antigens (Hurme et al 1978, von Boehmer and Haas 1979, Roopenian and Anderson 1988). The fact that both *H-3* and *H-4* are comprised of both HD and CD genes raises the possibility that both types of genes may have been selected for coincidentally during production of the minor H congenic strains because an H barrier caused as a consequence of TH-CTL collaboration be considerably stronger than that caused by either TH or CTL alone. This question can now be approached using mouse stocks established from mice carrying a recombination between HD and CD genes.

REFERENCES

Davis AP, and Roopenian, DC. Complexity at the mouse minor histocompatibility locus *H-4*. Immunogenetics, in press

Germain RJ. (1986) The ins and outs of antigen processing. Nature 332: 687-9

Morrison LA, Lukacher AE, Braciale VL, Fan D, and Braciale TJ. (1986) Differences in antigen presentation to MHC class I- and class II-restricted influenza virus-specific cytolytic T lymphocyte clones. J. Exp. Med. 163: 903-21

Roopenian DC, and Anderson PS. (1988) Generation of helper cell-independent cytotoxic T cells is dependent upon L3T4-positive helper T cells. J.Immunol. 141: 392-7.

Roopenian DC, and Davis AP. (1989) Responses against antigens encoded by the *H-3* histocompatibility locus: antigens stimulating class I MHC and class II MHC restricted T cells are encoded by separate genes. Immunogenetics , in press

von Boehmer H, Haas W. (1979) Distinct *Ir* genes for helper and cytotoxic T cells in the cytotoxic T cell response to *H-Y*. J Exp Med 150: 1134-42

Hurme M., C. M. Hetherington, P. R. Chandler, and E. Simpson. (1978) Cytotoxic T-cell responses to H-Y: Mapping of Ir genes. J. Exp Med147:758-67

Determination of Distinction Odortypes by Mutation of Class I and Class II MHC Genes

K. YAMAZAKI[1], G. K. BEAUCHAMP[1], L. THOMAS[2], J. BARD[3], and E. A. BOYSE[3]

[1] Monell Chemical Senses Center, Philadelphia, PA 19104, USA
[2] Cornell Graduate School of Medical Science and Memorial Sloan-Kettering Cancer Center, New York, NY 10021, USA
[3] Department of Microbiology and Immunology, University of Arizona, Tucson, AZ 85724, USA

Genetic polymorphism in the H-2:Qa:Tla region of chromosome 17 is associated with constitutive variation of bodily odor phenotypes which permit individual olfactory recognition among mice (Boyse et al.1987). We previously found that differences in the MHC genes affect the mating choices of mice (Yamazaki et al. 1976, 1978, 1988; Andrews and Boyse, 1978; Yamaguchi et al., 1978), and that mice can be trained to distinguish arms of a Y-maze scented by odors from mice differing only in this genetic region (Yamazaki et al.,1979, 1982; Yamaguchi et al.,1981). We also showed a raised incidence of blockage of pregnancy or pseudopregnancy in mated females exposed to the scent of alien males whose MHC type differs from that of the mate (Yamazaki et al., 1983). As no known gene complex exhibits such vast diversity as the MHC, its potential as a major source for genetically based signals of individual identity is great.

To determine whether known genes in the H-2:Qa:Tla complex are concerned in the constitution of odor phenotypes, we have studied the ability of trained mice to distinguish the odor of non-mutant strains from that of congenic mutant strains differing only by mutations of Class I or Class II MHC genes.

As in previous studies of the sensory discrimination of MHC phenotypes, mice were trained by reward in a Y-maze to distinguish the odors of urine samples from donors with various mutations. Successful training was confirmed by tests of transfer of training (generalization) with coded samples of urine from genetically equivalent urine donor mice which the trained mice had not previously encountered (for a description and rationale for this technique, see Yamaguchi et al., 1981).

RESULTS AND DISCUSSION

Experiment 1

Distinctions of B6 [C57BL/6 ($H\text{-}2^b$)] from the H-2K mutants, bm1, bm5, bm7 and bm8 were studied in the first series of tests. Distinction of B10.D2 [C57BL/10.D2 ($H\text{-}2^d$)] from the H-2D mutant, dm1 was demonstrated next. Finally, olfactory discriminations of B6 vs. the I-A (Class II) mutant congenic mouse strain $B6.C\text{-}H\text{-}2^{bm12}$ (bm12) were also made by trained mice .
In sum, the data demonstrate both Class I and Class II MHC genes are involved in individual odor constitution. The H-2- determined odor phenotype is thus a composite determined by several, perhaps many, genes within the MHC.

I.K. Egorov/C.S. David (Eds.)
Transgenic Mice and Mutants in MHC Research

In light of the possibility that $H-2^b$ (B6) might impart a particular odor not shared by $H-2^b$ mutants as a group, the next distinctions tested were between mutants. Table 1 shows that animals trained to discriminate B6 from each mutant generalized this training to B6 vs. other mutants. This suggests that the mutants as a class possess similar odors. However, there was no evidence in generalization tests of discrimination between mutants; that is for example, mice trained to discriminate B6 from bm1 responded randomly when offered a choice between bm1 and bm5, bm7 or bm8. Based on these data, B6 mice appear to have an odor distinguishing them from K mutant mice and Class II mutant mice. However, K mutants do not appear to be mutually discriminable under the Y-maze test conditions.

Experiment 2

It has previously been shown that MHC-congenic mouse odor can be discriminated by rats as well as mice (Beauchamp et al.,1985). In this experiment, rats' perception of MHC-determined odor differences were tested.
Rats were trained to detect and discriminate odors in a test chamber adapted from Slotnick and Katz (1974), a prototype of which is described in Beauchamp et al. (1985). Essentially, compressed filtered air is passed over urine samples and is briefly delivered to a tunnel-shaped test chamber housing the test animal. The test chamber is provisioned with a response bar connected to a touch circuit and a water reservoir for delivery of reinforcement. A discrete-trials "go, no go" successive discrimination training procedure was used.

Four female rats were first trained to discriminate panels of two inbred congenic strains, B6 and $B6-H-2^k$. Following demonstration of successful generalization to F_2 panels, two rats were trained to discriminate urine odors obtained from other panels of B6 vs. bm1 mice. The other two rats were trained to discriminate panels of B6 vs. bm12 mice.
All rats were able to discriminate urine of mutant (Class I and Class II) from non-mutant mice. Thus, trained rats can discriminate between very small genetic differences in two mouse strains, and in this regard do not differ from trained mice.

The next experiment was designed to determined whether rats would discriminate among Class I mutants or , like the trained mice in experiment 1, would fail to make this discrimination during generalization tests.
Two rats trained to discriminate B6 from bm1, were given generalization or probe trials with other Class I H-2K mutant mice (bm3, bm5, bm7 and bm8). For rat #1, generalization from S+ (bm1) to most test stimuli (bm3, bm5 and bm8) occurred. Thus, bm3, bm5 and bm8 mutant odors appear to be quite similar to the bm1 mutant, in agreement with mouse studies. Similarly, for rat #2, the bm3, bm5 and bm8 mutants elicited few bar touches and in this regard were responded to as if they smelled like bm1.

The current studies demonstrate that variation restricted to a few DNA base pairs of a single gene is reflected in variation in body odor. Each mouse in nature may have a unique odortype specified by its unique constellation of MHC genes.

Table 1 Distinction of K^b Mutant Pairs

Generalization Trials, Concordance % and (# of trials)

	B6	bm1	bm5	bm7	bm8
bm1	67* (212)	x			
bm5	70* (118)	57 (37)	x		
bm7	62* (143)	45 (58)	49 (82)	x	
bm8	78* (59)	57 (88)	53 (77)	46 (68)	x

*P < 0.01

REFERENCES

Andrews PW, Boyse EA (1978) Mapping of an H-2 linked gene that influences mating preference in mice. Immunogenetics 6: 265-268

Beauchamp GK, Yamazaki K, Wysocki CJ, Slotnick BM, Thomas L, Boyse EA (1985) Chemosensory recognition of mouse major histocompatibility types by another species. Proc Natl Acad Sci USA 82: 4186-4188

Boyse EA, Beauchamp GK, Yamazaki K (1987) The genetics of body scent. Trends Genet 3: 97-102

Slotnick BM, Katz BJ (1974) Olfactory stimulus control evaluated in a small animal olfactometer. Percept Mot Skills 39: 583-597

Yamaguchi M, Yamazaki K, Boyse EA (1978) Mating preferences tests with the recombinant congenic strain BALB.HTG. Immunogenetics 6: 261-264

Yamaguchi M, Yamazaki K, Beauchamp GK, Bard J, Thomas L, Boyse EA (1981) Distinctive urinary odors governed by major histocompatibility locus of the mouse. Proc Natl Acad Sci USA 78: 5817-5820

Yamazaki K, Boyse EA, Mike V, Thaler HT, Mathieson BJ, Abbott J, Boyse J, Zayas ZA, Thomas L (1976) Control of mating preferences in mice by genes in the major histocompatibility complex. J Exp Med 144: 1324-1335

Yamazaki K, Yamaguchi M, Andrews PW, Peake B, Boyse EA (1978) Mating preferences of F_2 segregants of crosses between MHC-congenic mouse strains. Immunogenetics 6: 253-259

Yamazaki K, Yamaguchi M, Baranoski L, Bard J, Boyse EA, Thomas L (1979) Recognition among mice: Evidence from the use of a Y maze differentially scented by congenic mice of different major histocompatibility types.J Exp Med 150: 755-760

Yamazaki K, Beauchamp GK, Bard J, Thomas L, Boyse EA (1982) Chemosensory recognition of phenotypes determined by the Tla and H-2K regions of chromosome 17 of the mouse. Pro Natl Acad Sci USA 79: 7828-7831

Yamazaki K, Beauchamp GK, Wysocki CJ, Bard J, Thomas L, Boyse EA (1983) Recognition of H-2 types in relation to the blocking of pregnancy in mice. Science 221: 186-188

Yamazaki K, Beauchamp GK, Kupniewski D, Bard J, Thomas L, Boyse EA (1988) Familial imprinting determines H-2 selective mating preferences. Science 240: 1331-1332

II. Mutants of MHC Genes as Probes of Immunological Function (HLA)

Analysis of HLA-B27-Specific T-Cell Epitopes with Site-Directed Mutants Mimicking HLA-B27 Polymorphism

S. Rojo, V. Calvo, D. López, B. Galocha, and J. A. López de Castro[1]

[1] Department of Immunology, Fundación Jiménez Díaz, Consejo Superior de Investigaciones Cientificas, Avenida Reyes Católicos, 2, 28040, Madrid, Spain

INTRODUCTION

Studies in several laboratories have demonstrated the existence of at least six subtypes of the HLA-B27 antigen in the human population. They are designated as B*2701 to B*2706 and differ from each other by just a few amino acid changes, that are located in the peptide binding site of the molecule (López de Castro 1989). Because of its location, subtype polymorphism induces significant, sometimes critical, differences in cytotoxic T lymphocyte (CTL) recognition, so that B27 subtypes are, in spite of their structural similarity, functionally distinct alleles.

Previous studies in our laboratory (Aparicio et al 1988; Jaraquemada et al 1988) have used the structurally characterized subtypes to analyse the clonal diversity of the alloreactive CTL response against HLA-B27 and to define some structural parameters of the determinants recognized by these CTL . An obvious limitation of these studies is that subtypes usually differ by more than one amino acid residue.

In this report we have focused on the immunologic relevance of individual positions, among those that are polymorphic in the HLA-B27 family. This was done by using site-directed mutagenesis to analyse their role in modulating CTL allorecognition at a clonal level. We have also examined the interactive effects of polymorphism at pairs of spatially close positions by the use of double mutations. The results show the extraordinary diversity of allospecific CTL epitopes on the HLA-B27 molecule and further define their topologic relationships and structural features.

MATERIALS AND METHODS

Site-Directed Mutagenesis of HLA-B27

A 6.5 Kb EcoRI DNA fragment containing the HLA-B*2705 gene (a gift of Dr. H. Coppin, Toulouse, France), cloned in pBR322 vector or recloned in pUC18, was used in this study. Mutations were introduced by oligonucleotide-mediated site-specific mutagenesis (Barbosa et 1987). Briefly, purified supercoiled plasmid DNA was digested separately with Hind III to linearize the plasmid in the vector sequence, and with Sma I (for mutations in the $\alpha 1$ coding sequence) or Kpn I (for mutations in $\alpha(2)$ to gap the region in the HLA-B27 gene

I.K. Egorov/C.S. David (Eds.)
Transgenic Mice and Mutants in MHC Research

to be mutated. The resultant molecules were mixed at equimolar amounts, denatured and renatured with the corresponding oligonucleotide at a molar ratio of 1:20 pmol, and at 10°C below its melting temperature, for 2 h. Polymerization and ligation was subsequentely carried out at 14°C for 15 h . The ligation mixture was used to transform competent E. Coli DH5. DNA from transformed colonies harbouring mutants was analysed by Southern blot hybridization to assure correct localization of the mutation to the appropriate region of the HLA-B27 gene. Primary colonies were retransformed and rescreened until colony homogeneity, to obtain purified mutant plasmid DNA. Confirmation of the intended mutation was obtained by DNA sequence analysis of at least the exon in which each mutation was located.

DNA-Mediated Gene Transfer.

Plasmids containing HLA-B*2705, B*2702 or the mutant B27 genes, as well as pSV2neo, were linearized by digestion with proper restriction enzymes and transfected into HMy2.CIR cells (a gift of Dr. P. Creswell, Durham, USA) by electroporation (Chu et 1987). Screening for HLA-B27 expression was done by fluorescence activated cell sorting (FACS) analysis with monoclonal antibodies ME1 (anti-HLA-B27+B7+Bw22) and W6/32 (anti-HLA class I monomorphic determinant) as described elsewhere (Aparicio et 1987). Populations showing a homogeneous FACS profile and high HLA-B27 expression, similar to those of lymphoblastoid cell lines (LCL), were selected for further use. Cell surface expression of HLA-B27 in all transfectants was stable as assessed by periodic FACS analysis.

T Cell Cloning.

Mixed lymphocyte cultures, T-cell cloning and ^{51}Cr-release cytotoxicity assays were performed as described elsewhere (Aparicio et 1987).

RESULTS AND DISCUSSION

Twenty alloreactive CTL clones were isolated by limiting dilution of peripheral blood mononuclear cells from various B27-negative responder individuals, stimulated in vitro with B*2705-positive LCL. A CTL clone, 64.8P (Jaraquemada et al 1988), derived from another B27-negative donor against B*2704, was also examined. These clones were analysed for their reactivity with the naturally occurring subtypes of this antigen, as well as with a set of mutants reproducing, individually or pairwise, the changes in the subtypes. The purpose of this study was threefold. First, to outline the diversity of epitope specificities generated in allogeneic CTL responses against HLA-B27. Second, to define some structural parameters of these epitopes. Third, to substantiate the role of subtype polymorphism in modulating HLA-B27 antigenicity.

Fine Specificity of HLA-B27-Specific CTL Clones

CTL clones were tested for their lytic ability towards a panel of LCL, including those expressing all six known HLA-B27 subtypes. As shown in Table 1, the clones raised against B*2705 could be classified, according to their reactivity with these subtypes, into six reaction patterns (A to F). Some heterogeneity was observed among CTL clones within a given group in: a) their lytic efficiency towards B*2705-positive targets; b) their relative cytotoxicity towards target cells expressing different B27 subtypes, and c) the occasional crossreactions of some CTL clones, with other HLA antigens.

Table 1. Reaction patterns of alloreactive CTL clones against B*2705 with HLA-B27 subtypes[a]

Reaction pattern	CTL Clone[b]	B*2705	B*2701	B*2702	B*2703	B*2704	B*2706
A	23, 40, 64DRF, 102DRF, 212DRF, 17A2, B54, G36	+	-	-	-	-	-
B	46, GM7, 5A2	+	-	-	+	-	-
C	2	+	+	-	-	-	-
D	28, 172DRF, 139DRD, 202DRD	+	-	+	-	-	-
E	7, 26, 29	+	-	+	+	-	-
F	72DRF	+	+	+	+	-	-
	64.8P	+	+	+	-	+	+

[a] Specific cytotoxicity towards target cells expressing the antigens shown in the Table is indicated. The reaction pattern of anti-B*2704 CTL 64.8P is also included for comparison.

[b] CTL 23 crossreacts with a HLA-DR2 subset; CTL 2 and 102DRF crossreact with HLA-B40*, a subtype of HLA-B40. The assignment of CTL 172DRF to reaction pattern D is tentative, as its lysis of B*2702-positive LCL is very close to background levels. However, it kills very efficiently B*2702-positive transfectant MHy2.CIR cells.

The main features of the reaction patterns shown in Table 1, for CTL clones raised against B*2705, are the following: 1) all twenty clones fail to recognize the B*2704 and B*2706 subtypes, with the same two changes at positions 77 and 152 (Table 2); 2) 19 of 20 clones fail to recognize either B*2701, B*2702 or both. These two alleles differ from B*2705 only within the segment 74-81 and 77-81 respectively (Table 2); 3) B*2703, differing from B*2705 only at position 59, is recognized by 7 (or 35%) of 20 CTL clones. Taken together, these results indicate that all subtype polymorphism influences, in a very drastic manner, recognition of HLA-B27 by a set of CTL clones from multiple individuals. In addition to this, virtually all epitopes involve residue 152 and one or more residues within the segment 74-81.

Effect of Subtype-Related Mutations on HLA-B27 Allorecognition

Nine site-directed HLA-B27 mutants were constructed to replace, individually or at two positions, the amino acid residues occurring in B*2705 by those in other subtypes (Table 2). A mutant at residue 63 was also constructed. This residue is invariant among HLA-B27 subtypes but is spatially close to residue 59, changed in B*2703, to which is related by an α-helical turn (Bjorkman et al 1987).

Table 2. Amino acid substitutions in HLA-B27 subtypes and mutants[a]

Molecule	Amino Acid Sequence Position							
Subtype	59	63	74	77	80	81	114	116	152
B*2705	Y	E	D	D	T	L	H	D	V
B*2701	-	-	Y	N	-	A	-	-	-
B*2702	-	-	-	N	I	A	-	-	-
B*2703	H	-	-	-	-	-	-	-	-
B*2704	-	-	-	S	-	-	-	-	E
B*2706	-	-	-	S	-	-	D	Y	E
Mutant									
N63	-	N	-	-	-	-	-	-	-
Y74	-	-	Y	-	-	-	-	-	-
N77	-	-	-	N	-	-	-	-	-
S77	-	-	-	S	-	-	-	-	-
I80	-	-	-	-	I	-	-	-	-
Y74N77	-	-	Y	N	-	-	-	-	-
N77I80	-	-	-	N	I	-	-	-	-
N77A81	-	-	-	N	-	A	-	-	-
E152	-	-	-	-	-	-	-	-	E

[a] Mutants were designated by the single-letter code of the amino acid(s) introduced followed by the corresponding position number(s).

The CTL clones described in Table 1 were tested for their lytic ability against a panel of HMy2.CIR transfectant target cells expressing HLA-B*2705, B*2702 and the mutant B27 molecules. The effect of each mutation on the anti-B*2705 CTL clones is shown in Fig.1. These results reveal the features of the B27 allospecific CTL epitopes listed below.

First, all twenty clones analysed have different epitope specificity. Even within a same reaction pattern group, as defined with the B27 subtypes (Table 1), no clones were found with identical specificity when tested against the panel of mutants. This result demostrates the extraordinary complexity of the allogeneic CTL response against HLA-B*2705, adding to studies suggesting similar diversity in responses against HLA-A2 (Santos-Aguado et al 1989; Hogan et al 1989).

Second, every single position that was mutated had important effects on CTL recognition (Table 3). The overall effect of any of these mutations on the particular set of CTL clones analysed was analogous. An exception to this was the mutation at residue 152 which totally abrogated recognition by all the CTL clones examined. On the basis of the three-dimensional model of HLA-A2 (Bjorkman et

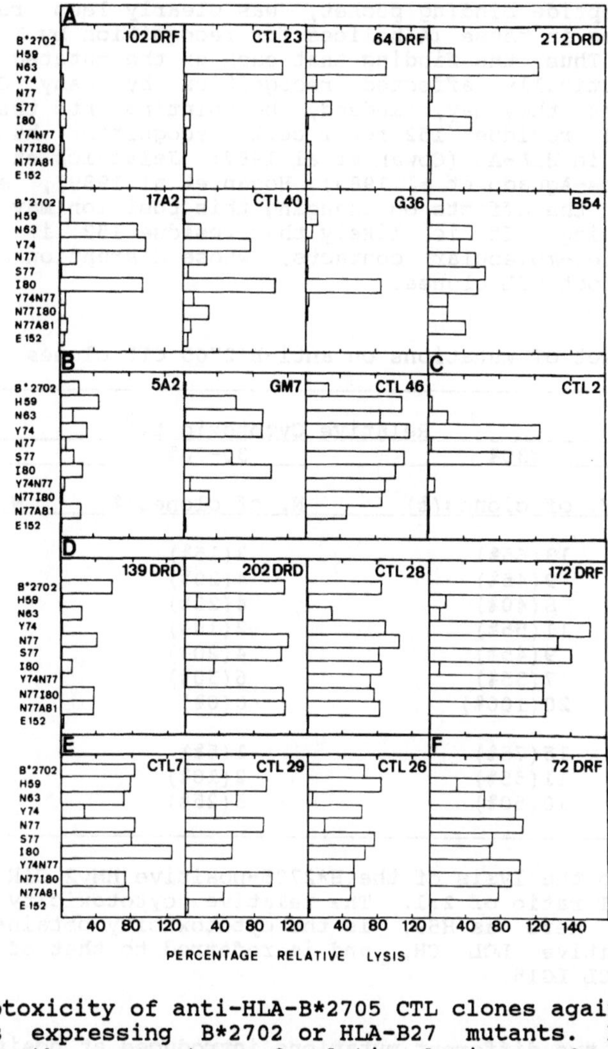

Fig. 1. Cytotoxicity of anti-HLA-B*2705 CTL clones against HMy2.CIR transfectants expressing B*2702 or HLA-B27 mutants. Results are expressed as the percentage of relative lysis, referred to the specific lysis of the B*2705-positive transfectant at the same E/T ratio. Results were obtained at E/T ratios of 0,5:1, 1:1 and 2:1 but only the latter one is presented here. CTL clones are grouped in panels A to F according to their reactivity with the B27 subtypes (Table 1). Specific lysis of the B*2705-positive transfectant cells at this E/T ratio was 40% to 100%, depending on each CTL clone. The relative lysis of the B*2703-positive LCL CH is also included for comparison. B*2703 is designated here as H59 to emphasize the nature of its single substitution. For CH cells, the relative lysis is referred to the specific lysis of the B*2705-positive LCL LG15 at an E/T ratio of 1:1. This was 13% to 86%, depending on the CTL clones

al 1987), all mutated positions are located in the α-helices of the foreign antigen binding site (Fig.2). They appear to be pointing into the groove, although the precise orientation of the side chains could be different in HLA-B27. In HLA-A2, it has been shown that residue 149, located in one of the α-helices but not pointing into the peptide binding pocket, was clearly less relevant than other residues in these α-helices for recognition by T cells (Hogan et al 1988). Thus, the finding that each of the mutated positions in HLA-B27 drastically affected recognition by many CTL clones, suggests that they may, indeed, be pointing into the site. The relevance of residue 152 for T cell recognition has also been demonstrated in HLA-A3 (Cowan et al 1987; Jelachich et al 1988) and HLA-A2 (Santos-Aguado et al 1989; Hogan et al 1989), although, for the latter, the effects on changing this position may be somewhat less overriding. It is likely that residue 152 is involved in critical inter-molecular contacts, whose disruption abrogate TCR binding for most CTL clones.

Table 3. Effect of mutations on anti-B*2705 CTL clones

	Relative Cytotoxicity[a]		
	<30%	30-70%	>70%
Mutation	N. of clones(%)	N. of clones(%)	N. of clones(%)
H59	13(65%)	3(15%)	4(20%)
N63	9(45%)	4(20%)	7(35%)
Y74	8(40%)	4(20%)	8(40%)
N77	11(55%)	2(10%)	7(35%)
S77	9(45%)	4(20%)	7(35%)
I80	7(35%)	6(30%)	7(35%)
E152	20(100%)	0(0%)	0(0%)
Y74N77	15(75%)	1(5%)	4(20%)
N77I80	11(55%)	2(10%)	7(35%)
N77A81	10(50%)	5(25%)	5(25%)

[a] Referred to the lysis of the B*2705-positive HMy2.CIR transfectant at an E/T ratio of 2:1. The relative cytotoxicity of B*2703, designated here as H59, is the cytotoxicity obtained with the B*2703-positive LCL CH, and is referred to that of the B*2705-positive LCL LG15.

Third, the two different mutations introduced at position 77 have different effects on recognition by some CTL clones. This result indicates that not only the location, but also the nature of the change, is relevant in modulating CTL recognition. Analogous observations have been recently made for HLA-A2 in a study involving other polymorphic positions (Hogan et al 1989).

Fourth, for most individual CTL clones, the effect of mutations on spatially close positions may be drastically different. Thus, changes at spatially close positions, related to each other by one α-helical turn , like positions 59 and 63, 74 and 77, 77 and 80, and 77 and 81 (Fig.2), have quite distinct effects for a given CTL clone. These effects may be different in different clones, suggesting that the alterations on CTL recognition induced by a given mutation are due more to a disruption of specific intermolecular contacts than to a more general disturbance of the local conformation in the HLA molecule.

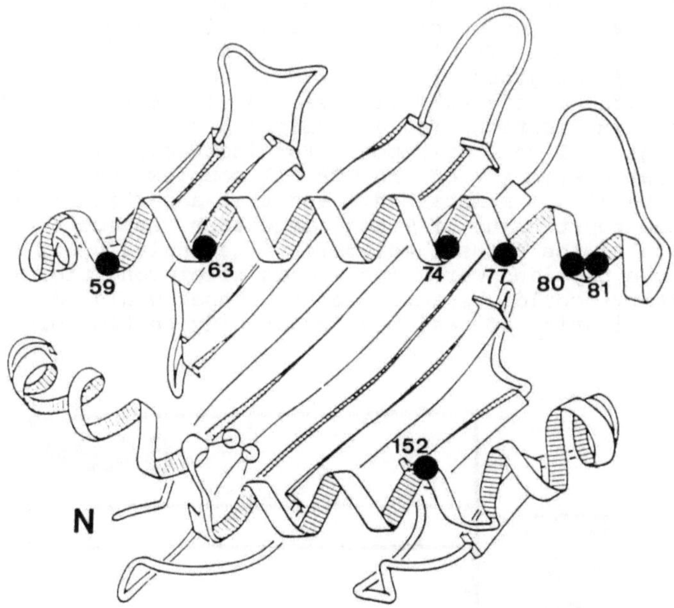

Fig. 2. Spatial location of the mutated residues in the HLA-B*2705 molecule. The figure is based on the three-dimensional structure of HLA-A2 (Bjorkman et al 1987) and shows a top view of the $\alpha 1$ and $\alpha 2$ regions

Fifth, the reaction pattern of CTL 172DRF reveals a heteroclytic effect of certain changes on recognition and lysis. The lytic efficiency of this clone towards the B*2702 transfectant and at least the Y74, N77 or S77 mutants is clearly higher than towards the transfectant expressing the stimulating B*2705 antigen. This suggests that certain mutations in the HLA-B27 antigen structure increase the affinity of its recognition by the corresponding T-cell receptor (TCR).

Sixth, the double mutants show that the negative effect of a mutation on recognition by some CTL clones can be diminished by the introduction of a second change in its vicinity. These compensatory effects are most clear in the reactivity patterns of CTL 202DRD and 172DRF. In both of these clones, the effect of the mutation at position 80 is compensated by the introduction of a second change at position 77. A molecular model to interpret the basis for these compensatory effects is still lacking, but they demonstrate the subtile interplay of polymorphic residues in modulating T-cell recognition and show that, for individual CTL clones, more changes may sometimes mean less antigenic distinctness.

The differential effects of various mutations at one or at spatially close positions, the existence of heteroclytic reactivity induced by some of the changes and the compensatory effects of multiple substitutions on recognition, all indicate the very strict stereochemical requirements of allospecific CTL epitopes.

Seventh, many individual CTL clones are affected by changes located at opposite ends of the peptide binding site of the class I molecule (Figs.1 and 2), suggesting that these epitopes may span the whole site. However, some CTL clones are only affected by changes located at one side of the groove (CTL 46, GM7, and 7). This observation, if confirmed with a more extensive series of mutants, should have significant implications for establishing the topographic diversity of alloantigenic T-cell epitopes. For instance, it is posible that the postulated symmetry of the TCR:HLA complex (Davis et al 1988) might not be always maintained, and that a certain diversity might exist in the positioning of T-cell receptors respect to the peptide binding site of the class I molecule. Alternatively, if presentation of endogenous peptides were involved in allorecongition (see below), such symmetry could be kept, but epitopes would differ in the spatial distribution of residues influencing peptide binding.

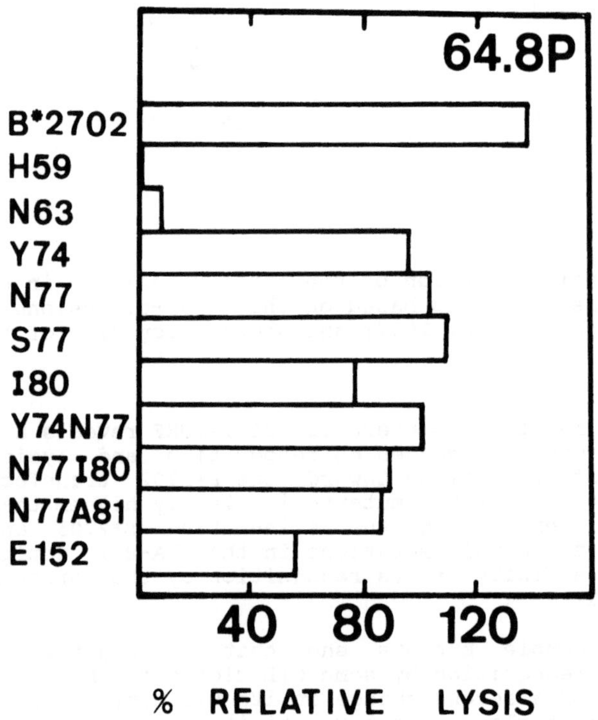

Fig. 3. Cytotoxicity of the anti-HLA-B*2704 CTL clone 64.8P. Experimental details are exactly as described in the legend to Fig. 1. The percent specific lysis of the B*2705-positive transfectant cell and of the B*2705-positive LCL LG15 by this CTL clone at the represented E/T ratio of 2:1 was 70% and 64% respectively

Figure 3 shows the effect of mutations on the reactivity of the single anti-B*2704 CTL clone, 64.8P, used in this study. This clone is unusual in that it is the only one crossreacting with all subtypes except B*2703 (Table 1). On this basis, it was suggested (Jaraquemada et al 1988) to recognize an epitope that would be largely independent from the area around residues 77 and 152, but would critically involve that around residue 59, at the other end of the peptide binding groove (Fig. 2). The result in Fig. 3 supports this suggestion as it shows that, besides the change at residue 59, only that at residue 63, abrogrates lysis.

In summary, the above observations further substantiate the significance of HLA-B27 subtype polymorphism. They indicate that every substitution, among those occurring in HLA-B27 subtypes, is relevant in modulating T-cell recognition of this antigen. This modulation is carried out through a complex interplay of effects between each polymorphic position and its molecular environment and acts on multiple clones that recognize different, but topographically related, epitopes.

The clonal diversity of the allogeneic recognition of HLA-B27 by T cells, the complexity of the epitopes involved and the existence of common structural features to all of them are all compatible with recognition of the alloantigen molecule alone by alloreactive T-cell clones. It could, however, be more easily explained by postulating that allorecognition of HLA-B27, as well as other class I antigens, involves a variety of peptides, perhaps of endogenous origin, bound to the alloantigen molecule. The complex effects observed with mutations could be, to a large extent, due to modulation on peptide binding, rather than to altering direct contact with the TCR. Confirmation of this hypothesis and, eventually, identification of the putative peptides involved, remains a major goal in current studies on alloreactivity.

REFERENCES

Aparicio P, Jaraquemada D, López de Castro JA (1987) Alloreactive cytolytic T cell clones with dual recognition of HLA-B27 and HLA-DR2 antigens. Selective involvement of CD8 in their class I-directed cytotoxicity. J Exp Med 165:428-443

Aparicio P, Jaraquemada D, Rojo S, López de Castro JA (1988) Clonal heterogeneity of HLA-B27 cellular allorecognition. Delineation of immunodominant sites. Eur J Immunol 18:203-209

Barbosa JA, Santos-Aguado J, Mentzer SJ, Strominger JL, Burakoff SJ, Biro PA (1987) Role of Glycosylation in surface expression and functional recognition. J Exp Med 166:1329-1350

Bjorkman PJ, Saper MA, Samraoui B, Bennett WS, Strominger JL, Wiley DC (1987) Structure of the human class I histocompatibility antigen HLA-A2. Nature 329:506-512

Chu G, Hayakawa H, Berg P (1987) Electroporation for the efficient transfection of mammalian cells with DNA. Nuc Acids Res 15:1311-1326

Cowan EP, Jelachich ML, Coligan JE, Biddison WE (1987) Site-directed mutagenesis of an HLA-A3 gene identifies amino acid 152 as crucial for major histocompatibility complex-restricted and alloreactive cytotoxic T-lymphocyte recognition. Proc Natl Acad Sci USA 84:5014-5018

Davis MM, Bjorkman PJ (1988) T-cell antigen receptor genes and T-cell recognition. Nature 334:395-402

Hogan KT, Clayberg C, Bernhard EJ, Walk SF, Ridge JP, Parham P, Krensky AM, Engelhard VH (1988) Identification by site-directed mutagenesis of amino acid residues contributing to serologic and CTL-defined epitope differences between HLA-A2.1 and HLA-A2.3. J Immunol 141:2519-2525

Hogan KT, Clayberg C, Bernhard EJ, Walk SF, Ridge JP, Parham P, Krensky AM, Engelhard VH (1989) A panel of unique HLA-A2 mutant molecules define epitopes recognized by HLA-A2-specific antibodies and cytotoxic T lymphocytes. J Immunol 142:2097-2104

Jaraquemada D, Galocha B, Aparicio P, Rojo S, Calvo S, López de Castro JA (1988) Modulation on immunogenicity by HLA-B27 subtype polymorphism. Eur J Immunol 18:1945-1950

Jelachich ML, Cowan EP, Turner RV, Coligan JE, Biddison WE (1988) Analysis of the molecular basis of HLA-A3 recognition by cytotoxic T cells using defined mutants of the HLA-A3 molecule. J Immunol 141:1108-1113

López de Castro JA (1989) HLA-B27 and HLA-A2 subtypes: Structure, Evolution and Function. Immunology Today. In press (July issue).

Santos-Aguado J, Crimmins MAV, Mentzer SJ, Strominger JL, Burakoff S (1989) Molecular characterization of allospecific cytotoxic T-lymphocyte recognition sites in the HLA-A2 molecule using oligonucleotide-generated site-specific mutants. In Immunobiology of HLA. Vol II: Immunogenetics and histocompatibility. Dupont, B. ed. Springer-Verlag. In press

The Role of Amino Acid Position and Side Chain Structure in Serological and CTL-Defined Epitopes on the HLA-A2.1 Molecule

K. T. HOGAN[1], C. CLAYBERGER[2], N. SHIMOJO[3], W. E. BIDDISON[3], A. M. KRENSKY[2], and V. H. ENGELHARD[4]

[1] Department of Microbiology, Medical College of Wisconsin, Milwaukee, WI 53226, USA
[2] Department of Pediatrics, Stanford University, Stanford, CA 94305, USA
[3] Neuroimmunology Branch, NINCDS, NIH, Bethesda, MD 20892, USA
[4] Department of Microbiology, University of Virginia, Charlottesville, VA 22908, USA

INTRODUCTION

A major focus of current immunological research is to determine the structure-function relationships of class I MHC molecules. The determination of the three-dimensional structure of HLA-A2.1 by Bjorkman et al. (1987a, 1987b) has significantly advanced our ability to achieve this goal. The α3 domain, which composes the membrane proximal region of the molecule, has a structure similar to that of the constant region of immunoglobulin, and primarily associates with β2-microglobulin. The membrane distal region consists of the α1 and α2 domains which interact to form a β-sheet platform, on top of which reside two antiparallel α-helices that are separated by a long groove. It was hypothesized that this groove acts as the binding site for peptides that are recognized by the TCR in association with class I molecules and in fact, an unidentified peptide(s) was associated with this groove in the crystal structure. An analysis of the crystallographic structure of HLA-A2.1 reveals that nearly all of the side chains of the polymorphic residues are oriented such that they point into the groove of the molecule and thus could be important in mediating contact with peptide, or they point upward from the α-helices and may thus be important in interacting with the TCR (Parham et al. 1988).

Our approach to defining the structure-function relationships of class I molecules has been to focus on the amino acid residues that distinguish HLA-A2.1 from the closely related but antigenically distinct molecules HLA-A2.2, and HLA-A2.3. HLA-2.2 differs from HLA-A2.1 by the substitutions of Tyr for Phe_{9}, Arg for Gln_{43}, Leu for Val_{95}, and Trp for Leu_{156}. HLA-A2.3 differs from HLA-A2.1 by the substitutions of Thr for Ala_{149}, Glu for Val_{152} and Trp for Leu_{156}. These substitutions occur in distinct regions of the molecule: amino acids 9 and 95 occur in the β-sheet of the peptide binding groove and are predicted to interact with peptide; amino acids 152 and 156 point into the peptide binding groove from the α2 helix and are also predicted to interact with peptide; amino acid 149 points up from the α2 helix and is predicted to interact with the TCR; and residue 43 occurs in a region of the molecule thought not to interact with the TCR or peptide. Although the variant molecules have been very useful in identifying particular residues in the class I molecule that may be important in CTL-defined and serological epitopes, it has not been possible to define the exact residues involved because the molecules differ by more than one substitution. In the work described here, we have used a combination of site-directed mutagenesis and hemi-exon shuffling to construct genes that encode novel HLA-A2 molecules that can be used to assess the importance of these residues in serological and CTL-defined epitopes.

I.K. Egorov/C.S. David (Eds.)
Transgenic Mice and Mutants in MHC Research

METHODOLOGY

Construction and Expression of Mutant HLA-A2 Molecules

The site-directed mutagenesis, hemi-exon shuffling and subsequent transfection of the mutant HLA-A2 genes has been previously described in detail (Hogan et al., 1988a,b). The recipient cell line used for transfections was Hmy2.C1R, a class I negative derivative of the human lymphoblastoid cell line, LICR.LON.Hmy2 (Storkus et al., 1987). The mutant HLA-A2 molecules utilized in this study are listed in Table 1.

Table 1. Protein Sequences of HLA-A2 Mutant Molecules

	Amino Acid Residues					
Molecule	9	43	95	149	152	156
A2.1	PHE	GLN	VAL	ALA	VAL	LEU
A2.3	-	-	-	Thr	Glu	Trp
TVL[a]	-	-	-	Thr	-	-
AEL	-	-	-	-	Glu	-
AVW	-	-	-	-	-	Trp
TEL	-	-	-	Thr	Glu	-
TVW	-	-	-	Thr	-	Trp
AEW	-	-	-	-	Glu	Trp
PVL	-	-	-	Pro	-	-
AAL	-	-	-	-	Ala	-
AQL	-	-	-	-	Gln	-
APL	-	-	-	-	Pro	-
AVG	-	-	-	-	-	Gly
AVS	-	-	-	-	-	Ser
A2.2	Tyr	Arg	Leu	-	-	Trp
A2.2/1	Tyr	Arg	Leu	-	-	-

[a] The three letters in the name of the mutant molecule correspond to the amino acids present at the positions 149, 152, and 156. Substitutions relative to HLA-A2.1 are indicated by underscoring.

Serological and CTL-Defined Epitope Analysis

Cellular radioimmunoassays (RIA) were done as previously described (Herman et al., 1983) using the mAbs MA2.1 (HLA-A2, -B17) (McMichael et al., 1980), PA2.1 (HLA-A2, -Aw69) (Brodsky et al., 1979), BB7.2 (HLA-A2, -Aw69) (Brodsky et al., 1979) and CR11-351 (HLA-A2, -A28) (Russo et al., 1983). HLA-A2.1-restricted, influenza matrix peptide-specific CTL were generated by culturing HLA-A2.1 positive PBL in the presence of M1 57-68 peptide, and the CTL were assayed on targets that had been preincubated with M1 55-73 peptide (Hogan et al., 1988c). Human CTL specific for HLA-A2.1 were derived and assayed as described by Clayberger et al. (1985). Murine CTL specific for HLA-A2.1 were derived from C57BL/6 mice (clones designated with the prefix AH or AX) and HLA-A2.1 transgenic C57BL/6 mice (clones designated with the prefix AT) and maintained as described (Bernhard et al., 1987, 1988).

RESULTS AND DISCUSSION

Identification of Residues Important in the CR11-351 Serological Epitope

The mAb CR11-351 recognizes an epitope that is present on HLA-A2.1 and HLA-A2.2, and is absent on HLA-A2.3 (Russo et al., 1983). As shown in Table 2, the epitope is absent on the TVL molecule and present on the AEL and AVW molecules. Furthermore, the substitution of both positions 152 and 156 in the AEW molecule does not destroy the epitope. This indicates that the substitution of Thr for Ala_{149} in HLA-A2.3 is responsible for the loss of the CR11-351 epitope. Based on sequence comparisons, Holmes et al. (1987) had previously proposed that residues 142 and 145 also composed a portion of the epitope recognized by CR11-351. An examination of the crystal structure of HLA-A2.1 reveals that residues 142, 145, and 149, all form a common face of an α-helix and would be accessible to binding by an antibody. Curiously, the epitope recognized by CR11-351 is present in the PVL molecule. The lack of effect of a Pro substitution at this position may be due to the fact that residue 149 occurs at a bend connecting two α-helices. Because Pro often occurs in such bends it may not significantly affect the secondary structure of the region. Furthermore, if the methyl group of Ala is part of the epitope that is recognized, it is possible that one of the similar methylene groups in Pro could substitute for it. In addition to the natural substitutions at positions 152 and 156, the

Table 2. Amino Acid Substitution Position and Side-Chain Structure Have Varying Effects on the Same Serological Epitope

	Monoclonal Antibody			
Molecule	MA2.1	PA2.1	BB7.2	CR11-351
A2.1	++++	++++	++++	++++
A2.3	++++	++++	++++	-
TVL	++++	++++	++++	-
AEL	++++	++++	++++	++++
AVW	++++	++++	++++	++++
TEL	++++	++++	++++	-
TVW	++++	++++	++++	-
AEW	++++	++++	++++	+++
PVL	++++	++++	++++	++++
AAL	++++	++++	++++	++++
AQL	++++	++++	++++	++++
APL	++++	++++	++++	-
AVG	++++	++++	++++	-
AVS	++++	++++	++++	++++

substitutions in the AAL, AQL, and AVS molecules, also did not lead to a loss of CR11-351 binding. However, the epitope was destroyed on the APL and AVG molecules. The fact that several substitutions at positions 152 and 156 have no effect on the epitope recognized by CR11-351 indicates that Val_{152} and Leu_{156} in HLA-A2.1 are not a part of the CR11-351 epitope. It seems most likely that the loss of the epitope that occurs when either Pro_{152} or Gly_{156} is present is mediated indirectly through conformation changes in the molecule. Such changes could be introduced through the propensity of both these amino acids to introduce bends into α-helices. Since both positions 152 and 156 occur within an a-helix, the introduction of a bend at either of these positions could result in an alteration of the epitope that is formed with residues 142, 145, and 149.

M1 Peptide Specific Recognition in the Context of HLA-A2 Mutant Molecules

The major component of the HLA-A2.1 restricted CTL response to influenza virus A/JAP/57 is the recognition of a single matrix protein peptide, M1 55-73 (Gotch et al., 1987). By studying CTL of this specificity it is possible to determine how amino acid substitutions in HLA-A2 affect the recognition of a single class I molecule-peptide complex. Of the 24 different CTL lines and clones that were obtained from individuals that expressed HLA-A2.1 and no other HLA-A2 allele, a single pattern of recognition referred to as "common" was observed as shown in Table 3. These CTL recognized the peptide when it was presented in association with the TVL molecule which suggests that residue 149 is not involved in the formation of the epitope recognized by M1 55-73 peptide-specific CTL. However, the epitope is not formed when the TVL transfectants are pulsed with peptide immediately before being assayed, although it is present following an overnight incubation with the peptide (Hogan et al., 1988c). These results indicate that the substitution of Thr for Ala_{149} decreases the rate at which the peptide can associate with the class I molecule and that the substitution does not alter the epitope once it has formed. These results also demonstrate that caution is needed in classifying residues as primarily being involved in either TCR or peptide interactions as residue 149 that is predicted to interact with the TCR can influence the rate of association of the peptide with the HLA-A2 molecule. These same CTL did not recognize the peptide-pulsed AEL transfectant. In fact, any substitution at position 152 destroyed the epitope recognized by these CTL. While the naturally occurring substitution of Trp for Leu_{156} completely destroyed the epitope recognized by these CTL, the peptide could be recognized to some extent in the context of the AVS and AVG molecules. These data indicate that residue 152, and to a lesser extent residue 156, are of major importance in determining whether or not the M1 55-73 peptide can be recognized in the context of the HLA-A2 molecule. Thus, in the case of HLA-A2.1-restricted, M1 55-73 peptide-specific CTL, it appears that the inability to recognize the peptide in association with HLA-A2.3 is mediated through the residues that interact primarily with the peptide and not with the TCR.

A second recognition pattern was also discernable among the tested CTL (Table 3). CTL demonstrating this pattern of recognition could only be obtained from donor Q66 and this person has subsequently been shown to express both HLA-A2.1 and -A2.2 (Shimojo et al., 1989). CTL from this individual were capable of recognizing the M1 55-73 peptide in association with the AAL and AVW molecule after a secondary stimulation. When the PBL of this individual were stimulated with the M1 55-73 peptide-pulsed AVW transfectant, CTL were generated that

Table 3. Recognition of the M1 55-73 Peptide in Association with the HLA-A2 Mutant Molecules

	M1 (55-73) Peptide Presented in the Context of:			
	A2.1			AVW
Transfectants	"common"[a]	2° Q66[b]	4° Q66[c]	Q66[d]
HLA-A2.1	+++	+++	+++	+++
TVL	+++	+++	+++	+++
AEL	-	-	-	-
AQL	-	-	-	++
APL	-	ND	-	+/-
AAL	-	++	-	++
AVW	-	++	-	++
AVS	+	ND	+	++
AVG	++	ND	++	++

[a]Reactivity pattern of the CTL derived from all donors with the exception of Q66.
[b]Reactivity pattern of CTL derived from secondary cultures of donor Q66.
[c]Reactivity pattern of CTL derived from tertiary cultures of donor Q66.
[d]Reactivity pattern of donor Q66 CTL when stimulated with M1 peptide in the context of AVW.

recognized the peptide in association with all of the mutants with the exception of AEL.

The fact that Q66 CTL are still capable of recognizing the M1 55-73 peptide in association with the AQL, APL, AAL, and AVW molecules indicates that the peptide is still capable of binding to the mutants. The inability of the "common" CTL clones to recognize the peptide in association with these molecules was not therefore, due to a failure of the peptide to bind to the molecule. Rather, the complex that forms must be altered in some fashion such that the TCR no longer recognizes it. This could be mediated through a change in peptide conformation or by direct recognition of residue Val_{152} and Leu_{156} themselves. Because no CTL have been found that recognize the peptide in association with the AEL molecule, it is possible that the substitution of Glu for Val_{152} prevents the binding of the peptide with the molecule. A resolution to this question will have to await the results of a direct binding assay, or the isolation of a clone that can recognize the peptide in the context of the AEL molecule. Perhaps most interesting, is the degree of flexibility that is demonstrated by the MHC-peptide interaction in tolerating changes within the MHC molecule itself. This is particularly true when one considers that the M1 peptide has only been shown to be recognized in association with HLA-A2 (Gotch et al., 1987). These results imply that there are at least two mechanisms to account for the inability to recognize a given peptide in association with a particular class I molecule. First, there could be a selective ability of the class I molecule to bind a particular peptide so that no recognition occurs in the absence of peptide binding. Secondly, the peptide could bind to the class I molecule, but in a conformation that is not recognized by any TCR.

Effect of Amino Substitutions in the β-sheet and the α2-Helix of HLA-A2.1 on Allogeneic and Xenogeneic CTL Recognition

The molecules HLA-A2.1, HLA-A2.2, and HLA-A2.3 are not only distinguishable by HLA-A2.1-restricted, M1 peptide-specific CTL, but can also be distinguished by allogeneic and xenogeneic CTL as well. To evaluate how the amino acid substitutions between HLA-A2.1 and HLA-A2.2 contribute to the epitope differences between the molecules, we have utilized both hemi-exon shuffling and site-directed mutagenesis to create two novel molecules. The first molecule, A2.1/2, differs from A2.1 at positions 9 and 95 in the β-sheet of the molecule and by a substitution at position 43 in a loop connecting two β-strands in the α2 domain. The second molecule created is AV<u>W</u> which differs from HLA-A2.1 only at residue 156. By studying the effects of the substitutions in both A2.1/2 and AV<u>W</u> it is possible to determine to what extent changes in the β-sheet and the α2-helix of the molecule affect the epitopes recognized by allogeneic CTL. The reactivity patterns of human allogeneic CTL clones that recognize epitopes present on HLA-A2.1 but not HLA-A2.2 were studied (Table 4). Two of the clones shown recognized epitopes that are influenced primarily by either residues 9, 43, and 95, or by residue 156. Thus the epitope recognized by AMSB.1 was present on the AV<u>W</u> molecule, but absent on the A2.1/2 molecule. Conversely, the epitope recognized by AMSB.8 was present on the A2.1/2 molecule and partially destroyed on the AV<u>W</u> molecule. The epitope recognized by clone AMSB.3 was not present on either the A2.1/2 or AV<u>W</u> molecules, while the epitope recognized by clone ALI1.1 was present on both the A2.1/2 and AV<u>W</u> molecules. Thus, both of these clones define epitopes that are sensitive to changes in both regions of the molecule. However, they are different in that clone AMSB.3 is sensitive to changes in either region, while clone ALI1.1 is sensitive only when changes are made in both sets of residues. It is possible therefore, to define four different epitopes that are recognized among these CTL clones. Similar patterns of recognition were also found for xenogeneic CTL clones as well (Hogan et al., 1988b). These data indicate that at least one of residues 9, 43, and 95, as well as residue 156 are important in defining the epitopes recognized by these CTL. The results further demonstrate that while some epitopes can be mapped to a particular region of the molecule, other epitopes are formed through a complex interaction of residues in different regions of the molecule.

Table 4. Effect of Substitutions in the β-Sheet and the α2 Helix of HLA-A2.1 on the Epitopes Recognized by Allogeneic CTL

	% specific ^{51}Cr-release from:				
CTL Clones	A2.1	A2.2	A2.1/2	AV<u>W</u>	C1R
AMSB.1	40	7	0	29	0
AMSB.8	41	3	40	21	0
AMSB.3	37	2	0	2	0
ALI1.1	28	3	45	29	0

Effect of Amino Acid Substitutions at Residues 149, 152, and 156 on Allogeneic and Xenogeneic CTL Recognition

The epitopes recognized by the allogeneic and xenogeneic CTL clones utilized in this study are not present on the HLA-A2.3 molecule. An extensive analysis was therefore performed to determine the overall importance of residues 149, 152, and 156 in defining the epitopes recognized by these CTL. The panel of mutants included both substitutions that occur naturally at these positions in HLA-A2.3, as well as additional substitutions at these same positions (Table 1). In light of the fact that the substitutions of Pro for Val_{152} and Gly for Leu_{156} might be expected to have significant effects on the secondary structure of the molecule as reflected in their effects on the CR11-351 epitope (Table 2), it was important to establish that at least some allogeneic epitopes were still present on these molecules. As shown in Figure 1, the epitopes recognized by at least one allogeneic clone are still present for each of the mutant molecules tested. This result, combined with the fact that most of the serological epitopes on these molecules are still intact (Table 2 and Hogan et al., 1989), argues that the overall structure of the molecule has not been altered to such an extent that it is no longer recognizable by any CTL clone.

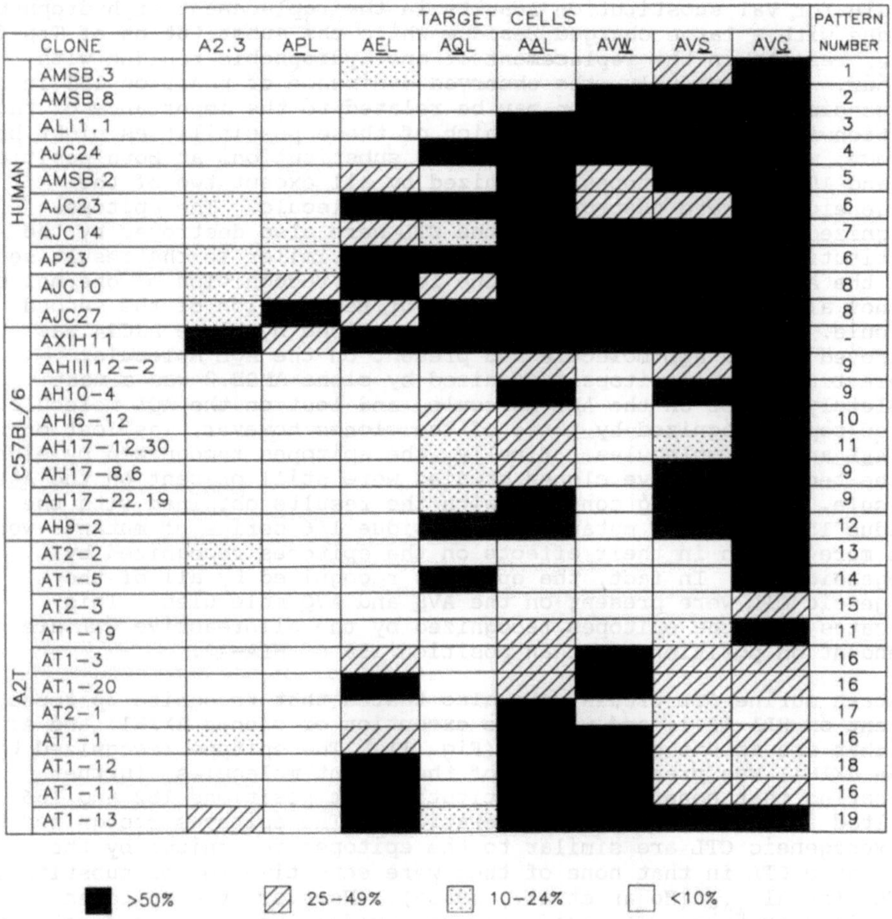

Figure 1. Reactivity patterns of allogeneic and xenogeneic CTL clones. The results are presented as relative lysis compared with reactivity on HLA-A2.1. Only clones that did not recognize an epitope present on HLA-A2.3 were included in the epitope pattern count

In comparing the effects of the naturally occurring substitutions on epitopes recognized by allogeneic CTL, it was found that the epitopes were always present on the TVL molecule (Hogan et al., 1988a), and were often lost on the AEL and AVW molecules (Fig. 1). Based on reactivity patterns with the AEL and AVW molecules, a minimum of four different epitopes could be discerned. One type of epitope was absent when either substitution was made, as exemplified by clone AMSB.3. The second pattern of recognition is illustrated with clone AMSB.8 which recognizes an epitope that is absent on the AEL molecule and present on the AVW molecule. In contrast clone AJC23 recognizes an epitope that is present on the AEL molecule and largely destroyed on the AVW molecule. Finally, clones AP23 and AJC10 recognize an epitope that is present when any single amino acid substitution is made. These results indicate that residue 149 is relatively unimportant in defining allogeneic CTL epitopes and that residues 152 is more important than residue 156.

The apparent importance of position 152 may be related to the fact that the Glu for Val substitution results in the replacement of hydrophobic residue with a large charged residue while the substitution of Trp for Leu_{156} results in the replacement of one hydrophobic residue with another. Alternatively, the observed dominance of position 152 in allogeneic epitope formation may be related to the importance of the position itself. To determine which of these possibilities might be correct, we have introduced additional substitutions at both positions 152 and 156. The epitopes recognized by all except two of the allogeneic CTL were destroyed on the APL molecule. The epitopes recognized by about one half of the CTL were also destroyed by the substitution in the AQL molecule. This is similar to the result seen with the AEL molecule, although the loss of recognition of one molecule did not always correspond to the loss of recognition of the second molecule. For example, the epitope recognized by clone AJC24 was destroyed on the AEL molecule and present on the AQL molecule. Interestingly, the epitope recognized by clone AMSB.2 was almost completely intact on the AEL molecule, and lost on the AQL molecule. The epitope recognized by three of the clones however, was lost on both the AEL and AQL molecules. Finally, the epitopes recognized by eight of the ten alloreactive clones studied were still present on the AAL molecule. In striking contrast with the results obtained with the residue 152 series of mutants, the residue 156 series of mutants were much more benign in their effects on the epitopes recognized by allogeneic CTL. In fact, the epitopes recognized by all of the allogeneic CTL were present on the AVS and AVG molecules. This indicates that the epitopes recognized by the alloreactive CTL are more dependent on position 152 than position 156 of HLA-A2.

Nineteen murine CTL clones were also tested that recognize epitopes present on HLA-A2.1, and with the exception of clones AXIH11 and AT1-13, that are absent on HLA-A2.3 (Fig. 1). The epitope recognized by clone AXIH11 was present on all of the mutant molecules, further indicating that none of the substitutions at positions 152 and 156 resulted in gross structural alterations. The epitopes recognized by the xenogeneic CTL are similar to the epitopes recognized by the allogeneic CTL in that none of them were sensitive to the substitution of Thr for Ala_{149} (Hogan et al., 1988a). However, the epitopes recognized by these CTL could only be divided into two broad categories based on their reactivity with the AEL and AVW molecules. The majority of the clones were sensitive to both the substitution of Glu for Val_{152} and Trp for Leu_{156}. However, a number of clones recognized an epitope that was insensitive to any single amino acid substitution at these positions. The xenogeneic clones that recognized an epitope present on both the AEL and AVW molecules were further tested on a panel of double mutant molecules that included TEL, TVW, and AEW (Hogan et al., 1988a). The results indicated that when residues 149 and 152, or residues 152 and 156 were substituted in the same molecule, the epitope was destroyed. However, the epitope was not affected in the molecule substituted at positions 149 and 156. These results indicate two things. First, position 152 is most important in defining the epitope recognized by these CTL, and secondly, position 149 can sometimes be demonstrated to affect CTL-defined epitopes.

The epitopes recognized by the xenogeneic clones were all sensitive to the substitution of Pro for Val_{152}. The substitutions in the AEL and AQL molecules were comparable to one another in that they both affected the epitopes recognized by the majority of the CTL clones. There were only three clones that had their epitope destroyed by the substitution of Ala at this position. Thus, the epitopes recognized by the xenogeneic CTL were similar to those recognized by the allogeneic CTL in that they were affected in the same hierarchical fashion by the substitutions at position 152 in the HLA-A2 molecule. Although non-naturally occurring substitutions at position 156 had a minimal effect on the epitopes recognized by allogeneic CTL, they had a substantially greater effect on the epitopes recognized by xenogeneic CTL. Thus the epitopes recognized by 8 CTL clones were destroyed when Ser was substituted for Leu_{156} and the epitopes recognized by 2 CTL clones were destroyed when Gly was substituted at the same position.

Overall Role of Residues in the β-Sheet and α2-Helix in Defining Serological, M1 55-73 Peptide-Specific and Allogeneic CTL Epitopes

Amino acid position 149, with its side chain extending up and away from the HLA-A2 molecule, is important in the formation of the CR11-351 serological epitope. Interestingly, even though this residue has been predicted to interact with the TCR (Bjorkman et al., 1987b), it was not found to affect epitopes recognized by either M1 55-73 peptide-specific or allogeneic CTL. This contrast sharply with the results of Ajitkumar et al. (1988) who studied a series of $H\text{-}2K^b$ mutant molecules that differed at residues predicted to interact with the TCR. By using a large series of anti-K^b CTL clones these authors demonstrated that these positions were important in defining the epitopes recognized by allogeneic CTL. However, the panel of mutant $H\text{-}2K^b$ molecules did not include a substitution at position 149 so it is not possible to determine what effect such a substitution would have had on $H\text{-}2K^b$-specific CTL. These results may indicate that the residue at position 149 is not recognized by the TCR or that if it is, the TCR of the CTL are of sufficient affinity that the substitution of Thr for Ala_{149} does not affect the interaction.

Residues 9 and 95 in the β-sheet and residues 152 and 156 in the α2-helix of the molecule all have side chains that extend into the peptide binding pocket of the HLA-A2 molecule, and are important in the formation of CTL-defined epitopes but not in the formation of serological epitopes. This finding, along with the fact the residue 149 does not appear to be involved in the formation of epitopes recognized by the TCR, suggests that differences in peptide binding are responsible for the epitope differences among HLA-A2 variant molecules. This is particularly true of residues 9 and 95 which occur on the floor of the "peptide binding pocket" and that would probably be inaccessible to direct TCR interaction. This suggests that differences in peptide binding rather than differences in class I molecule residues that interact with the TCR, may be more critical in defining class I variant molecule differences. The peptide mediated effects may either occur through the binding of a different subset of peptides, or alternatively, many of the same peptides may bind in a different conformation. In either case, the ultimate result would be the display of a new set of epitopes that would distinguish the variant and prototypic molecules. This does not preclude the possibility however, that in at least some circumstances the direct recognition of residues 152 or 156 by the TCR is important.

When the substitution of Glu for Val_{152} or Trp for Leu_{156} leads to the loss of a CTL-defined epitope, one possible explanation for the loss is that the epitopes are dependent upon Val_{152} or Leu_{156}. However, in many of these cases it is possible to substitute additional amino acids at these sites and maintain the presence of a particular epitope. This indicates that the epitopes recognized by most CTL do not require Val_{152} or Leu_{156}, but rather that the nature of the substitution at these positions is important. These same results also indicate that charge substitutions are not an important force in altering CTL-defined epitopes as proposed by Monos et al. (1984). This is evidenced by the fact that the neutral substitution of Gln for Val_{152} is sometimes more destructive for CTL-defined epitopes than is the negatively charged Glu. In addition, the substitution of Ala at the same position is often as destructive as the substitution of Glu.

Based on the fact that position 156 displays a higher degree of polymorphism than does position 152, it might be expected that position 156 would be more important in CTL-defined epitope formation than position 152. However, the results demonstrate that position 152 is of greater importance than position 156. For example, a conservative substitution at position 152 has a stronger influence on CTL-defined epitopes than does a nonconservative substitution at position 156. This suggests that the degree of polymorphism of a position may not be a good indicator of the significance of a position in CTL-defined epitopes. Interestingly, position 152 appears to be of great importance in CTL-defined epitopes on class I molecules other than HLA-A2 including HLA-A3 (Cowan et al., 1987; Jelachich et al., 1988), HLA-B27 (Vega et al., 1985), and $H\text{-}2K^b$ (Nathenson et al, 1986). The reason for the significance of residue 152 in defining epitope differences among class I molecules is not presently clear as there is nothing in the structure of the molecule to suggest that this position should be more important than any others in either peptide binding or TCR interactions.

HLA-A2 Mutant Molecules Define a Large Number of Epitopes Recognized by CTL

In comparing the reactivity patterns of the 28 different allogeneic and xenogeneic CTL utilized in this study, a total of 19 different patterns of reactivity can easily be discerned. These include 8 recognition patterns among 10 allogeneic CTL clones and 11 recognition patterns among 18 xenogeneic CTL clones. Each of these recognition patterns defines a separate epitope that is recognized on the HLA-A2 molecule. Because there is little duplication in the epitopes that are observed it is likely that this represents a lower limit for the number of epitopes that can be recognized on the HLA-A2 molecule. In fact, if reactivity on HLA-A2.2 and the double mutants is taken into account, then there are a minimum of 22 different epitopes that can be discerned among the 28 CTL clones. Furthermore, because we are considering a subset of clones that recognize HLA-A2.1 and not HLA-A2.3, it is likely that a large number of other epitopes could be further defined. These results demonstrate that there is considerable diversity in the TCR repertoire that is utilized to recognize HLA-A2 as an alloantigen. This presents a strong contrast to the results that have been obtained with the HLA-A2.1-restricted, M1 55-73 peptide-specific CTL in which only a single pattern of recognition can be detected when the donor expresses only the HLA-A2.1 allele. This suggests that a limited TCR repertoire is available to respond against a defined class I molecule-peptide complex. Because a defined class I molecule-peptide complex can only be recognized by a very limited portion of the TCR repertoire it seems unreasonable that a class I variant differing by only a few residues from the prototypic molecule should be recognized by a very diverse TCR response. Because this is in fact the case, it suggests that the epitopes recognized by allogeneic CTL may in fact consists of a diverse collection of self-peptides.

ACKNOWLEDGMENT

This work was supported by Grants AI20963 (VHE) and CA47609 (AMK) from the United States Public Health Service (USPHS). AMK is supported by an American Heart Association Clinician-Scientist Award. During a portion of this work, KTH was a USPHS trainee, supported by Grant CA09109.

REFERENCES

Ajitkumar P, Geier SS, Kesari KV, Borriello F, Nakagawa M, Bluestone JA, Saper MA, Wiley DC, Nathenson SG (1988) Evidence that multiple residues on both the α-helices of the class I MHC molecule are simultaneously recognized by the T cell receptor. Cell 54:47-56

Bernhard EJ, Le AT, Barbosa JA, Lacy E, Engelhard VH (1988) Cytotoxic T cells from HLA-A2 transgenic mice specific for HLA-A2 expressed on human cells. J Exp Med 168:1157-1162

Bernhard EJ, Le AT, Yannelli JR, Holterman MJ, Hogan KT, Parham P, Engelhard VH (1987) The ability of cytotoxic T cells to recognize HLA-A2.1 or HLA-B7 antigen expressed on murine cells correlates with their epitope specificity. J Immunol 139:3614-3621

Bjorkman PJ, Saper MA, Samraouri B, Bennet WS, Strominger JL, Wiley DC (1987a) Structure of the human class I histocompatibility antigen, HLA-A2. Nature 329:506-512

Bjorkman PJ, Saper MA, Samraouri B, Bennet WS, Strominger JL, Wiley DC (1987b) The foreign antigen binding site and T cell recognition regions of class I histocompatibility antigens. Nature 329:512-518

Brodsky FM, Parham P, Barnstable CJ, Crumpton MJ, Bodmer WF (1979) Monoclonal antibodies for analysis of the HLA system. Immunological Rev 47:3-61

Clayberger C, Holmes N, Wang PL, Koller TD, Parham P, Krensky AM (1985) Determinants recognized by human cytotoxic T cells on a natural hybrid class I HLA molecule. J Exp Med 162:1707-1714

Cowan EP, Jelachich ML, Coligan JE, Biddison WE (1987) Site-directed mutagenesis of an HLA-A3 gene identifies amino acid 152 as crucial for major histocompatibility complex-restricted and alloreactive cytotoxic T-lymphocyte recognition. Proc Natl Acad Sci USA 84:5014-5018

Gotch F, Rothbard J, Howland K, Townsend A, McMichael, A. (1987) Cytotoxic T lymphocytes recognize a fragment of influenza virus matrix protein in association with HLA-A2. Nature 326:881-882

Herman A, Parham P, Weissman SM, Engelhard VH (1983) Recognition by xenogeneic cytotoxic T lymphocytes of cells expressing HLA-A2 or HLA-B7 after DNA-mediated gene transfer. Proc Natl Acad Sci USA 80:5056-5060

Hogan KT, Clayberger C, Bernhard EJ, Walk SF, Ridge JP, Parham P, Krensky AM, Engelhard VH (1988a) Identification by site-directed

mutagenesis of amino acid residues contributing to serologic and CTL-defined epitope differences between HLA-A2.1 and HLA-A2.3. J Immunol 141:2519-2525
Hogan KT, Clayberger C, Bernhard EJ, Walk SF, Ridge JP, Parham P, Krensky AM, Engelhard VH (1989) A panel of unique HLA-A2 mutant molecules defines epitopes recognized by HLA-A2 specific antibodies and CTL. J Immunol 142:2097-2104
Hogan KT, Clayberger C, Le AT, Ridge JP, Walk SF, Parham P, Krensky AM, Engelhard VH (1988b) CTL-defined epitope differences between HLA-A2.1 and HLA-A2.2 map to two distinct regions of the molecule. J Immunol 141:4005-4011
Hogan KT, Shimojo N, Walk SF, Engelhard VH, Maloy WE, Coligan JE, Biddison WE (1988c) Mutations in the α-2 helix of HLA-A2 affect presentation but do not inhibit binding of influenza virus matrix peptide. J Exp Med 168:725-736
Holmes N, Ennis P, Wan AM, Denney DW, Parham P (1987) Multiple genetic mechanisms have contributed to the generation of the HLA-A2/A28 family of class I MHC molecules. J Immunol 139:936-941
Jelachich ML, Cowan EP, Turner RV, Coligan JE, Biddison WE (1988) Analysis of the molecular basis of HLA-A3 recognition by cytotoxic T cells using defined mutants of the HLA-A3 molecule. J Immunol 141:1108-1113
McMichael AJ, Parham P, Rust N Brodsky F (1980) A monoclonal antibody that recognizes an antigenic determinant shared by HLA-A2 and B17. Hum Immunol 1:121-129
Monos DS, Tekolf WA, Shaw S, Cooper ML (1984) Comparison of structural and functional variation in class I HLA molecules: The role of charged amino acid substitutions. J Immunol 132:1379-1385
Nathenson SG, Geliebter J, Pfaffenbach GM, Zeff RA (1986) Murine major histocompatibility complex class-I mutants: Molecular Analysis and Structure-Function Implications. Ann Rev Immunol 4:471-502
Parham P, Lomen CE, Lawlor DA, Ways JP, Holmes N, Coppin HL, Salter RD, Wan AM, Ennis PD (1988) Nature of polymorphism in HLA-A, -B, and -C molecules. Proc Natl Acad Sci USA 85:4005-4009
Russo C, Ng A-K, Pellegrino MA, Ferrone S (1983) The monoclonal antibody CR11-351 discriminates HLA-A2 variants identified by T cells. Immunogenetics 18:23-35
Shimojo N, Cowan EP, Engelhard VH, Maloy WL, Coligan JE, Biddison WE (1989) A single amino acid substitution in HLA-A2 can alter the selection of the CTL repertoire that responds to influenza virus matrix peptide 55-73. J Immunol 143:558-564
Storkus W, Howell DN, Salter RD, Dawson JR, Cresswell P (1987) NK susceptibility varies inversely with target cell class I HLA antigen expression. J Immunol 138:1657-1659
Vega MA, Ezquerra A, Rojo S, Aparicio P, Bragado R, Lopez de Castro JA (1985) Structural analysis of an HLA-B27 functional variant: Identification of residues that contribute to the specificity of recognition of cytolytic T lymphocytes. Proc Natl Acad Sci USA 82:7394-7398

Two Amino Acids on the α Helical Region of the α_1 Domain Form Multiple Epitopes Recognized by HLA-Bw52 Specific Human Cytotoxic T Cell

M. TAKIGUCHI, J. YAMAMOTO, H. HAYASHI, and K. KANO[1]

[1] Department of Immunology, Institute of Medical Science, University of Tokyo, Minato-ku, Tokyo 108, Japan

INTRODUCTION

We have previously demonstrated that amino acid substitutions of Asn →Glu at position 63 and Ser→Phe at position 67 are only differences between HLA-B51 and HLA-Bw52 (Hayashi 1989). As these substitutions are located on the helical region of the α_1 domain, both T and B cells are suspected to discriminte one from the other. Several studies (Ajitkumar 1988, Hogan 1988) using mutant HLA or H-2 antigens recently demonstrated that the repertoire of allogeneic CTL was extremely large. We focused on only two residues of the helical region on the α_1 domain, in which the amino acid substitutions between HLA-B51 and Bw52 are found, to analyze the allorecognition of CTL. We attempted to generate HLA-Bw52 specific CTL clones and constructed the chimeric molecules between HLA-B51 and Bw52 to investigate the epitope formation of alloantigen which T cells recognize and the repertoire of allogeneic CTL against only two amino acid substitutions.

RESULTS AND DISCUSSIONS

Generation of HLA-Bw52 specific CTL clones

Five HLA-Bw52 specific CTL clones from an individual with HLA-B51 and six from that without HLA-B5 CREG were generated and their allospecificity was determined by using pannel cells with HLA-B51, Bw52 or B35 and Hmy2CIR cells expressing HLA-B51, Bw52 or B35. These results indicate that two amino acid substitutions at these residues form the dominant epitopes that T cells recognize.

Construction of chimeric molecules between HLA-B51 and HLA-Bw52.

Two chimeric genes, HLA-B51/52 (Asn^{63}, Ser^{67}) and HLA-B52/51 (Glu^{63}, Phe^{67}) were constructed between HLA-B51 and Bw52. These hybrid genes as well as parental genes were tansfected into HLA-A, B negative Hmy2CIR cells. The clones expressing high levels of these gene products were selected by flowcytometry.

The epitopes recognized by HLA-Bw52 specific human CTL clones.

Killing of Hmy2CIR cells expressing HLA-B51, Bw52 or chimeric moleucles by six HLA-Bw52 specific CTL clones from the individual without HLA-B5 CREG revealed four different pattens of recognition (Fig.1). These results indicate that HLA-Bw52 specific CTLs recognize the multiple epitopes which are formed by the combination of two residues at positions 63 and 67 on the helical region.

I.K. Egorov/C.S. David (Eds.)
Transgenic Mice and Mutants in MHC Research

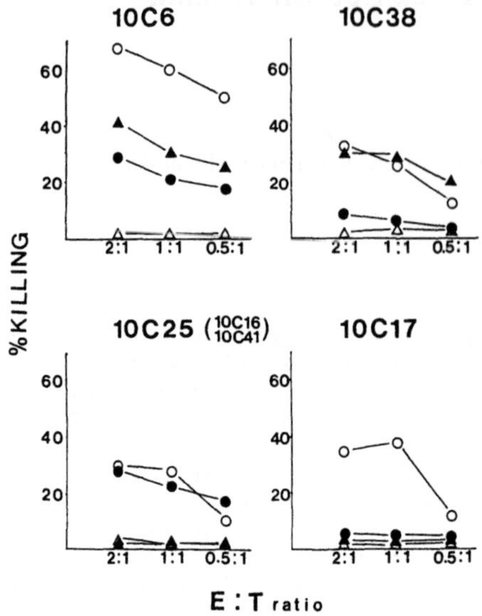

Fig.1. Recognition of chimeric molecules between HLA-B51 and Bw52 by HLA-Bw52 specific CTL clones from an individual without HLA-B5 CREG. Hmy2CIR cells expressing HLA-B51(△), HLA-Bw52(○), HLA-B51/52(▲), HLA-B52/51(●) were used as target cells.

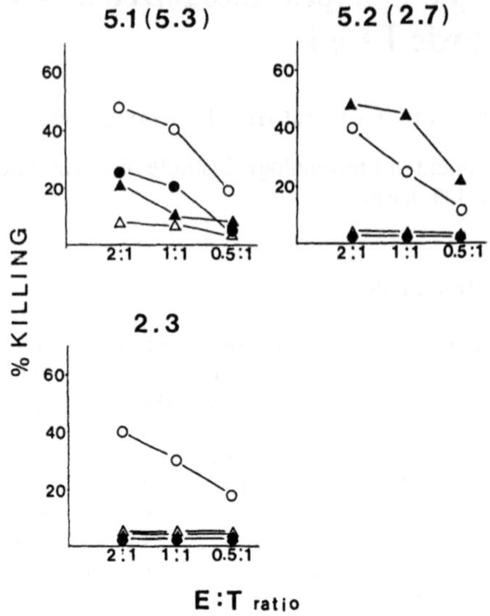

Fig.2. Recognition of chimeric molecules between HLA-B51 and Bw52 by HLA-Bw52 specific CTL clones from an individual with HLA-B51. Hmy2CIR cells expressing HLA-B51(△), HLA-Bw52 (○), HLA-B51/52(▲), HLA-B52/51 (●) were used as target cells.

The results shown in Fig.1 might be influenced by HLA phenotype of the CTL donor. To investigate this possibility, five HLA-Bw52 speific human CTL clones from the individual with HLA-B51 were tested against Hmy2CIR cells expressing chimeric molecules (Fig.2). Killing of Hmy2CIR cells expressing chimeric molecules by these clones also revealed the different paterns of recognition. Three different recognition patterns analogous to those shown in Fig.1, except that of the clones recognizing B52/51 but not B51/52, were observed.

Both results described in Figs.1 and 2 demonstrated that the epitopes of HLA-Bw52 discriminated from HLA-B51 by the T cell clones were not single but multiple, suggesting a diverse repertoire of allogeneic T cells against only two amino acid substitutions.

REFERENCES

Ajitkumar P, Geier SS, Kesari KV, Borriello F, Nakagawa M, Bluestone JA, Saper MA, Wiley DC, Nathenson SG (1988) Evidence that multple residues on both the α-helices of the class I MHC molecule are stimulataneously recognized by the T cell receptor. Cell 54:47-56.

Hayashi H, Ennis PD, Ariga H, Salter RD, Parham P, Kano K, Takgicuhi M (1989) HLA-B51 and HLA-Bw52 differ by only two amino acids which are in the helical region of the α_1 domain. J. Immunol. 142:306-311.

Hogan KT, Clayberger C, Bernhard EJ, Walk SF, Ridge JP, Parham P, Krensky AM, Engelhard VH (1988) Identification by site-directed mutanogenesis of amino acid residues contributing to serologic and CTL-defined epitope differences between HLA-A2.1 and HLA-A2.3. J. Immunol. 141:2519-2525.

III. Regulation of MHC Gene Expression

Mouse Mammary Tumor Virus-Directed Gene Expression in Transgenic Mice

S. R. Ross[1], Y. Choi[1], C.-L. L. Hsu[2], and J. Dudley[2]

[1] Department of Biochemistry, University of Illinois, Chicago, IL, USA
[2] Department of Microbiology, University of Texas, Austin, TX, USA

INTRODUCTION

Transgenic mice have proven to be useful for the study of many different biological phenomenon, such as gene regulation, oncogenesis, and immunological function. The expression of novel gene products in transgenic mice, such as the major histocompatibility antigens, has allowed investigators to more easily study the role of these antigens in tissue transplantation and rejection, in organ development and tumorigenesis. A useful vector for use in such studies would be one that is capable of directing expression of a variety of gene products to specific cell types in different tissues. We have been studying the tissue-specific expression of the mouse mammary tumor virus (MMTV), with the goal of developing such vectors.

MMTV is an endogenous retrovirus in mice that is expressed at high levels in the epithelial cells of the lactating mammary gland and at lower levels in the epithelial cells of the salivary gland, seminal vesicles, lung and kidney and in the lymphoid cells of the spleen and thymus (Henrard and Ross, 1988). Transcription of MMTV is stimulated by glucocorticoid hormones, the result of a glucocorticoid response element (GRE) encoded in the long terminal repeat (for review, see Yamamoto, 1985) of the virus. The MMTV LTR has been used by several groups to drive the expression of marker and oncogenes in transgenic mice (Stewart et al., 1984; Ross and Solter, 1985; Leder et al., 1986; Choi et al., 1987, 1988). We have begun to dissect the LTR to determine whether it encodes multiple regulatory elements that are responsible for the tissue-specific expression of MMTV. We have also used the LTR to direct the expression of the SV40 early region genes, large T and small t antigen, to these different tissues in order to study their oncogenic potential.

RESULTS

Expression of MMTV in Transgenic Mice

We used the MMTV LTR linked to two different marker genes to make transgenic mice (Fig.1). The full-length LTR (1200 to +101) was ligated upstream of the SV40 early region to create LTag (Choi et al., 1987). To examine the expression of both the endogenous virus and the transgene, we subcloned the fragment of LTag containing the MMTV-SV40 junction into a Gemini vector (Promega Biotech, Inc.) to use for RNAseT1 protection assays (Fig.1). We also used immunohistochemical staining with antibody against the SV40 large

I.K. Egorov/C.S. David (Eds.)
Transgenic Mice and Mutants in MHC Research

tumor antigen to determine the cell type in the various tissues in which expression occurred (Choi et al., 1987, 1988). Expression of the transgene in LTag mice was identical to that of the endogenous provirus (Table I, Fig.1). Transgene expression was detected in the epithelial cells of the mammary and salivary gland, lung, kidney, prostate/seminal vesicle, and in the lymphoid cells of the spleen and thymus.

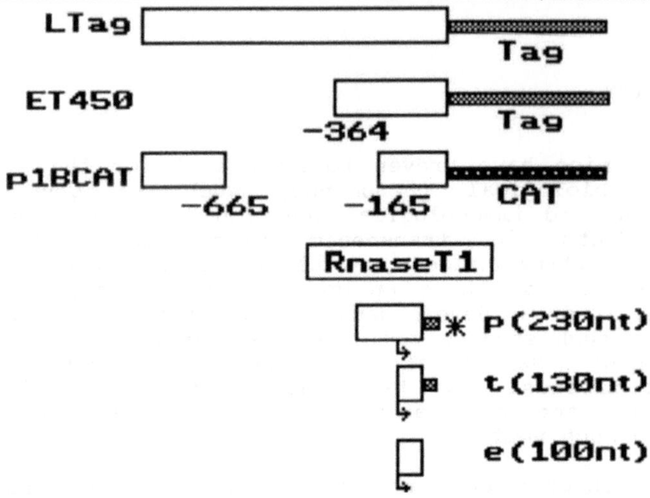

Figure 1. Diagram of LTag, ET450 and p1BCAT transgenes and the probe used for RNaseT1 mapping. Abbreviations: p, probe; t, transgene; e, endogenous

To determine whether the expression that we detected in the lymphoid organs was the result of expression in B and/or T cells, we removed the spleens of CD1 mice and cultured cell suspensions made from these organs in the presence of lipopolysaccharide or Concanavalin A for 48 hours. This treatment results in populations of cells that are about 75% enriched for B-cells or T-cells, respectively. When RNAseT1 protection assays were carried out on RNA isolated from these populations, we saw protection of an 130nt band, corresponding to transcription from the endogenous proviruses (Fig.2). The amount of protection from these populations was approximately equal, indicating that expression of endogenous provirus was occurring in both B and T cells. This is consistent with previous results in which transgenic mice which had the LTR linked to a variety of oncogenes got both B- and T-lymphomas (Stewart et al., 1984; Leder et al., 1986; Choi et al., 1987, 1988).

The Role of the SV40 Large and Small Tumor Antigens in Tumorigenesis

Because the MMTV LTR can be used to direct the expression of linked genes to several different cell types and tissues, we decided to test the role of the SV40 tumor antigens in the transformation of these tissues. We made transgenic mice containing the SV40 early region genes under the control of the MMTV LTR to determine whether the

Table 1. Endogenous MMTV and transgene expression in LTag, ET450 and p1BCAT transgenic mice.

		Tissue										
Strain	#[a]	MG	SG	LU	KI	SV	SP	TH	TE	BR	SM	HE
LTag	>20	e	e	e	e	e	e	e	e	-	-	-
		t	t	t	t	t	t	t	t	-	-	-
ET450	3	e	e	e	e	e	e	e	e	-	-	-
		t	t	t	t	t	t	t	t	t	t	t
p1BCAT	2	e	e	e	e	e	e	e	e	-	-	-
		t	t	t	t	t	t	t	t	t	t	t

[a] Number of independently generated transgenic strains.

Abbreviations: MG, mammary gland; SG, salivary gland; LU, lung; KI, kidney; SV, seminal vesicle; SP, spleen; TH, thymus; TE, testes; BR, brain; SM, skeletal muscle; HE, heart; e, endogenous; t, transgene; -, no expression.

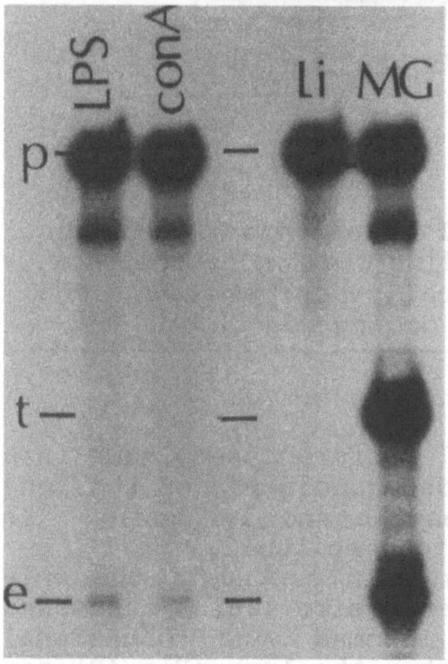

Figure 2. RNaseT1 protection assay of RNA isolated from LPS- and Concanavalin A-treated spleens from normal mice and the liver and mammary gland from LTag transgenic mice. Abbreviations: p, probe; t, transgene; e, endogenous; Li, liver; MG, mammary gland

SV40 tumor antigens showed tissue-specificity in their transforming ability and to examine the role of the SV40 small tumor antigen in transformation. Although it has long been known that the large tumor antigen of SV40 is required for transformation, the role of the small tumor antigen is less clear (Tegtmeyer, 1980).

Three different transgenes were used to make transgenic mice (Fig.3). The LTag transgene contained the entire SV40 early region under the control of the MMTV LTR, while the LTM transgene contained these same sequences, except that the small t coding region was disrupted by the insertion of a fragment of the MMTV coding region; this insertion resulted in a loss of stable small t antigen mRNA (Choi et al., 1988). The LST transgene, which codes only for a functional small t protein, was made by deleting a HpaI fragment in the large T coding region (Fig.3). Six different LTag transgenic mice, five different LTM transgenic mice and two strains of LST transgenic mice were produced. All of the transgenic mice expressed the transgene in the lung, kidney, mammary and salivary gland, prostate, seminal vesicles, and lymphoid tissue (Choi et al., 1987, 1988; Choi and Ross, unpublished).

Figure 3. Diagram of LTag, LTM, and LST transgenes. Open box, MMTV LTR; dotted box, small t antigen coding region; hatched box, large T antigen coding region; diagonally-striped box, MMTV coding region genes inserted in small t coding region in LTM

The LTag transgenic mice developed several different neoplasms in a variety of tissues (Table 2), including adenocarcinomas of the lung (Fig.4A), kidney and prostate/seminal vesicle and lymphomas. In addition to the neoplasms, there was atypical epithelial cell proliferation in the mammary gland ducts. In contrast, the only neoplasms that occurred in the LTM transgenic mice, which expressed only the large T antigen, were lymphomas and occasional prostate/seminal vesicle carcinomas. Neither the lung (Fig.4B) nor the kidney exhibited any neoplasia in the LTM mice. Not surprisingly, LST mice, which express only the small t tumor antigen, were completely normal histologically in all tissues, including the lung (Fig.4C) and survived to normal life expectancy.

To show that the large T and small t tumor antigens could cooperate to cause lung and kidney adenocarcinomas when they were inherited on different chromosomes, we bred LTM 54.3 transgenic mice with LST 5 transgenic mice and compared histological sections of the lung, kidney and lymphoid tissues of their offspring. The mice that inherited both the LTM and the LST transgenes developed adenocarcinomas of the lung (Fig.4D) and kidney and lymphomas, whereas the animals that inherited the LTM and LST transgenes alone

Table 2. Tumorigenesis in LTag, LTM and LST transgenic mice

Strain	Age (days)	Sex	Tissue						# mice
			LU	KI	LY	MG	SG	SV	
LTag1	75	M	+	+	-	-	-	+	1
LTag2	75	F	+	+	-	D	-	NA	1
LTag3	150	F	D	-	+	D	-	NA	1
LTag4	1	?	+	NA	+	NA	-	NA	1
LTag5	120	F	+	D	+	D	-	NA	1
LTag6	1	?	-	-	-	-	-	NA	1
LTM21	120	M	-	-	+	-	-	+	1
LTM57	195	M	-	-	+	-	-	+	>20
LTM54.1	150	M/F	-	-	+	D	-	+	2
LTM54.2	150	M/F	-	-	+	D	-	+	2
LTM54.3	180	M/F	-	-	+	D	-	+	>20
LST5	>200	M/F	-	-	-	-	-	-	>20
LST2	>200	M/F	-	-	-	-	-	-	>20
LST5 x LTM54.3	200	M/F	+	+	+	-	-	+	2

Abbreviations and symbols: +, adenocarcinoma in lung, kidney, mammary gland or seminal vesicle, lymphoma in lymphoid (LY) tissue; -, no tumor seen; NA, not available for analysis; D, dysplastic; additional abbreviations, see Table 1 legend.

were identical to their parents, developing only lymphomas or remaining free of tumors, respectively. Thus, the small t and large T tumor antigens can function in trans.

Negative Regulation is Involved in Tissue-Specific Expression

To determine whether there were different elements within the LTR which specified expression in the different organs and whether the GREs had any role in this expression, we created transgenic mice with two different deletions in the LTRs, ET450 and p1BCAT (Fig.1). Transgene ET450 has a deletion of all of the LTR sequences 5' of -364 and contains the SV40 early region as a marker gene; Northern blot and RNAseT1 protection analysis of RNA, as well as Western blot analysis and immunocytochemistry was used to examine expression in ET450 transgenic mice, as previously described (Choi et al., 1988). Transgene p1BCAT has an LTR deletion from -165 to -665 and contains the bacterial chloramphenicol acetyltransferase (CAT) gene (Fig.1); CAT enzymatic activity and RNaseT1 protection assays were both carried out to analyze transgene expression in p1BCAT mice.

We found that both of these transgenes are expressed in the same tissues as the endogenous proviruses (Table I). In addition, the ET450 and p1BCAT transgenes are expressed in the heart, brain and skeletal muscle of these animals, tissues that express neither endogenous virus nor transgenes containing the entire LTR. Thus, deletion of a region between -364 and -665 resulted in expression of MMTV-directed transgenes in novel tissues, the heart, brain and

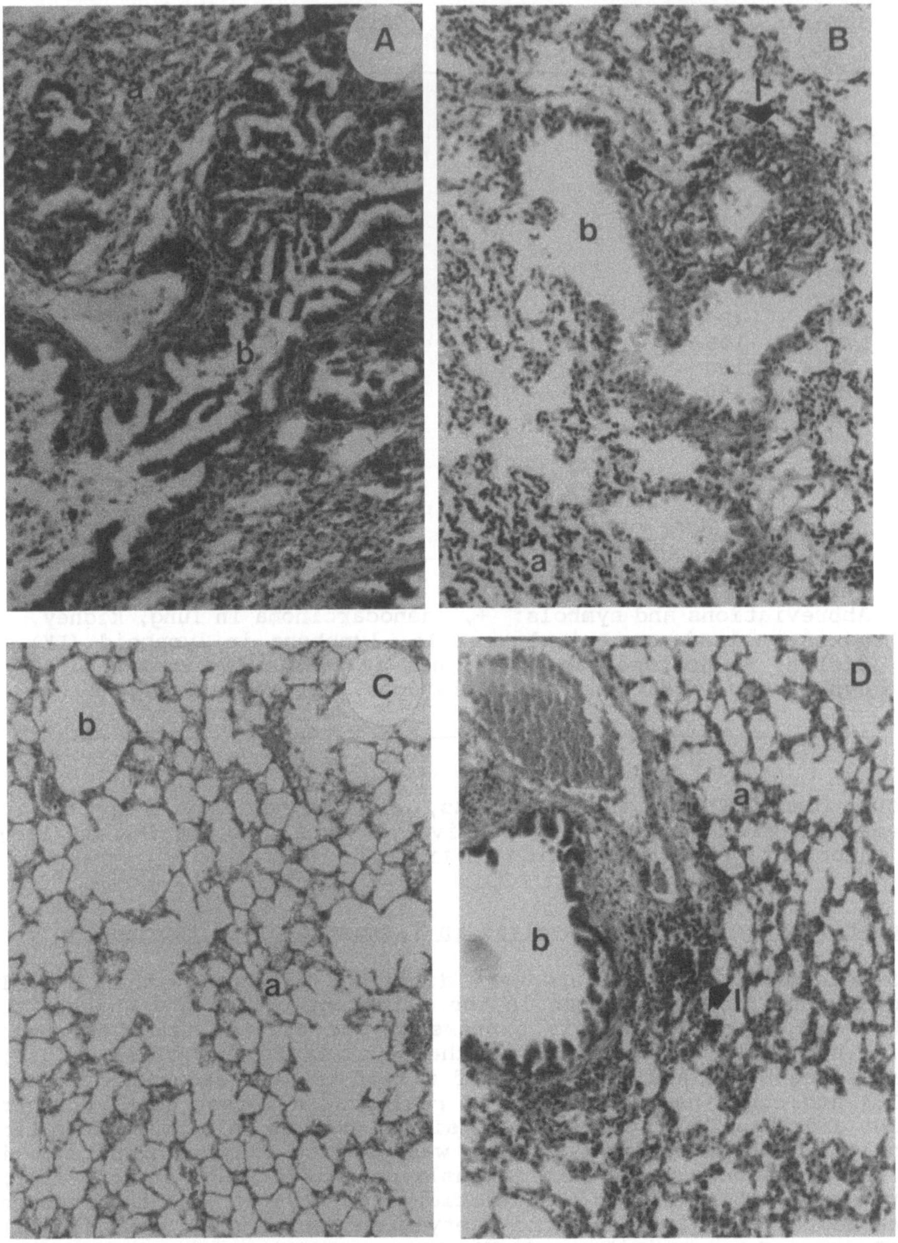

Figure 4. H & E staining of sections of the lung from LTag, LTM, LST and LTM/LST mice. The bronchial epithelial cells (b) of the LTag (A) and LST/LTM (D) mouse show malignant proliferation, while those of the LTM (B) and LST (C) mice are normal. Alveoli (a) are normal in all four sections. Lymphoma cells (l) are also visible in (B) and (D)

skeletal muscle. ET450 mice, which express both the large T and small t antigens, develop adenocarcinomas of the lung, kidney and prostate/seminal vesicle, as well as lymphomas (not shown), but do not appear to develop tumors in heart, brain or skeletal muscle. Preliminary in situ hybridization studies (not shown), indicate that transgene expression is occurring at low levels in all cells in these tissues; the level of the SV40 tumor antigens in these cells may not be high enough to mediate transformation.

The 5' end of the LTR encodes an enhancer

We also compared the relative levels of expression of the transgenes to that of endogenous virus in the tissues in which endogenous MMTV is expressed. The level of expression of endogenous MMTV varies from tissue to tissue, with mammary gland having very high levels, salivary gland and prostate/seminal vesicle intermediate levels and lung, kidney and lymphoid tissue the lowest levels of MMTV RNA (Henrard and Ross, 1988). The expression of transgenes containing the complete LTR, such as LTag, LTM, or LST in mice, mimics that of the endogenous virus in this pattern (Choi et al., 1987).

When we examined expression in the ET450 mice, however, a different pattern of expression was seen. Although all of the same tissues that transcribe endogenous MMTV provirus expressed the transgene (Table 1), the relative level of transgene RNA in the lactating mammary glands and salivary glands of these mice was lower or equal to that detected in the other organs (Table 3). This indicated that the region from -364 to -1185 of the LTR contained an element that caused increased expression in the mammary and salivary glands. This deletion had no apparent effect on the relative level of expression in the prostate/seminal vesicle.

Table 3. Comparison of the levels of expression of endogenous MMTV to transgene in LTag, ET450 and p1BCAT mice.

	MG	SG	SV	LU	KI	LY
Endogenous MMTV	+++	++	++	+	+	+
LTag	+++	++	++	+	+	+
ET450	+	+/-	++	++	++	++
p1BCAT	+++	++	++	+	+	+

Multiple + represent highest levels of expression; +/- represents very low expression; - represents no expression.

To further map this putative enhancer element, we examined expression in the p1BCAT mice. The relative level of expression of the p1BCAT transgene in these mice again mimicked that of the endogenous virus (Table 3). Thus, since the p1BCAT transgene contains the region from -665 to -1185 of the LTR, a putative enhancer for mammary gland and salivary gland maps to this region.

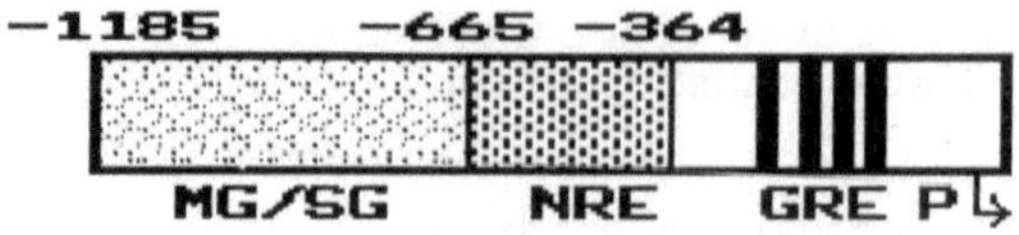

Figure 5. Diagram of the MMTV LTR

DISCUSSION

We have shown that the MMTV LTR is capable of directing the expression of marker genes to the ductal epithelial cells of the mammary gland, salivary gland, lung, kidney, prostate/seminal vesicle and to lymphoid cells. In addition, we have begun to define the elements responsible for this expression. Our results suggest that there is a negative regulatory element(s) (NRE) within the LTR, which maps between -365 and -664, relative to the start of transcription and that in addition, an enhancer for mammary and salivary gland expression lies upstream of -664 (Fig.5). This NRE may act to prevent transcription in tissues that do not normally express virus or transgenes containing the full-length LTR, perhaps because this element represents a binding site for a factor(s) that suppresses transcription in these cells.

Since the ET450 and p1BCAT transgenic mice all express the transgene in the same tissues as endogenous virus or transgenes containing the full-length LTR, it appears that sequences responsible for expression in mammary and salivary gland, thymus, spleen, lung and seminal vesicle are present downstream of -165. The region downstream of -165 encodes both the MMTV promoter and the glucocorticoid regulatory elements (Fig.5); tissue specificity could be encoded by either or both of these regions. Studies to determine whether the GRE or the promoter or both mediate tissue-specificity are underway. This dissection of the elements present in the LTR allows us to manipulate the LTR sequences in order to direct expression of linked transgenes to various tissues.

We have used this ability to direct tissue-specific expression of LTR-linked transgenes to study the ability of the SV40 tumor antigens to transform different cell types. We showed that small t antigen cooperates with the large T antigen in the transformation of ductal epithelial cells in the lung and kidney and that the large T antigen alone is capable of transforming lymphoid cells and prostate/seminal vesicle epithelial cells. Ductal epithelial cells in the lung and kidney are slowly dividing in the adult, while lymphoid stem cells and the epithelial cells of the prostate continue to divide. Our results suggest that the SV40 small t antigen is necessary for rapid transformation of slowly dividing ductal epithelial cells. For the lymphoid tissue and the prostate, and perhaps for the mammary gland, which also divides postnatally, the factors which induce cellular division may obviate the need for small t function in the transformation process.

In conclusion, we have shown that the MMTV LTR represents a useful control element for directing the expression of linked genes, such as oncogenes, to the tissues of transgenic mice. Additionally, since the LTR contains a glucocorticoid-hormone inducible GRE, that also functions in transgenic mice (Ross and Solter, 1985), the level of expression of linked genes can be manipulated by administration of hormone to transgenic animals. The use of this viral element to direct expression should prove useful in many studies.

REFERENCES

Choi Y, Henrard D, Lee I, Ross SR (1987) The mouse mammary tumor virus long terminal repeat directs expression in epithelial and lymphoid cells of different tissues in transgenic mice. J Virol 61:3013-3019

Choi Y, Lee I, Ross SR (1988) Requirement for the Simian Virus 40 small tumor antigen in tumorigenesis in transgenic mice. Mol and Cell Biol 8:3382-3390

Henrard D and Ross SR (1988) Endogenous mouse mammary tumor virus is expressed in several organs in addition to the lactating mammary gland. J Virol 62:3046-3049

Leder A, Pattengale PK, Kuo A, Stewart TA, Leder P (1986) Consequence of widespread deregulation of c-myc gene in transgenic mice: multiple neoplasms and normal development. Cell 45:485-495

Ross SR and Solter D (1985) Glucocorticoid regulation of mouse mammary tumor virus sequences in transgenic mice. Proc Natl Acad Sci USA 82:5880-5994

Tegtmeyer P (1980) Molecular biology of tumor viruses, 2nd ed. Cold Spring Harbor Laboratory, Cold Spring Harbor, New York, p. 297.

Yamamoto KR (1985) Steroid receptor regulated transcription of specific genes and gene networks. Ann Rev Genet 19:209-252.

Regulation of Class I MHC Expression: In Vivo Function of Regulatory DNA Sequence Elements in Transgenic Mice

J. E. MAGUIRE[1], R. EHRLICH[1], W. I. FRELS[2], and D. S. SINGER[1]

[1] Experimental Immunology Branch, Building 10, Room 4B-17, NIH, Bethesda, MD 20892, USA
[2] Reproduction Lab., USDA-ARS, BARC-E, Beltsville, MD 20705, USA

INTRODUCTION

The class I MHC genes constitute a multigene family consisting of a set of highly homologous DNA sequences (Kindt, Singer, 1987). Despite the relatedness of the class I genes, their products have been divided into three sub-families, based on their function, serological characteristics, and tissue distribution (Klein, 1975). One of these subfamilies, containing the classical class I genes, encodes transplantation antigens which are the primary restriction elements recognized in a cytolytic T cell response (Singer, Maguire, 1989). The other two families, containing the non-classical class I genes such as Qa and TL in the mouse, are less well functionally characterized, but are distinguished by the tissue distribution of their gene products (Singer, Maguire, 1989). Early serological studies of the transplantation antigens indicated that they are ubiquitously expressed, but that their level of expression varies among the tissues (Amos, Kostyu, 1980; Bodmer, 1981). Subsequent analyses of class I RNA levels with molecular probes has revealed that steady state levels of class I mRNA vary, suggesting that class I gene expression is transcriptionally regulated (Singer, Maguire, 1989; Fahrner, Hogan, Flavell, 1987; Ehrlich et al 1989). Analyses from our laboratory, as well as others, have identified DNA sequence elements which regulate class I gene expression in transfected cell lines (Ehrlich et al 1989; Satz et al 1985; Israel et al 1986; Korber, Hood, Stroynowski, 1987; Singer et al 1988; Miyazaki, Appella, Ozato, 1986; Kimura et al 1986; Ehrlich, Maguire, Singer, 1988). However, none of these has been shown to function in vivo. In the present report, we used transgenic mice to study the role of DNA sequence elements in _in vivo_ expression. Using a series of derivatives of a class I MHC

I.K. Egorov/C.S. David (Eds.)
Transgenic Mice and Mutants in MHC Research

gene, we find that a silencer element associated with the class I gene (Ehrlich, Maguire, Singer, 1988) contributes to the establishment of tissue-specific levels of expression.

RESULTS

A series of postive and negative regulatory elements are associated with the 5′ flanking sequences of class I MHC genes (Ehrlich, Maguire, Singer, 1988; Kimura et al 1986; Miyazaki, Appella, Ozato, 1986) (Figure 1). In addition, an interferon response element has been identified in the classical transplantation antigen genes (Ehrlich et al 1989; Satz et al 1985; Israel et al 1986; Korber, Hood, Stroynowski, 1987; Singer et al 1988). To assess the role of these elements, as well as others directly linked to a class I gene, we constructed a transgenic mouse with a 9 kb segment of swine DNA containing 1.1 kb of 5′ flanking sequence, 3.5 kb of a class I gene (PD1), and 4.4 kb of 3′ flanking sequences (Frels et al 1985). All of the regulatory elements diagrammed in Figure 1 were contained within this piece of DNA. PD1 expression in

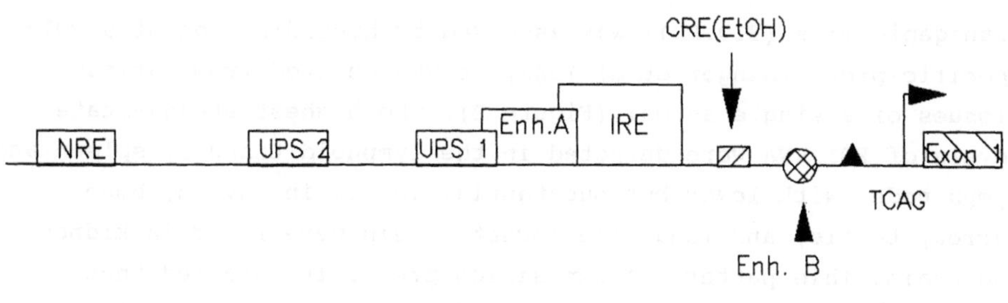

Figure 1. Regulatory DNA sequence elements in a class I MHC gene. Contained within the 5′ flanking sequences of class I MHC genes are a series of regulatory elements. Including among these are the promoter (▲) consisting of the TATA and CCAAT boxes, enhancers (enhancer A, enhancer B), a cAMP responsive element and a silencer (NRE). In addition, two conserved sequences (UPS1 and UPS2) have been identified in all swine class I genes

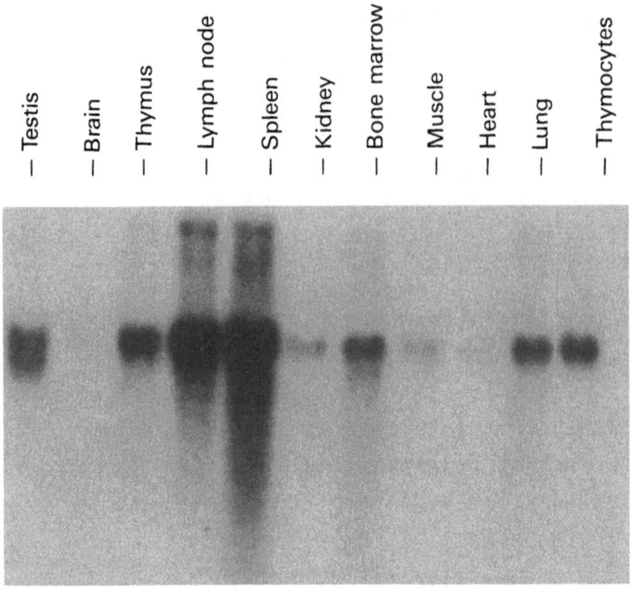

Figure 2. PD1 is expressed in transgenic mice. Total RNA (10ug) samples from tissues of a single animal were resolved on formaldehyde-agarose gels and hybridized with a specific PD1 probe

transgenic mice (B10.PD1) was assessed by hybridization of a PD1-specific probe (Singer et al 1982) to RNA derived from various tissues of a single animal (Figure 2). The highest steady-state levels of PD1 RNA were detected in the lymphoid tissues, spleen and lymph node, with lower but substantial levels in thymus, bone marrow, testis, and lung. The lowest levels were found in kidney and brain. This pattern of expression precisely paralled that observed for PD1 in situ in the pig (Ehrlich et al 1989). Thus the elements necessary for normal PD1 expression are contained within the 9 kb swine DNA segment introduced into the transgenic mouse.

To directly analyze the role of 5′ flanking sequences in the regulation of PD1 expression, a series of nested deletions was generated from the 1.1 kb of 5′ flanking DNA sequences linked to

PD1. These deletions, all sharing a common 3′ boundary which spanned a basic promoter, were then ligated to a reporter gene, chloramphenicol acetyl transferase (CAT). The ability of each of these constructs to direct CAT synthesis was assessed both in transfected L cells and in transgenic mice. The results for two of these constructs are summarized in Figure 3. In transfected L cells, an alternating array of positive and negative regulatory elements was identified (Figure 4). In particular, a strong silencer element was identified in the DNA segment between -516 and -894. The ability of this silencer to function in vivo was demonstrated by the finding of consistently higher levels of CAT activity in the tissues of transgenic mice containing the -516 construct than in those containing the -894 construct (Figure 3). In addition, its level of activity appears to be tissue specific. Thus, the silencer is not only active in vivo, but also functions to determine tissue levels of expression.

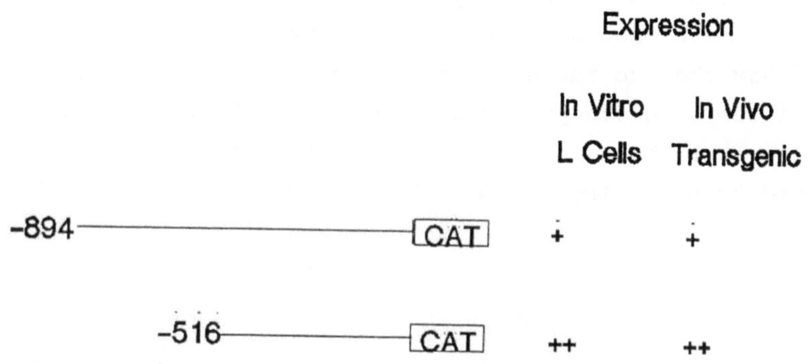

Figure 3. PD1 contains a silencer element which functions both in transgenic mice and transfected cells. Each of the deletion mutants was introduced into either L cells or transgenic mice. Levels of expression were assessed by assaying for CAT activity. -894 and -516 constructs terminate at 894 bp and 516 bp upstream of transcriptional initiation

DISCUSSION

Molecular dissection of the 5′ flanking sequences of class I DNA has identified a series of regulatory elements. From the studies in our laboratory, there are at least four negative or positive elements, as well as an interferon response element and a cAMP response element. In addition, at least one other enhancer element (enhancer B) has been identified associated with the mouse H-2K^b gene (Kimura et al 1986). Each of these elements, individually, has been shown to modulate class I transcriptional levels in transfected tissue culture cells. In the present study, we establish for the first time that at least one of these elements, a silencer, also functions in vivo in transgenic mice. Indeed, its level of activity varies among the different tissues such that it is most active in tissues which express the least amount of class I mRNA (data not shown). These data suggest that the silencer contributes to the determination of the final level of class I expression in a given tissue.

Given the complex array of regulatory elements occurring in the 5′ flanking sequences, it is unlikely that the silencer element alone, or any one of the elements alone, is sufficient to regulate class I expression. How then do these various elements act in concert to establish tissue specific patterns of expression? Previous studies from our laboratory have demonstrated that the silencer element has an associated trans acting factor (Ehrlich, Maguire, Singer, 1988).

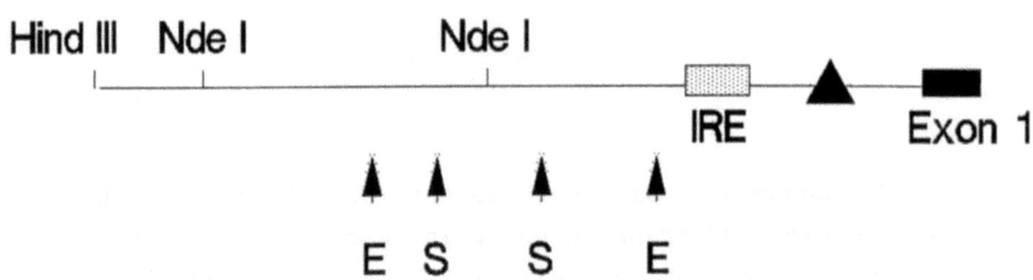

Figure 4. PD1 contains an array of positive and negative regulatory elements. S= silencer, E= enhancer

The ability of this trans acting factor to function is dependent upon its interaction with an enhancer and its cognate binding factor. Thus, we speculate that class I regulation is a dynamic process in which levels of expression are determined by an equilibrium between complexes consisting of either a silencer and enhancer (leading to reduced expression) or a set of enhancers (leading to increased expression). Studies are currently underway to identify these complexes and their substituent cognate factors.

References

Amos, D. and Kostyu, D. HLA - a central immunological agency of man.In: Advances in human genetics, edited by Harris, H. and Hirschhorn,K. New York: Plenum Press, 1980, p. 137-157.

Bodmer W (1981) HLA structure and function: a contemporary view.Tissue Antigens 17:9

Ehrlich R, Maguire J, Singer D (1988) Identification of negative and positive regulatory elements associated with a class I major histocompatibility complex gene. Mol Cell Biol 8:695

Ehrlich R, Sharrow S, Maguire J, Singer D (1989) Expression of a class I MHC transgene: Effects of in vivo a/b-interferon treatment. Immunogenet In press:

Fahrner C, Hogan B, Flavell R (1987) Transcription of H-2 and Qa genes in embryonic and adult mice. EMBO J 6:1265

Frels W, Bluestone J, Hodes R, Capecchi M, Singer D (1985) Expression of a microinjected porcine class I major histocompatibility complex gene in transgenic mice. Science 228:577

Israel A, Kimura A, Fournier A, Fellous M, Kourilsky P (1986) Interferon response sequence potentiates activity of an enhancer in the promoter region of a mouse H-2 gene. Nature 322:743

Kimura A, Israel A, LeBail O, Kourilsky P (1986) Detailed Analysis of the Mouse H-2Kb Promoter: Enhancer-like Sequences and Their Role in the Regulation of Class I Gene Expression. Cell 44:261

Kindt T, Singer D (1987) Class I Major Histocompatibility Complex Genes in Vertebrate Species: What is the Common Denominator?. ImmunolRes 6:57

Klein, J. Biology of the mouse histocompatibility complex, New York:Springer-Verlag, 1975.

Korber, B., Hood, L. and Stroynowski, I. Regulatory Elements in the Promoter of the H-2D[d] Class I Gene. In: H-2 Antigens, edited by David, C.S. : Plenum Publishing Corp., 1987, p. 201-208.

Miyazaki J, Appella E, Ozato K (1986) Negative regulation of the MHC class I gene in undifferentiated embryonal carcinoma cells. ProcNatlAcadSciences(USA) 83:9537

Satz M, Wang L, Singer D, Rudikoff S (1985) Structure and expression of two porcine genomic clones encoding class I MHC antigens. JImmunol 135:2167

Singer D, Camerini-Otero D, Satz M, Osborne B, Sachs D, Rudikoff S (1982) Characterization of a porcine genomic clone encoding a major histocompatibility antigen: expression in mouse L cells. Proc Natl Acad Sci USA 79:1403

Singer, D., Ehrlich, R., Golding, H., Satz, L., Parent, L. and Rudikoff, S. Structure and expresion of class I MHC genes in the miniature swine. In: The Molecular Biology of the Major Histocompatibility Complex of Domestic Animal Species, edited by Warner, C., Rothschild, M. and Lamont, S. Ames: Iowa State University Press, 1988, p.53-62.

Singer D, Maguire J (1989) Regulation of the Expression of Class I MHC Genes. CRC Critical Reviews in Immunology (In Press)

Expression of a Non-Classical Class I Gene in Transgenic Mice

C. Bordallo[1], W. I. Frels[2], J. Lunney[3], A. Rosenberg[4], S. Rudikoff[5], and D. S. Singer[1]

[1] Experimental Immunology Branch, Bldg. 10, Room 4B-17, NIH, Bethesda, MD 20892, USA
[2] Reproduction Lab. and [3] HDL, ARS, USDA, Beltsville, MD 20705, USA
[4] CBER, FDA, Bethesda, MD 20892, USA
[5] Laboratory of Genetics, NCI, NIH, Bethesda, MD 20892, USA

Introduction

The class I MHC genes can be divided into two major categories: classical and non-classical (Singer, Maguire, 1989). The classical class I genes encode the major transplantation antigens. The non-classical class I genes encode highly related cell surface molecules, which associate with β_2 -microglobulin, but which do not function as major transplantation antigens. Indeed, the function, if any, of these gene products is not known. Early serological studies suggested that cell-surface expression of the non-classical gene products was tissue-restricted (Flaherty, 1976). However, more recent studies have detected a more general distribution of non-classical class I transcripts (Fahrner, Hogan, Flavell, 1987; Singer, Maguire, 1989). Nevertheless, their patterns of expression are still distinct from those of the classical class I genes (Ehrlich et al 1987). In order to elucidate the molecular mechanisms which regulate such non-classical class I genes, we have been studying their patterns of expression and comparing them with those of the classical class I genes.

The seven class I MHC sequences in the genome of the miniature swine can be divided into three sub-families by DNA sequence homology (Singer et al 1988). One of the families contains three members; two of these members, PD1 and PD14, encode the major transplantation antigens of the swine. The remaining member, PD7, appears to be a non-classical class I gene. In this study, we report that although PD7 grossly parallels the classical class I genes in its pattern of expression, its level of expression is much lower. This reduced level of expression is mediated by at least two distinct regulatory DNA sequence elements, both of which are demonstrated to function in vivo in transgenic mice.

Results

PD7 was isolated from pig genomic DNA in a clone (PD7.7) which contained a 7.7 kb DNA fragment (Figure 1). DNA sequence analysis demonstrated that PD7 is over 80% homologous to the classical class I genes, PD1 and PD14, but only about 50% homologous to the remaining SLA class I genes which have been sequenced (Singer et al1988). PD7 consists of eight exons with open reading frames and legitimate splice sites. It is thus capable of encoding a class I molecule. In addition to the PD7.7 clone, another clone (PD7.4) containing the entire PD7 gene was isolated. Unlike PD7.7 which extends approximately 3 kb 5′ of the PD7 gene, PD7.4 is truncated 600 bp upstream of the transcriptional initiation and thus contains only a 4.3 kb DNA insert (Figure 1).

I.K. Egorov/C.S. David (Eds.)
Transgenic Mice and Mutants in MHC Research

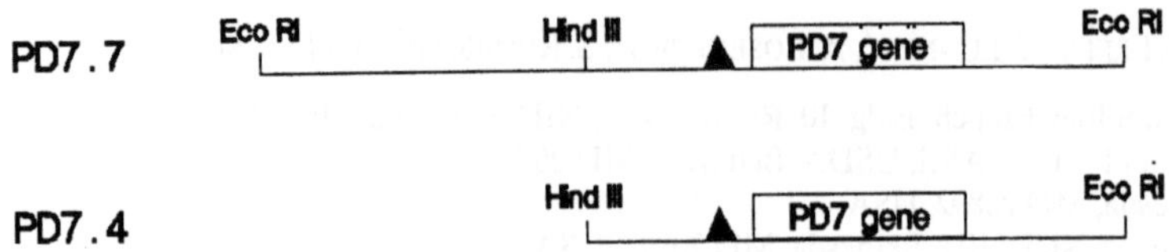

Figure 1.Two genomic clones, PD7.7 and PD7.4, both contain the entire coding region of PD7 and 3′ flanking sequences, but differ in the extent of 5′ flanking sequence. Partial restriction maps indicate regions of overlap

Analysis of pig tissues revealed that PD7 is expressed in vivo in nearly all tissues (Frels et al., ms. in prep). The overall pattern of expression of PD7 parallels that of the classical class I genes, PD1 and PD14, although its level of expression appears to be consistently lower (Ehrlich et al1989). To determine if either of the isolated clones of PD7 could be expressed, each one was transfected into mouse L cells (Figure 2). PD7 specific transcripts are found in both cell lines; however, their level of expression is very low relative to that of the classical transplantation antigens. Furthermore, the expression of PD7.7 is consistently lower than that of the 5′ truncated clone, PD7.4. Quantitation of the relative magnitude of expression, following corrections for differences in copy number in the different transfectants, reveals a 5-6 fold better expression of PD7.4. Thus, it is concluded that there are at least two regulatory elements associated with the PD7 gene. One of these is shared between PD7.7 and PD7.4 which governs the low level of expression of both of these clones. The other, a negative regulatory element occurring in the 5′ flanking region, is unique to PD7.7 and has been deleted from PD7.4.

The in vivo role of the PD7.7-specific negative element was investigated by constructing transgenic mice from PD7.7 and PD7.4. Total RNA samples from tissues of these transgenics were assayed for the presence of PD7 transcripts. Both constructs of PD7 were able to direct PD7 expression in vivo. However, as summarized in Table 1, mice containing the PD7.4 construct displayed higher levels of PD7 RNA in a number of tissues. Thus, the PD7 negative regulatory element functions in vivo, and in a tissue-specific fashion.

No PD7 gene product has been identified previously using a variety of serological reagents (H.Golding, J.Lunney, and D.Singer, unpublished observations). Since PD7 transcripts were found in the skin of transgenic animals, normal littermates were grafted with transgenic skin in order to determine whether PD7 encoded a product which could function as a transplantation antigen. Both sets of transgenic mice containing the PD7 gene were able to mediate graft rejection, although PD7.4 transgenic skin was consistently rejected more rapidly and completely than PD7.7 transgenic skin (Frels et al, in preparation). Furthermore, serum from engrafted animals which had rejected their grafts was cytotoxic to pig peripheral blood lymphocytes (PBL), whereas normal mouse serum was not (Table 2). Thus, PD7 encodes a product which is expressed on the cell surface

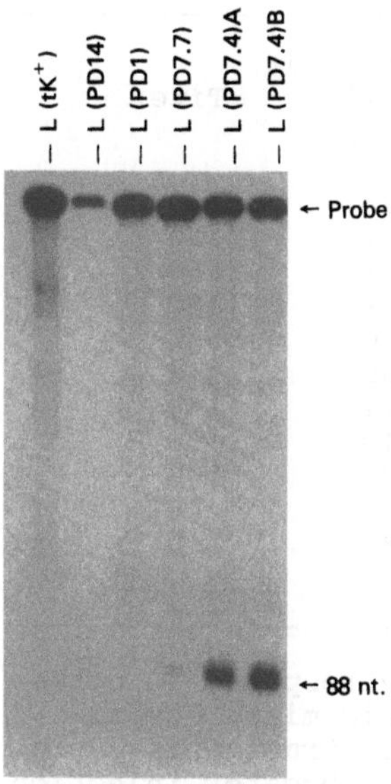

Figure 2. PD7 is expressed in L cell transfectants containing either PD7.7 or PD7.4. Mouse L(tk-) cells were cotransfected with HSV tk and either PD7.7 or PD7.4. RNA from such stable transfectants was assayed for the presence of PD7 transcripts by S1 nuclease analysis, using a PD7 probe which spanned exons 2 and 3. L(PD1) and L(PD14) are L cells lines stably transfected with PD1 or PD14, respectively, and serve to demonstrate the specificity of the PD7 probe

Tissue	Relative Level of PD7 Expression (PD7.4/PD7.7)
Liver	0.9
Spleen	1.8
Skin	4.9

Table 1. Relative Levels of PD7 Expression in PD7.7 and PD7.4 Transgenic Mice. Samples of total RNA were sequentially hybridized with a PD7 probe and an actin probe. Intensity of hybridization was quantitated by densitometry and normalized to actin. Values represent the average of two experiments.

and is able to elicit both cellular and humoral immune responses, as assessed by graft rejection and antibody production, respectively.

Discussion

Despite the high homology of PD7 with the two class I genes encoding major transplantation antigens, PD7 expression is lower than that of the other two class I genes, both in transfected cells and in vivo in various tissues. At least two DNA sequence elements regulate this pattern of expression. One of these is contained within the DNA sequences shared between the two clones PD7.7 and PD7.4, since L cells transfected with either clone accumulate much less class I RNA than comparable L cells transfected with the other class I genes. All class I gene encoding major transplantation antigens contain a DNA sequence element which has been identified as an enhancer (enhancer A)(Korber, Hood, Stroynowski, 1987; Israel et al1986; Kimura et al1986; Miyazaki, Appella, Ozato, 1986; Ehrlich, Maguire, Singer, 1988). Enhancer A has been shown to bind a variety of nuclear proteins (Korber, Hood, Stroynowski, 1987; Korber et

Animal	Post-Graft Serum Lysis (%)	Titer	Post-Boost Serum Lysis (%)	Titer
1	103	160	106	160
2	100	80	107	160
3	82	80	101	160
4	63	20	92	160
5	60	10	100	80
6	94	40	105	160
7	82	20	111	160
8	117	80	110	160
9	75	20	103	160
10	77	20	70	40
11	63	10	87	40
Normal	0	-	1	-

Table 2. The PD7 gene product elicits an antibody response in normal mice receiving skin grafts from PD7.4 transgenic mice. Normal animals were engrafted with skin from the PD7.4 transgenic. Sera were collected after graft rejection or after a booster injection with PD7.4 transgenic mouse cells and were assayed by complement-mediated cytotoxicity on PBL from miniature swine (SLA^{dd}). Lysis was calculated relative to a standard anti-class I SLA monoclonal antibody.

al 1988; Israel et al 1987; Shirayoshi et al 1988). In PD7, this enhancer element is altered in a residue which has been demonstrated to be an important binding site for all trans acting factors. It is thus possible to speculate that normal levels of class I transcription are regulated by the effective binding of trans acting factors to enhancer A and that the mutation of enhancer A in PD7 has led to its reduced transcription. Enhancer A may control class I RNA levels in vivo, since PD7 expression in either transgenic line is lower than that of a PD1 transgene, as evidenced by the observation that skin from either PD7 transgenic line is rejected more slowly than skin from a PD1 transgenic (data not shown).

The present studies have also identified a negative regulatory element contained within the 5′ flanking DNA sequences unique to PD7.7, since expression of PD7 in PD7.4 is consistently higher than in PD7.7. Similar studies of PD1, which encodes a major transplantation antigen, have identified a silencer element which functions both in vivo and in transfected cells (Ehrlich, Maguire, Singer, 1988). Comparison of the PD7 sequence with that of PD1 reveals the presence of a highly related element in PD7.7. This element has been deleted from PD7.4. Thus, it is possible that the deletion of the silencer from PD7.4 enables its higher level of expression. The function of the PD7 silencer also appears to be tissue specific, since the magnitude of effect of its removal varies among the tissues.

Due to the combined effects of these two regulatory elements, PD7 expression in vivo is normally low (Frels et al, in prep). Thus, the PD7 antigen does not ordinarily stimulate either cellular or humoral responses. As a result, no antibody reagents had been previously generated by standard techniques, and there was no evidence that the PD7 gene encoded a cell surface antigen. The ability of skin from

PD7 transgenics to mediate graft rejection demonstrates that PD7 does encode a cell surface molecule which is able to function as a transplantation antigen. We speculate that its failure to do so in situ in the pig simply reflects its low level of expression. Indeed, it is possible to further speculate that PD7 may have at one time functioned as a third transplantation antigen, prior to the mutation in enhancer A which reduced its level of expression. Studies are currently in progress to evaluate the role of enhancer A in determining the extent of PD7 transcription and the role of the silencer element in establishing tissue specific patterns of expression.

References

Ehrlich R, Lifshitz R, Pescovitz M, Rudikoff S, Singer D (1987) Tissue-specific expression and structure of a divergent member of a class I MHC gene family. JImmunol 139:593

Ehrlich R, Maguire J, Singer D (1988) Identification of negative and positive regulatory elements associated with a class I major histocompatibility complex gene. Mol Cell Biol 8:695

Ehrlich R, Sharrow S, Maguire J, Singer D (1989) Expression of a class I MHC transgene: Effects of in vivo a/b-interferon treatment. Immunogenet In press:

Fahrner C, Hogan B, Flavell R (1987) Transcription of H-2 and Qa genes in embryonic and adult mice. EMBO J 6:1265

Flaherty L (1976) The Tla region of the mouse: identification of a new serologically defined locus, Qa-2. Immunogenet 3:533

Israel A, Kimura A, Fournier A, Fellous M, Kourilsky P (1986) Interferon response sequence potentiates activity of an enhancer in the promoter region of a mouse H-2 gene. Nature 322:743

Israel A, Kimura A, Kieran M, et al (1987) A common positive trans-acting factor binds to enhance sequences in the promoters of mouse H-2 and b2-microglobulin genes. ProcNatlAcadSci(USA) 84:2653

Kimura A, Israel A, LeBail O, Kourilsky P (1986) Detailed Analysis of the Mouse H-2Kb Promoter: Enhancer-like Sequences and Their Role in the Regulation of Class I Gene Expression. Cell 44:261

Korber, B., Hood, L. and Stroynowski, I. Regulatory Elements in the Promoter of the H-2Dd Class I Gene. In: H-2 Antigens, edited by David, C.S. : Plenum Publishing Corp., 1987, p. 201-208.

Korber B, Mermod N, Hood L, Stroynowski I (1988) Regulation of Gene Expression by Interferons: Control of H-2 Promoter Responses. Science 239:1302

Miyazaki J, Appella E, Ozato K (1986) Negative regulation of the MHC class I gene in undifferentiated embryonal carcinoma cells. Proc Natl Acad Sciences (USA) 83:9537

Shirayoshi Y, Burke P, Appella E, Ozato K (1988) Interferon-induced transcription of a major histocompatibility class I gene accompanies binding of inducible nuclear factors to the interferon consensus sequence. Proc Natl Acad Sci USA 85:5884

Singer, D., Ehrlich, R., Golding, H., Satz, L., Parent, L. and Rudikoff, S. Structure and expresion of class I MHC genes in the miniature swine. In: The Molecular Biology of the Major Histocompatibility Complex of Domestic Animal Species, edited by Warner, C., Rothschild, M. and Lamont, S. Ames: Iowa State University Press, 1988, p. 53-62.

Singer D, Maguire J (1989) Regulation of the Expression of Class I MHC Genes. CRC Critical Reviews in Immunology (In Press)

Regulation of Expression of Human MHC Class I Heavy (HLA-B7) and Light (hβ_2-m) Chain Genes in Transgenic Mice

J. W. CHAMBERLAIN, P. J. CONRAD, and S. M. WEISSMANN[1]

[1] Departments of Human Genetics and Pathology, Yale University School of Medicine, 333 Cedar Street, New Haven, CT 06510, USA

INTRODUCTION

The class I transplantation antigens encoded in the major-histocompatibility complex (MHC) (HLA-A, -B, -C in man, H-2K, -D, -L in mouse) are expressed at the surface of most somatic cells of the organism as heterodimers of a polymorphic 43,000 to 45,000 M.W. glycoprotein non-covalently associated with an invariant 12,000 M.W. light chain called β_2-microglobulin (β_2-m)(Ploegh et. al., 1981; Klein, 1986). Class I gene expression has been demonstrated to be regulated during embryonic development (Ozato et. al., 1985), between tissues in the adult (Harris et. al., 1986; Halloran et. al., 1986), and in response to various lymphokines and viruses (Harris et. al., 1986; Halloran et. al., 1986). The primary function of these molecules is in the presentation of foreign antigens to CD8-positive cytotoxic T lymphocytes (Klein, 1986).

The multiplicity of class I molecules expressed on the cell surface and the large number of alleles for each locus present within the population complicate analyses of expression and function of individual class I products. The recent development of the technology for introducing cloned DNA into the germ-line of mice (reviewed in Palmiter and Brinster, 1986) provides a powerful approach for studying the regulation and functions of specific MHC genes in the whole organism. In order to map and characterize the role(s) played by specific cis-active sequences in the regulation of class I MHC gene expression in vivo, as well as to study the effects of specific class I genes on the development of immune responsiveness, we have generated a number of lines of transgenic mice carrying various fragments of the human class I MHC gene HLA-B7 (Chamberlain et. al., 1988a, 1988b). This report describes our analyses of constitutive and interferon -γ (IFN-γ)- enhanced tissue-specific expression of this HLA class I transgene. The results obtained indicate that a 6.0 kb HLA-B7 transgene with 0.66 kb of 5' and 2.0 kb of 3' flanking sequence apparently contains all cis-regulatory information to drive appropriate tissue-specific levels of expression. Further dissection of this fragment indicates the presence of a dominant regulatory element(s) within a 550 bp 5' flanking region which is involved in conferring position-independent, copy-number-dependent expression. We also present results showing that in addition to the probable involvement of 5' sequences in IFN-γ-responsiveness, regulatory mechanisms involving sequences downstream of the known interferon -α/β consensus/response sequence (ICS/IRS) play an important role in both broad tissue-specific constitutive expression and IFN-γ- enhanced class I expression.

MATERIALS AND METHODS

All procedures used were either standard (Maniatis et. al., 1982; Hogan et. al., 1986) or have been described previously (Chamberlain et. al., 1988a, 1988b).

I.K. Egorov/C.S. David (Eds.)
Transgenic Mice and Mutants in MHC Research

RESULTS AND DISCUSSION

A series of 16 independent transgenic lines were derived from microinjection of 3 HLA-B7 DNA fragments containing varying amounts of 5' and 3' flanking sequence. These fragments (a, b and c) are indicated below the restriction map shown in Fig. 1. Fragment (a) contained 0.66 kb and 2.0 kb of 5' and 3' sequence, while fragment (c) contained about 12 kb and 2.0 kb of 5' and 3' sequence. Analysis of DraI-restriction endonuclease digests of tail-skin DNA of founder mice and DNA-positive progeny for each line by Southern blot hybridization allowed an accurate determination of those founder mice carrying one or more intact copies of the B7 transgene. As indicated in the restriction map in Fig. 1, DraI sites are located just upstream and downstream of the B7 gene and thus one or more intact copies of the integrated B7 transgene would be clearly identified by the presence of a 5.0 kb DraI fragment. Mice carrying rearranged B7 transgenes were identified by the inappropriate size of the detected band(s). Approximately 1/4 of the lines carrying single or multi-copy integrations were found to contain B7-hybridizing bands of inappropriate size. As expected, no expression of B7 RNA was detected in any tissues tested from mice of these lines and thus these lines were not analyzed further. In contrast, as shown in Table 1, all other lines contained intact copies of the B7 transgene with from 1 to about 30 copies per cell. Two founders (18 and 35) were found to carry 2 integrations (A and B) of distinct copy number which could be clearly identified by segregation of integration-specific junction fragments in progeny DNA (not shown). In addition, 4 of the founders were found to be mosaic for the transgene (18, 179, 7, 17) by the reduced frequency of transmission to offspring (germline mosaicism) and/or by the reduced intensity of B7-specific bands detected on Southern blots of parental compared to offspring tail DNA (somatic tissue mosaicism) (not shown). For these reasons, transgene copy number estimates were made from progeny rather than founder DNA samples.

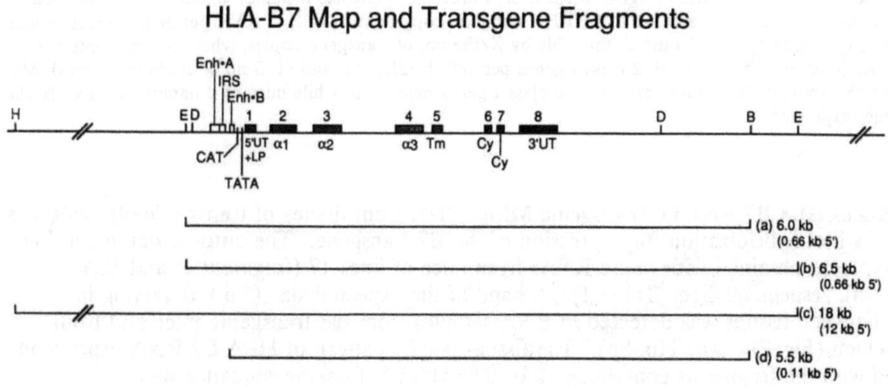

Fig. 1.
Structure of HLA-B7 DNA fragments used for microinjection. The uppermost line is a partial restriction map (B, BamHI; D, DraI; E, EcoRI; H, HindIII). The stippled boxes represent exons 1 to 8 encoding the 5' untranslated region and leader peptide (5' UT + LP), α1, α2, α3, transmembrane (Tm) and cytoplasmic (Cy) domains and the 3' untranslated region (3'UT). Lines a, b, c, and d represent the fragments microinjected into eggs. a, b, and d were 6.0-kilobase (kb) EcoRI-BamHI, 6.5-kb EcoRI, and 5.5-kb MluI-BamHI fragments, respectively, and contained no vector DNA; c was an 18 kb HindIII fragment and contained 24 bp of vector DNA. The open boxes represent the corresponding positions of known cis-regulatory elements identified in the H-2K^b gene (Kimura et al., 1986). Probes for detecting HLA-B7 DNA and RNA were described previously (Chamberlain et al., 1988b). Scale is approximate

Table 1. **Transgene copy numbers and expression in *HLA-B7* transgenic mice.**

HLA-B7 DNA fragment	Line no. *	Copy no. §	Expression of HLA-B7 RNA ratio Φ (B7/H2)
a	18A	30	5.0
a	18B	25	3.6
a	34	3	0.4
a	35A	7	ND
a	35B	1	0.1
a	65	1	0.2
b	176	7	1.9
b	179	10	2.4
c	2	10	1.5
c	7	7	ND
c	17	30	5.0

ND: Not determined

* The strains of the founder mice of each line were as follows: 18-$(B6/SJL)F_1$; 34,65,176,179-$(B6/D2)F_2$; 35,2,7,17-$(B6/SJL)F_2$.

§ Transgene copy number was determined by dot- and Southern blot analysis of offspring tail DNA. When more than one integration was identified in a founder, the individual integrations were represented as A, B, etc.

Φ The ratio of the level of HLA-B7 RNA (B7 RNA) to H-2 RNA in spleen was determined by quantitating autoradiographic hybridization signal intensities. A ratio of 1 indicates a similar amount of B7 and total H-2 RNA. An estimate of the amount of B7 RNA per gene copy relative to H-2 RNA per gene copy is obtained by multiplying the B7/H-2 ratio in the table by X/the no. of transgene copies, where X represents the number (four, five, or six) of H-2 class I genes per cell (H-2K, -D, and -L from both chromosomes). Mice of b or s haplotype have four conventional class I genes expressed, while mice of d haplotype have six class I genes expressed.

Expression of HLA-B7 RNA in Transgenic Mice. RNA from tissues of transgenic offspring was analyzed by blot hybridization for expression of the B7 transgene. The autoradiograms in Fig. 2 A and C were obtained with tissue RNAs from mice of lines 17 (fragment c) and 18A (fragment a), respectively (see Table 1). A band of the expected size (1.6 kb) varying in intensity between tissues was detected in RNA samples from the transgenic mice and from human spleen (Fig. 2A, lane Hu. Sp). The tissue-specific pattern of HLA-B7 RNA expression generated with the fragments containing 12 or 0.66 kb of 5' flanking sequence was indistinguishable. Lymphoid tissues and lung contained the highest level of B7 RNA. Relative to spleen, small intestine contained 20-30%; liver, heart and kidney contained 5-10%; skeletal muscle and brain contained 1-3%; and pancreas contained <1% of the amount of B7 RNA. A similar pattern was observed for all mice examined from all lines carrying three or more copies of any of the three microinjected fragments. Corresponding tissues from mice of independent lines carrying a different number of copies of the same fragment contained levels of B7 RNA that varied over a 30- to 50-fold range (Table 1). Multiple offspring within a line (lines 18A, 34 and 65 were examined) expressed similar levels of B7 RNA with a similar tissue-specific distribution.

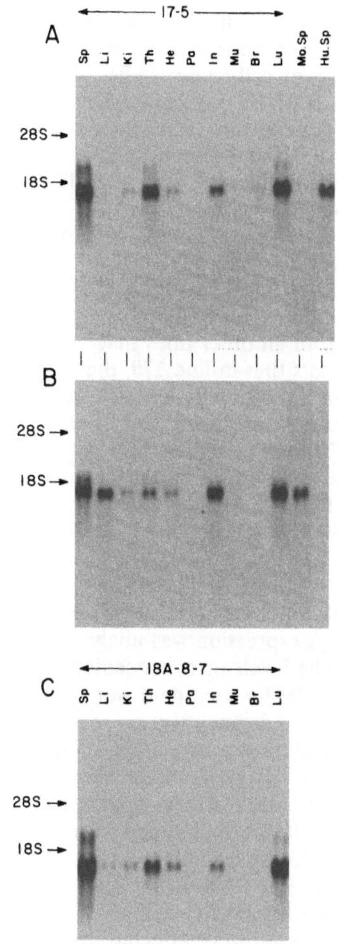

Fig. 2.

To compare HLA-B7 expression to endogenous H-2 class I genes, the blot of Fig. 2A was stripped of the HLA-B7 cDNA probe and rehybridized with a H-2 class I probe from exon 8 of an H-$2K^k$ cDNA conserved among mouse class I genes (Fig. 2B) (Lew et. al., 1986; Chamberlain et. al., 1988b). The highest levels of 1.6 kb H-2 transcripts were detected in lymphoid tissues, liver, lung and small intestine. Relative to spleen, thymus, liver and small intestine contained 50-90%, heart and kidney contained 10-15%, brain and skeletal muscle contained 1-2%, and pancreas contained <1% of the amount of H-2 RNA. Similar results were obtained for the nine lines analyzed (Table 1) and for non-transgenic siblings (not shown). There was less than a 2-fold difference in the ratios of the RNA blot signal intensities for HLA-B7 relative to H-2 (i.e., B7 RNA/H-2 RNA) for most tissues from mice for a given line. However, B7 RNA was about 10-fold under-represented relative to H-2 RNA in liver compared with other tissues and 3-fold under-represented relative to H-2 RNA in small intestine. The B7 RNA/H-2 RNA ratios for spleen RNA for most lines are given in Table 1. There was a strong correlation between the number of intact copies of the HLA-B7 transgene and the level of B7 RNA detected. If all classical H-2 class I genes contribute approximately equally to the hybridization signals detected with the H-2 probe, then the level of B7 RNA in spleen from most lines is similar to the level of H-2 RNA on a per gene copy basis.

These results indicate that HLA-B7 transgene fragments (a), (b) and (c) are relatively insensitive to position effects as expression appeared to be largely independent of the integration site both with regard to the quantitative level of expression and with regard to tissue-specific expression. However, further analyses have identified two instances where expression seems to be

---

Blot-hybridization analysis of HLA-B7 RNA in transgenic mice. RNA samples were from tissues of transgenic mice 17-5 (fragment c) (A and B) and 18A-8-7 (fragment a) (C). Lanes: Sp, spleen; Li, liver; Ki, kidney; Th, thymus; He, heart; Pa, pancreas; In, small intestine; Mu, skeletal muscle; Br, brain; and Lu, lung. Nontransgenic mouse spleen (lane Mo.Sp) and human spleen (lane Hu.Sp) RNAs were also analyzed (A and B). The autoradiograms in A and C were obtained by hybridization with an HLA-B7 cDNA exon 8-derived probe (Chamberlain et al., 1988b). The autoradiogram in B was obtained by stripping the blot in A and rehybridizing with the H-2 class I cDNA probe. The exposure times were 4 hr for all samples in A except Hu.Sp. RNA, which was for 10 hr; exposure times were 16 hr for samples in B and 4 hr for samples in C.

---

influenced by neighboring DNA. In the first case, a series of transgenics were generated with an 8.5 kb Hinc II - Hind III fragment of the HLA-B7 clone pJY150. This fragment contained the 6.5 kb EcoRI fragment similar to fragment (b) but was flanked on either end by about 0.87 kb and 1.25 kb of pBR328 vector DNA. In 2 out of 3 lines analyzed, the relative level of RNA detected in specific tissues and/or the tissue specific pattern of expression was noticeably altered in comparison to results obtained with transgene fragments (a), (b) and (c) (data not shown). These mice have not been analyzed in sufficient detail to determine whether the presence of vector DNA on the B7 transgene fragment has a tissue- or cell-type-specific negative effect on expression as was reported for the murine class II MHC transgene I-Eα (Widera et. al., 1987) or whether the negative effect is of a more general repressive nature as seemed to be the case for both human β-globin (Townes et. al., 1985) and α-fetoprotein transgenes (Hammer et. al., 1987). In the second case, analysis of peripheral T cells of male or female offspring of line 179 by flow cytometry with anti-HLA mAbs (Chamberlain et. al., 1988a, 1988b) identified two distinct populations of cells (data not shown). About 15-25% of T cells were negative for B7 expression while the remaining 75-85% were uniformly positive. B cells from the same animal gave a single positive peak of fluorescence for B7, and T and B cells gave the expected single peak of fluorescence for H-2K or H-$2D^b$. This bimodal distribution is in contrast to the single peak observed for T cells from all other lines that express surface B7 protein (Chamberlain et. al. 1988b). This suggests that in line 179, the transgene integration site has a negative effect on B7 expression in a sub-population of T cells. It is unclear at what level this effect is exerted, but is currently being investigated. In summary, the HLA-B7 transgene with no more than 0.66 kb 5' and 2.0 kb 3' of native flanking DNA functions largely, although not exclusively, independently of its site of insertion into the mouse genome. This relative independence of integration site position effects is seen both in quantitative and qualitative terms.

The effects of *in vivo* treatment with murine γ-interferon (IFN-γ) (Skoskiewicz et. al., 1985) on HLA-B7 and H-2 RNA expression was analyzed for line 65 (Fig. 3). The levels of enhancement of B7 and H-2 RNA after IFN-γ treatment were similar within a factor of 2 in each tissue examined except skeletal muscle and brain. The low level of expression of B7 RNA in this single copy line is likely responsible for these apparent exceptions. Similar results were obtained with offspring of line 34, including a similar enhancement of B7 and H-2 RNA in skeletal muscle (not shown) (see legend to Fig. 3). Although all tissues analyzed were affected, tissues which constitutively expressed lower levels of class I RNA were affected to a greater degree (e.g. kidney, heart and pancreas) than those which

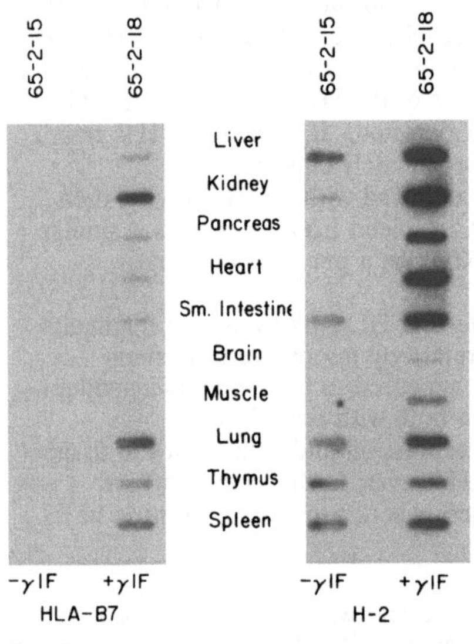

Fig. 3.

Slot-blot analysis of HLA-B7 RNA expression after in vivo treatment of transgenic mice with IFN-γ. Transgenic siblings of line 65 were injected i.p. daily for 7 days with phosphate-buffered saline (PBS) containing 5 x 10^4 units of recombinant murine IFN-γ (Genentech, South San Francisco, CA) (mouse 65-2-18, +IFN-γ) or PBS without IFN-γ (mouse 65-2-15, -IFN-γ). On day 8, RNA was prepared from the indicated tissues. (Left) Hybridization with the HLA-B7 probe (film exposure time 24 hr). (Right) Hybridization with the H-2 class I probe (film exposure time 8 hr). The average enhancement of HLA-B7 RNA relative to H-2 RNA caused by IFN-γ in various tissues,

[{B7 RNA (+IFN-γ)/B7 RNA (-IFN-γ)}/ {H-2 RNA (+IFN-γ)/H-2 RNA (-IFN-γ)}]

for lines 65 and 34 offspring were as follows: liver (6.5/5.8), kidney (31.7/42.9), pancreas (28.3/32.2), heart (24.5/44.1), small intestine 6.0/8.3), brain (-/4.7), skeletal muscle (13.2/14.0), lung (7.5/5.4), thymus (1.3/1.8), and spleen (2.6/2.8). For RNAs from nontreated mice with no apparent hybridization signal, a value was assigned based on longer exposures of the same blot (H-2), or on the relative level of expression in the same tissue(s) of higher-expressing lines (HLA-B7).

constitutively expressed higher levels of class I RNA (i.e. lymphoid tissues). As the tissue specific degree of enhancement of B7 and H-2 in multiple lines were similar, this suggests that the level of regulation, which may include both transcriptional and post-transcriptional mechanisms (Yoshie et. al., 1984; Israel et. al., 1986; Korber et. al., 1987), is largely independent of integration site-specific position effects.

The thymus is composed of a number of different cell types including stromal cells (epithelial and dendritic cells) in the medulla and cortex, and developing cortical and medullary thymocytes (Adkins et. al., 1987). Expression studies of RNA from fractionated thymocytes and stromal cells from line 18A transgenic mice are shown in Fig. 4A. The level of B7 and H-2 RNA in thymocytes (lane 1) was 5-10% of that detected in stromal cells (lane 2) or whole thymus (lane 3). This low level of class I transgenic and endogenous RNA is consistent with our earlier estimate of surface expression of HLA-B7 and H-2K/H-2D^b protein on thymocytes compared to peripheral T cells (Chamberlain et. al., 1988b). We are currently analyzing B7 expression in thymic stomal cells in more detail relative to H-2. It will be important to relate the results of these studies to functional analyses of the human B7 molecule in tests of immunological tolerance and restriction.

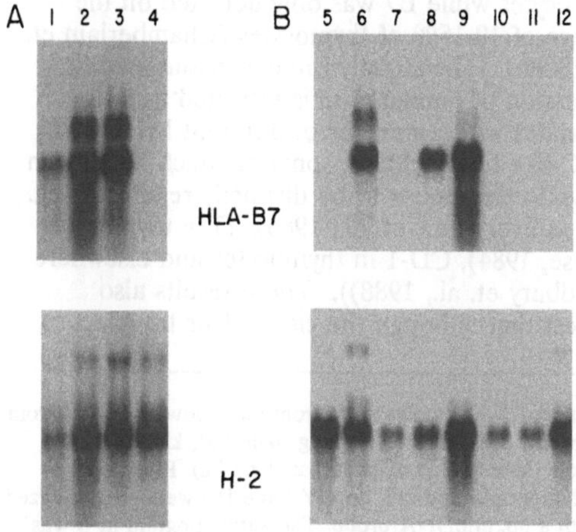

Fig. 4.

Blot hybridization analysis of HLA-B7 and H-2 RNA expression in transgenic thymocytes and stromal cells (A) and PEC (B). Thymocytes (lane 1) were separated from thymic stromal components (lane 2) of line 18A HLA-B7 mice as described previously (Chamberlain et al., 1988b). Transgenic (lane 3) and nontransgenic (lane 4) whole thymus RNAs were analyzed in parallel. Thioglycollate-induced PEC (Mishell et al., 1980) from line 18A HLA-B7 mice (lanes 7-9) and control (B6/SJL) F1 mice (lanes 10-12) were analyzed directly (lanes 7 and 10) or after adherence and culture for 48 hours in the absence (lanes 8 and 11) or presence (lanes 9 and 12) of murine IFN-(200 u/ml). Nontransgenic liver (lane 5) and line 18A HLA-B7 whole thymus (lane 6) RNAs were analyzed in parallel. In both A and B, the upper panel was obtained from hybridization with the probe for HLA-B7 and the lower panel was obtained (after removal of the B7 probe) from hybridization with the probe for H-2.

In order to examine B7 and H-2 expression in macrophage-like cells, northern blot hybridization studies were carried out with RNA prepared from thioglycollate-induced peritoneal exudate cells (PEC) (Mishell et. al., 1980) from line 18A (fragment (a)) (Fig. 4B), line 17 (fragment (c)) (not shown), and control (B6/SJL) F1 mice (Fig. 4B). Relative to H-2, B7 was under-expressed several-fold in PEC analyzed directly (lane 7, line 18A) compared to after culture of adherent cells for 48 hours (lane 8) when the level of B7 RNA increased significantly with no change in H-2 RNA levels (lanes 7, 8 and 10, 11 - upper panel, HLA-B7; lower panel, H-2). The slight apparent increase in H-2 RNA levels in lane 8 compared to lane 7 is due to incomplete removal of the B7 probe. Culture in the presence of IFN-γ had a similar enhancing effect on the levels of both B7 and H-2 (lane 9). There was no difference detected in the expression pattern of B7 RNA in PEC from mice of lines 18A and 17 under these various treatments. These results suggest that in PEC the mechanisms governing constitutive levels of B7 and H-2 RNA differ and therefore indicate that multiple mechanisms regulate class I RNA expression.

Class I MHC heavy chains are expressed at the cell surface in association with β_2 microglobulin (β_2-m). We and others have generated and crossed human β_2-m (hβ_2-m) transgenic mice with HLA transgenic mice to ensure efficient surface expression of the human heavy chain (Chamberlain et. al., 1988b; Krimpenfort et. al., 1987). Northern blot analyses were carried out with tissue RNAs from line 14 hβ_2-m transgenic mice (which carried about 30 copies of the hβ_2-m gene) (Chamberlain et. al., 1988b) (Fig. 5). A band of the expected size of about 1 kb was detected in human JY cell RNA (JY) and in transgenic tissue RNAs. The tissue-specific pattern of endogenous mouse β_2-m (mβ_2-m) expression was very similar to that of the human β_2-m transgene (Fig. 5). It should also be noted that the overall tissue-specific pattern of hβ_2-m expression was very similar, though not identical, to that of HLA-B7 expression. In addition to the much higher relative level of hβ_2-m RNA in liver compared to HLA-B7, FACS analysis showed that hβ_2-m was detectable on the surface of all thymocytes while B7 was only detected on the surface of 10-15% of thymocytes (Chamberlain et. al., 1988b). The slightly broader tissue-specific expression of human β_2-m is expected as β_2-m associates with a number of different MHC and MHC-like heavy chains, some of which have been shown in the mouse to be distinctly regulated (eg. Q10 in liver (Lew et. al., 1986), Tl in thymocytes (Boyse, 1984), CD-1 in thymocytes and elsewhere (Bradbury et. al., 1988)). These results also suggest that although the cis- and/or trans-acting regulatory mechanisms controlling HLA-B7 and β_2-m expression may be similar, they are not identical.

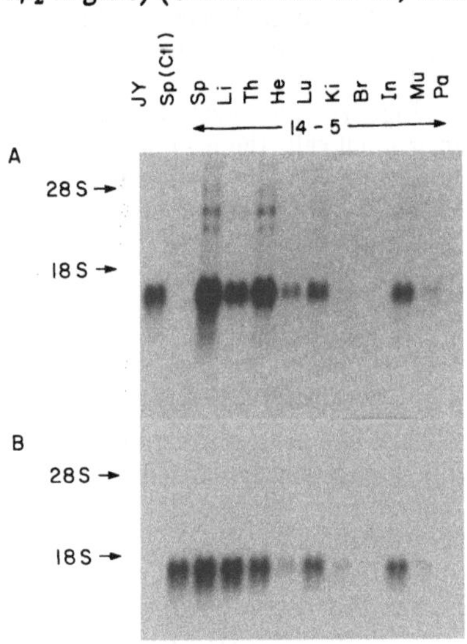

Fig. 5.

Blot hybridization analysis of hβ_2-m RNA in transgenic mice. RNA samples were from the following tissues from mouse 14-5: spleen (lane Sp), liver (lane Li), thymus (lane Th), heart (lane He), lung (lane Lu), kidney (lane Ki), brain (lane Br), small intestine (lane In), skeletal muscle (lane Mu), and pancreas (lane Pa). RNA from nontransgenic mouse spleen [lane Sp (Ctl)] and human B lymphoblastoid cell line JY (lane JY) were also analyzed. The autoradiogram in A was obtained hy hybridization with a hβ_2m cDNA probe. The autoradiogram in B was obtained by stripping the blot in A and rehybridizing with a mβ_2m cDNA probe. Exposure times were 6 hr for samples in A and 12 hr for those in B.

The effect of *in vivo* IFN-γ treatment on hβ_2-m expression was analyzed as above for hβ_2-m line 8 siblings 8-9-3 (-IFN-γ) and 8-9-4 (+IFN-γ). Line 8 contained 5-10 copies of the hβ_2-m transgene per cell. As shown in Fig. 6, the overall pattern and fold-enhancement of RNA levels was approximately the same for hβ_2-m and mβ_2-m. This pattern was similar, though not identical, to that seen for HLA-B7 (Fig. 3). The high constitutive level of hβ_2-m and mβ_2-m RNA in liver was unaffected by IFN-γ while HLA-B7 and H-2 RNA levels were both increased about 6-fold in liver, HLA-B7 from very low constitutive levels and H-2 from much higher constitutive levels (Fig. 3). This indicates that the hβ_2-m and mβ_2-m genes are closely co-regulated both constitutively and in response to IFN-γ, apparently in all tissues, presumably a result of conserved cis- and trans-regulatory mechanisms. In contrast, HLA-B7 and hβ_2-m may

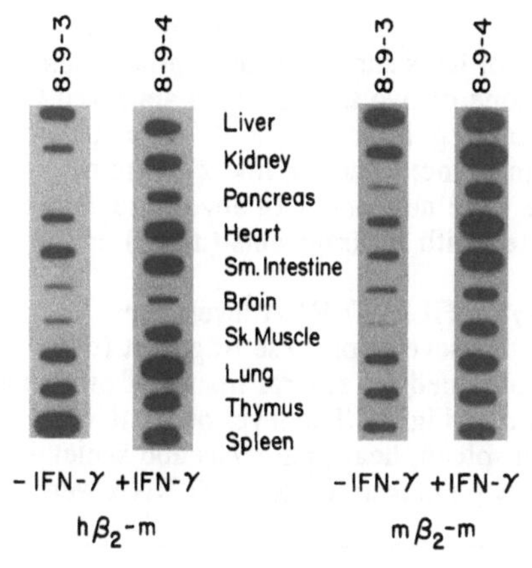

Fig. 6.

Slot blot analysis oh hβ_2-m RNA expression after in vivo treatment of transgenic mice with IFN-γ. hβ_2-m transgenic siblings of line 8 were treated as in the legend to figure 3. (Left) Hybridization with a hβ_2-m cDNA-derived probe. (Right) Hybridization with a mβ_2-m cDNA-derived probe.

share some, but apparently not all, regulatory mechanisms involved in constitutive and IFN-γ-enhanced expression.

HLA-B7 transgenes with only 0.66 kb of 5' sequence and 2.0 kb of 3' sequence gave what appeared to be appropriate tissue-specific levels of expression on a per-copy basis relative to H-2. To begin to delineate the relevant cis-sequences involved in the regulation of class I genes, transgenic mice were generated with HLA-B7 transgenes containing 0.11 kb of 5' sequence and 2.0 kb of 3' sequence (i.e., the 5.5 kb MluI-BamHI fragment (d) in Fig. 1). This 5' truncation deletes the region of the HLA-B7 gene corresponding to the "enhancer A" and IRS/ICS sequence elements

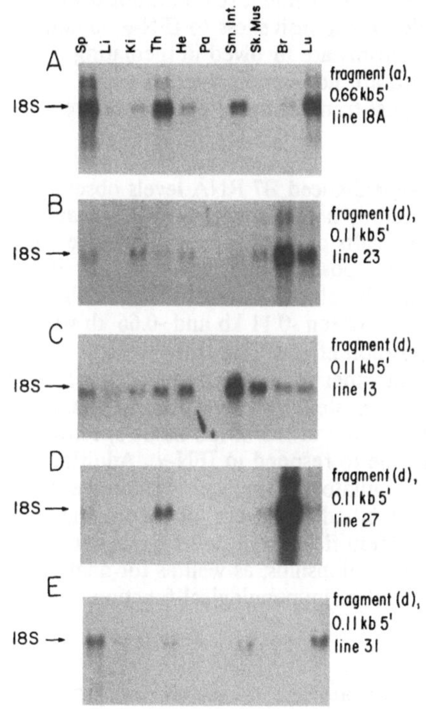

Fig. 7.

located in the H-2K^b gene between about -137 bp and -213 bp (Kimura et. al., 1986) (about -151 bp to -228 bp of the B7 gene) plus all DNA 5', as well as DNA 3' to -110 bp but leaves "enhancer B" (Kimura et. al., 1986) and any regulatory sequences further 3' (Korber et. al., 1987; Ganguly et. al., 1989) intact. Ten independent fragment (d) lines were produced and tissue RNAs of transgenic offspring of seven of these were analyzed by Northern blot hybridization (Fig. 7). For comparison, panel A shows the tissue-specific expression of HLA-B7 fragment (a) (line 18A - fragment (a), 0.66 kb 5'). Panels B through E show the results obtained for 4 independent lines. Several points can be made: **i)** HLA-B7 RNA was detected in at least several tissues of all lines analyzed, albeit at levels ranging from about 5-30 times lower overall than fragment (a) lines. There was no obvious correlation between the transgene copy number and the level of expression for fragment (d) lines (not shown); **ii)** in most lines, B7 RNA was detected in almost all appropriate tissues, although in most instances the high ratio of B7 RNA levels in lymphoid tissues compared to most non-lymphoid tissues was significantly reduced (i.e. compare spleen and kidney lanes

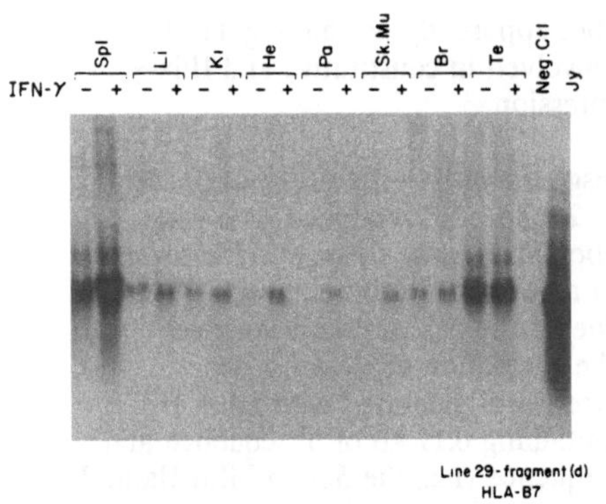

Fig. 8.

Blot-hybridization analysis of HLA-B7 RNA expression in fragment (d) (line 29) transgenic siblings after in vivo treatment with IFN-γ. The tissues analyzed (Te-testis) and whether or not the mouse received IFN-γ (mouse 29-7-9, -IFN-γ; mouse 29-7-10, +IFN-γ) are indicated above each lane. The negative and positive controls were nontransgenic mouse spleen RNA and human JY cell RNA respectively.

for Panels A to E); **iii)** superimposed on the above pattern, about 50% of lines showed inappropriate high level expression in one or two tissues which appeared to be specific to the line (eg. Panel B, line 23 - brain; Panel C, line 13 - small intestine; Panel D, line 27 - brain). Effects of this type were not seen with any of the transgenics generated with B7 fragments (a), (b) or (c).

The effect of IFN-γ on HLA-B7 RNA levels was examined as above for several of these fragment (d) lines. The results obtained for several tissues of offspring of line 29 are shown in Fig. 8. The level of B7 RNA is clearly enhanced in spleen, heart, pancreas and skeletal muscle. Significantly enhanced levels of B7 RNA were also seen for the other lines examined although there was some line-to-line variability in the tissue-specificity and magnitude of the effect (not shown). As fragment (d) does not contain the conserved IRS/ICS sequence believed to play a role in responsiveness to IFN-α, -β and possibly-γ, this suggests that additional regulatory mechanisms are involved in mediating increased levels of class I RNA to exogenous IFN-γ. We are not certain whether these mechanisms act at the level of transcription or involve post-transcriptional effects but are currently investigating this.

The line-to-line variability in both constitutive and IFN-γ-enhanced B7 RNA levels observed for fragment (d) lines is likely to be due to effects exerted by neighboring mouse DNA at the site of transgene integration (i.e. position effects). The results demonstrate both positive and negative position effect influences on B7 RNA levels. Such position effects were rarely (if at all) observed for B7 transgene fragments (a), (b) or (c). Therefore our results suggest the presence of a regulatory mechanism involving sequences between -0.11 kb and -0.66 kb which play a strong role in normally preventing such strong position effects. Whether these sequences involve the "enhancer A"/IRS/ICS elements or other regulatory sequences will have to await further experiments. Our results also suggest the existence of regulatory mechanisms involving sequences downstream of -0.11 kb which are involved in the broad tissue-specific expression of the HLA-B7 gene and the ability of this gene to respond to IFN-γ. Additional experiments are in progress to identify these important regulatory elements. Finally, the novel patterns of tissue-specific expression of HLA-B7 RNA observed for several of the fragment (d) transgenic lines should make these animals valuable for identification of novel tissue-specific cis-active regulatory elements, for studies of cell lineage relationships, as well as for analyses of the tissue-specific expression requirements for development of immunological function.

Acknowledgements:

We would like to thank Dr. H.M. Shepard (Genentech) for generously providing recombinant murine IFN-γ; H. Vasavada, L. Klein, and B. Broughton for technical assistance; and L. Madsen for assembling this manuscript.

REFERENCES

Adkins, B., Mueller, C., Okada, C., Reichardt, R., Weissman, I. and Sprangrude, G. (1987) Early events in T-cell maturation. Annu. Rev. Immunol. 5: 325-365.

Boyse, E. (1984) The biology of Tla. Cell 38: 1-2.

Bradbury, A., Belt, K., Neri, M., Milstein, C. and Calabi, F. (1988) Mouse CD-1 is distinct from and co-exists with Tl in the same thymus. EMBO J. 7: 3081-3086.

Chamberlain, J.W., Nolan, J.A., Gromkowski, S.H., Kelly, K., Eisenstadt, J., Herrup, K., Janeway, C.A. and Weissman, S.M. (1988a) Cell surface expression and alloantigenic function of a human class I MHC heavy chain gene (HLA-B7) in transgenic mice. J. Immunol. 140: 1285-1292.

Chamberlain, J.W., Nolan, J.A., Conrad, P.J., Vasavada, H.A., Vasavada, H.H., Ploegh, H., Ganguly, S., Janeway, C.A., Jr. and Weissman, S.M. (1988b) Tissue-specific and cell surface expression of major histocompatibility complex class I heavy (HLA-B7) and light ($h\beta_2$-m) chain genes in transgenic mice. Proc. Natl. Acad. Sci. USA 85: 7690-7694; 86: 3792.

Ganguly, S., Vasavada, H. and Weissman, S.M. (1989) Multiple enhancer-like sequences in the HLA-B7 gene. Proc. Natl. Acad. Sci. USA (In press).

Halloran, P., Wadgymar, A. and Autenried, P. (1986) The regulation of expression of major histocompatibility complex products. Transplantation 41: 413-420.

Hammer, R., Krumlauf, R., Camper, S., Brinster, R. and Tilghman, S. (1987) Diversity of α-fetoprotein gene expression in mice is generated by a combination of separate enhancer elements. Science 235: 53-58.

Harris, H., and Gill, T. (1986) Expression of class I transplantation antigens. Transplantation 42: 109-117.

Hogan, B., Costantini, F. and Lacy, E. (1986) Manipulating the Mouse Embryo--A Laboratory Manual. Cold Spring Harbor Laboratory, Cold Spring Harbor, N.Y.

Israel, A., Kimura, A., Fournier, A., Fellous, M. and Kourilsky, P. (1986) Interferon response sequence potentiates activity of an enhancer in the promoter region of a mouse H-2 gene. Nature 322: 743-746.

Kimura, A., Israel, A., LeBail, O. and Kourilsky, P. (1986) Detailed analysis of the mouse H-$2K^b$ promoter: enhancer-like sequences and their role in the regulation of class I gene expression. Cell 44: 261-272.

Klein, J. (1986) Natural History of the Major Histocompatibility Complex. John Wiley & Sons, New York.

Korber, B., Hood, L. and Stroynowski, I. (1987) Regulation of murine class I genes by interferons is controlled by regions located both 5' and 3' to the transcription initiation site. Proc. Natl. Acad. Sci. USA 84: 3380-3384.

Krimpenfort, P., Rudenko, G., Hochstenbach, F., Guessow, D., Berns, A. and Ploegh, H. (1987) Crosses of two independently derived transgenic mice demonstrate functional complementation of the genes encoding heavy (HLA-B27) and light (β_2-m) chains of HLA class I antigens. EMBO J. 6: 1673-1676.

Lew, A., Lillehoj, E., Cowan, E., Maloy, W., VanSchravendijk, M. and Coligan, J. (1986) Class I genes and molecules: an update. Immunology 57: 3-18.

Maniatis, R., Fritsch, E., and Sambrook, J. (1982) Molecular Cloning--A Laboratory Manual. Cold Spring Harbor Laboratory, Cold Spring Harbor, N.Y.

Mishell, B.B. and Shiigi, S.M. (1980) Selected Methods in Cellular Immunology. Freeman, San Francisco.

Ozato, K., Wan, Y., and Morrison, B. (1985) Mouse major histocompatibility class I gene expression begins at midsomite stage and is inducible in earlier stage embryos by interferon. Proc. Natl. Acad. Sci. USA 82: 2427-2431.

Palmiter, R. and Brinster, R. (1986) Germline transformation of mice. Annu. Rev. Genet. 20: 465-499.

Ploegh, H., Orr, H. and Strominger, J.L. (1981) Major histocompatibility antigens: the human (HLA-A, -B, -C) and murine (H-2K, H-2D) class I molecules. Cell 24: 287-299.

Skoskiewicz, M., Colvin, R., Schneeberger, E. and Russell, P. (1985) Widespread and selective induction of major histocompatibility complex-determined antigens in vivo by γ-interferon. J. Exp. Med. 162: 1645-1664.

Townes, T., Lingrel, J., Chen, H., Brinster, R. and Palmiter, R. (1985) Erythroid-specific expression of human β-globin genes in transgenic mice. EMBO J. 4: 1715-1723.

Widera, G., Burkly, L., Pinkert, C., Bottger, E., Cowing, C., Palmiter, R., Brinster, R. and Flavell, R. (1987) Transgenic mice selectively lacking MHC class II (I-E) antigen expression on B cells: An in vivo approach to investigate Ia gene function. Cell 51: 175-187.

Yoshie, O., Schmidt, H., Lengyel, P., Reddy, E., Morgan, W. and Weissman, S.M. (1984) Transcripts of human HLA gene fragments lacking the 5' terminal region in transfected mouse cells. Proc. Natl. Acad. Sci. USA 81: 649-653.

Cloned Trans-Acting Factors that Bind to the Regulatory Elements of the Major Histocompatibility Complex Class I Gene

K. OZATO[1], S. L. GLEASON[1], B.-Z. LEVI[1], S. HIRSCHFELD[1], P. H. DRIGGERS[1], J. R. FLANAGAN[1], D. L. ENNIST[1], K. HAMADA[1], and E. APPELLA[2]

[1] Laboratory of Developmental and Molecular Immunity, National Institute of Child Health and Human Development, National Institutes of Health, and the [2] Laboratory of Cell Biology, National Cancer Institute, Bethesda, MD 20892, USA

ABSTRACT

Transcription of MHC class I genes is controlled by the two conserved cis sequences, the class I regulatory element (CRE) and the interferon consensus sequence (ICS). The CRE binds two nuclear factors that are involved in enhancing class I transcription. The enhancer activity is observed in cells expressing the class I genes at high levels. The ICS binds interferon inducible and constitutive factors, and controls induction of the class I genes in response to interferons. The CRE/ICS also controls repression of class I gene transcription in embryonal carcinoma cells which, like early embryos, do not express class I genes.

By screening phage expression libraries with the CRE and ICS as probes, we isolated cDNA clones encoding CRE- and ICS- binding proteins. One of the cloned factors, H-2RIIBP binds to region II of the CRE and has a Zn finger domain very similar to that of steroid/thyroid hormone receptors. The overall structure of the N- and C-terminal domains of H-2RIIBP resembles that of the hormone receptor superfamily. Another clone encoding a CRE binding factor, obtained as a collaboration with other laboratories, binds region I and highly homologous to a human cDNA clone described previously. We also isolated a clone encoding an ICS binding protein (ICSBP). This gene is expressed in cells of macrophage/lymphocyte lineages and inducible by interferon.

These cloned factors show binding activity not only to the cis sequences of the class I genes, but also regulatory sequences of other genes, indicating that cis regulatory sequences and the trans-acting factors are shared by many genes.

INTRODUCTION

Classical class I antigens encoded by genes in the MHC are expressed in the majority of somatic cells in adult. However, cells in the central nervous system and in the germ line are negative for class I gene expression, illustrating a clear tissue specificity for class I gene expression. This tissue specificity is characteristic of classical class I genes, and is not seen with the less polymorphic class I genes encoded by the Qa and Tla regions. The class I gene expression is developmentally controlled: The levels of classical class I transcripts are extremely low in early embryonic stage and begin to be detected after mid-somite stage of mouse embryogenesis, when primordial organogenesis is well under way (Ozato et al, 1985). Class I gene expression remains low during the entire gestation, and a sharp increase takes place in the first two weeks of neonatal stage (Kasik et al, 1987). MHC class I genes are induced by interferons (all types of interferons α, β, γ) and tumor necrosis factor.

Regulatory Sequences of the MHC Class I Genes

In the past several years our laboratory (Miyazaki et al, 1986; Sugita et al, 1987; Silverman et al, 1988) among other groups (Kimura et al, 1986; Israel et al, 1987; Baldwin and Sharp, 1987; Korber et al, 1988) identified two cis acting sequences that control transcription of classical class I genes. These sequences reside between -139 bp and -203 bp relative from the RNA start site, and designated the MHC class I regulatory element (CRE) and the interferon consensus sequence (ICS) (see Fig. 1). The CRE acts as an enhancer and elevates

I.K. Egorov/C.S. David (Eds.)
Transgenic Mice and Mutants in MHC Research

transcription of class I genes in various tissue culture cells expressing class I genes. The ICS, partially overlapped and juxtaposed with the CRE, induces transcription of MHC class I genes in response to IFNs. These cis sequences are well conserved in mouse and human classical class I genes (Burke and Ozato, 1989). However, the genes encoded by the Qa/Tla show some diversification in this region. Besides the CRE/ICS complex, there may be additional cis elements involved in regulating of class I genes. For example, the enhancer B (Kimura et al, 1986) residing downstream may have a significant functions. The presence of a cyclic AMP response element in the region close to the CAAT and TATA boxes has also been noted.

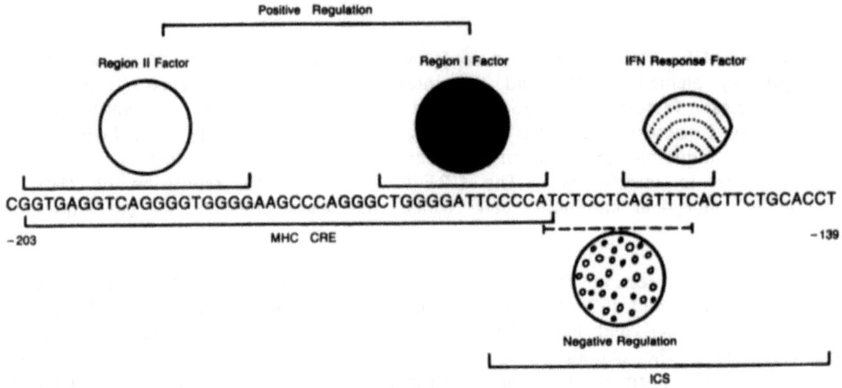

Fig. 1. Factor binding to the cis-acting sequence of the MHC class I gene. A summary. The CRE has two positive regulatory sequences, region I and region II, both of which bind separate nuclear factors. The ICS binds IFN-inducible factors and is responsible for IFN mediated induction of the class I gene. A negative regulatory factor, uniquely expressed in undifferentiated EC cells, binds toward the 5' side of the ICS

Trans-acting Factors that Bind to the CRE and ICS

By using gel mobility shift assays and methylation interference experiments, binding of nuclear factors to the CRE and ICS was studied in detail (Shirayoshi et al, 1987; Shirayoshi et al, 1988; Burke et al, 1989). Within the CRE there are at least three sequences that bind separate nuclear factors. These three regions were designated region I, II and III. As seen in Fig. 1, region I is composed of a small inverted repeat, and corresponds to the binding site for KbF1 or H-2TF1 as defined by Israel et al, (1988) as well as Baldwin and Sharp (1987). Furthermore, a factor cloned by Singh et al, (1988) binds region I. Additionally, this region binds NF-kB which binds to the Ig enhancer and the regulatory regions of the IL2 receptor and the IFN genes. We found that the region I binding factor is present in tissues expressing class I genes at high levels, but not in adult brain and fetal tissues in which class I gene expression is absent or low (Burke et al, 1989). Handy et al showed that region I of the Q10 gene that contains two base changes, failed to bind a nuclear factor. Similarly, region I of the Q7 gene that contains a single base deletion shows a greatly reduced factor-binding activity. The region II binding activity was found in all tissues tested regardless of class I gene expression. Region II of the Q10 and Q7 genes, although there are nucleotide alterations, showed factor-binding activity comparable to that of the classical CRE. A third factor that binds to region III which correspond to a large inverted repeat present in the CRE, was detected abundantly in tissue culture cells. This binding activity, dependent on Mg^{2+} was, however, not readily detected in tissues in vivo. To assess functional roles of the factor binding to the CRE, we performed CAT assays in fibroblasts using CAT constructs containing a 1.4 Kb H-$2L^d$ promoter. Single two bp substitutions were introduced into each of the three binding site of the CRE by site directed mutagenesis. When factor binding to region I or region II was abolished, the enhancer activity was significantly decreased. On the other hand, a mutation in region III did not reduce enhancer activity (Burke et al, 1989). These results indicate that region I and region II, by binding to specific factors, elicit enhancer activity.

The ICS also binds nuclear factors. We previously showed that the ICS binds factors inducible by IFN-α/β. In addition, the ICS binds a constitutive factor present in cells regardless of IFN treatment (Shirayoshi et al, 1988). These binding activities were seen at a ~10 bp sequence within the ICS (Fig. 1). There appears to be at least two IFN-inducible factors, as the inducible bands in gel shift assyas show differential sensitivity to cycloheximide. Similar ICS-binding activities inducible by IFN-γ have been reported (Blaner et al, 1989). We showed that mutations that eliminate factor binding to the ICS eliminate the ability to induce MHC class I gene transcription by IFN, which indicated that factor binding seen in vitro reflects a functional mechanism of transcription in vivo (Shirayoshi et al, 1988). It is noteworthy that sequence motifs very similar to the binding site of the ICS occur in a number of genes regulated by IFNs. This includes β-2 microglobulin, 2-5A synthetase, complement factor B, ISG 15, and 6-16 genes. Some of these genes show factor binding similar to that seen with the ICS (Reid et al, 1989). In addition, the IFN genes which are autoregulated by IFNs, also have the motif (Fan and Maniatis, 1989). Further, binding of mouse nuclear factors to the ICS is competed by the sequence motifs found in human genes. Thus, it is most likely that the same or similar factors control transcription of genes regulated by IFNs, and that these factors are conserved in human and mouse.

Repression of Class I Gene Transcription in Embryonal Carcinoma Cells

Embryonal carcinoma (EC) cells, extensively used as a model for early embryos, do not express MHC class I transcripts at appreciabe levels (Morrello et al, 1982; Wan et al, 1987). Previously we reported that transcription of class I genes is negatively regulated in undifferentiated F9 EC cells, and that the negative regulation, observed as repressed transcription activity, is mediated by the CRE/ICS complex (Miyazaki et al, 1986). This transcriptional repression was relieved when cells were differentiated and class I expression induced following retinoic acid treatment. We tested CAT activities in undifferentiated F9 cells using constructs driven by the class I promoter that contained mutations in various parts of the CRE and ICS. Results showed that mutations in the 5' side of the ICS, but not those in the CRE, markedly elevated CAT activity in F9 cells. These results suggest that the repression is mediated by the ICS, and the mutations in this region abrogate binding of a negative factor present in undifferentiated EC cells. Gel mobility shift assays carried out with probes corresponding to this region identified a factor-binding activity present in undifferentiated F9 cells, but much less or none in differentiated F9 cells. An equivalent binding activity was detected in other EC cells, but not in fibroblasts, lymphocytes and other cells expressing class I genes. Moreover, cotransfection of F9 EC cells with oligonucleotides corresponding to the binding sites of this factor relieved repression of CAT activity. This relief occurred presumably because of the competitive removal of a negative regulatory factor due to the co-introduced oligonucleotides. These results indicate that the ICS-binding factor, uniquely expressed in undifferentiated EC cells controls represssion of class I transcription in these cells.

Isolation of cDNA Clones Encoding CRE or ICS-binding Proteins

To study structure and function of trans-acting factors, it is necessary to obtain the proteins in sufficient amounts. However, because in most cells these factors are minute in quantity, their purification and biochemical characterizations have been difficult to achieve. Screening of a phage expression library with a cis DNA sequence as a probe circumvents this difficulty (Singh et al, 1988). This approach has been successfully used for some DNA binding proteins, although it would not be applicable to all trans-acting factors. This method relies on in situ DNA-protein interactions, and allows to isolate a gene for a DNA-binding protein even when no information is available for protein structure of the factor. We screened λgt11 and λ ZapII libraries using CRE and ICS as probes by following the procedure described by Vinson et al (1988). IPTG induced phage plaques were transferred onto nitrocellulose filters and subjected to a cycle of a guanidine-HCl treatment to denature and renature phage proteins. The filters were then incubated with [^{32}P]-labeled and concatenated probes, and then autoradiographed. With this method we have so far cloned two cDNAs encoding a region II and ICS binding factors. In addition, as a collaboration with other laboratories in NIH, a cDNA clone encoding a region I-binding protein has been isolated. Sequence data of this clone indicate that this region I binding factor is a mouse homologue of a human cDNA clone isolated by Singh et al (1988).

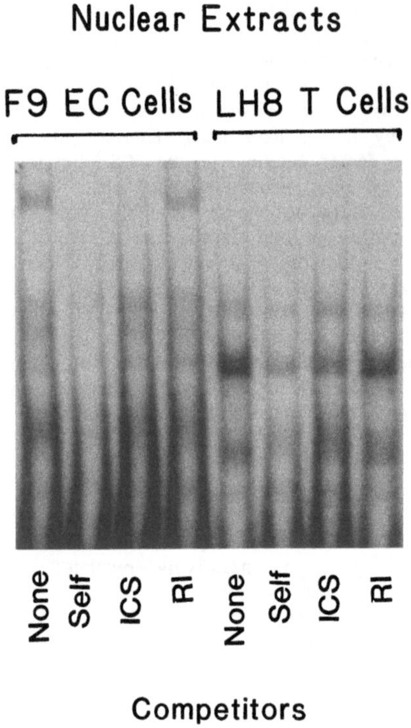

Fig. 2. ICS-binding activity detected in undifferentiated F9 EC cells. Gel mobility shift experiments were performed using a probe corresponding to the ICS and region I of the CRE. The uppermost band present in undifferentiated F9 cells was competed by excess unlabeled ICS oligomer and unlabeled probe itself. This binding activity was distinct in mobility and in competition profiles from the ICS binding activity seen in other cells (See Fig. 1 also).

Region II Binding Protein (H-2RIIBP)

This cDNA clone was obtained from a λ gt11 library prepared from mouse liver, and the fusion protein expressed in the lambda clone showed binding to region II, but not to region I or region II of the CRE, nor to the ICS. Mutant region II sequences either failed to bind or bound only weakly to H-2RIIBP in the λ clone. The specificity pattern of H-2RIIBP was very similar to that of the natural region II factor seen previously in mouse cells. DNA sequencing of the 2.2Kb H-2RIIBP insert indicates that it has a long open reading frame with a stop codon and putative poly adenylation sites in the 3' region. Deduced amino acid sequence of H-2RIIBP indicates that this clone has a Zn finger domain consisting of two Zn fingers of the C_4C_5 type. This Zn finger domain closely resembles the DNA binding region of the steroid/thyroid hormone receptor superfamily (Evans, 1988; Beato, 1989). The Zn finger region of H-2RIIBP is greater than 40% homologous to that of the nuclear hormone receptor superfamily. This superfamily includes glucocorticoid, thyroid, estrogen, progestrone, vitamin D, retinoic acid- receptors as well as an immediate early gene, Nur77, the c-erb oncogene and Drosophila developmental genes. The Zn finger region is flanked by the N-terminal and C-terminal halves where similarity with the hormone receptor superfamily is only marginal. However, this general domain structure parallels the distinct modular domains characteristically seen in the superfamily (Fig. 3). In agreement with the sequence data, in vitro translation of H-2RIIBP insert RNA produced a translated material of m.w. 50Kd. H-2RIIBP mRNA is expressed ubiquitously in many mouse tissues. Sequences similar to H-2RIIBP is conserved widely even in species as distant as Drosophila.

The 2.2 Kb insert of this cDNA was recloned into pBS vector, which expressed a fusion protein much smaller than that in the λ clone in the β-galactosidase moiety.

This insert protein was used for further studying binding specificity. As seen in Fig. 4, H-2RIIBP protein bound not only to region II of the MHC CRE, but also the estrogen response element (ERE) (Kumar and Chambon, 1988) of the vitellogenin gene. The ERE of the vitellogenin gene responds to estrogen by binding to the estrogen-receptor complex and induces transcription of the vitellogenin gene. H-2RIIBP did not bind to other hormone response elements. The natural region II factor also showed binding activity to the ERE but not to other hormone response elements. Thus, it is likely that this clone represents the natural region II factor previously detected in mouse cells.

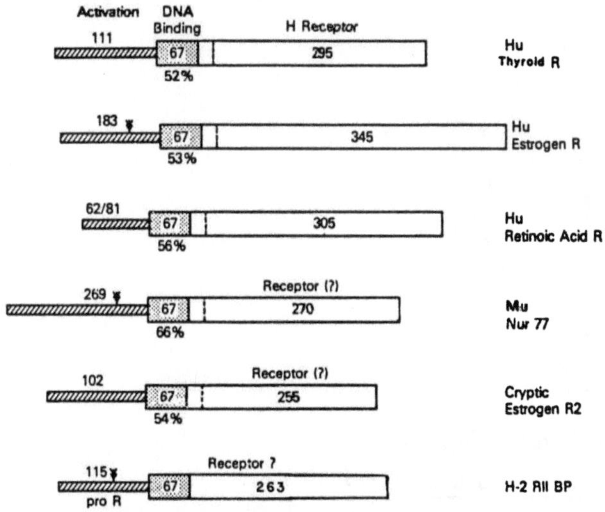

Fig. 3A. A structural comparison of H-2RIIBP with the nuclear hormone receptor superfamily. The hatched N-terminal region with variable amino acids sequences corresponds to the transcription activation domain. The Zn finger region of 67 amino acids corresponds to the DNA binding domain. The open C-terminal region represents the hormone binding domain. Note that H-2RIIBP has modular domains similar to the members of the superfamily

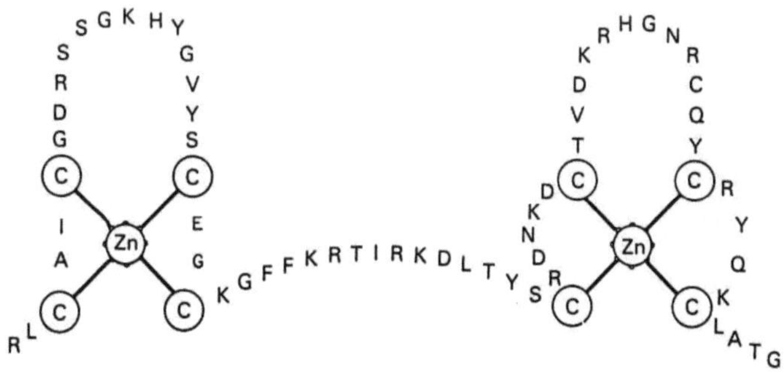

Fig. 3B. The Zn finger domain of H-2RIIBP. Nur 77 (Hazel et al, 1988) shows the greatest similarity in the Zn finger region

ICS Binding Protein (ICSBP)

This clone was obtained from a λ ZapII library. Northern blot hybridization with the insert fragment (Fig. 5) shows that the gene encodes a ~3Kb message. An additional mRNA species of 1.8 Kb is also detected. These mRNAs are expressed only in lymphoid tissues *in vivo*, and in tissue culture cells of lymphocyte/macrophage lineages. Southern blot hybridization indicates that human, mouse and chicken, but not lower animal species have a hybridizable sequence. The partial sequence information available for this clone indicates that ICSBP does not have a sequence corresponding to a known DNA binding domain, nor a leucine-zipper noted in some transcription factors. However, the N-terminal region of the ICSBP shows a considerable similarity with the murine interferon response factor (IRF), cloned by Miyamoto et al (1988). This similarity most likely reflects shared target specificity of the two genes: The IRF binds not only the regulatory sequence of the IFN-β gene, but <u>cis</u> sequences of various IFN-regulated genes, including the ICS of MHC class I genes.

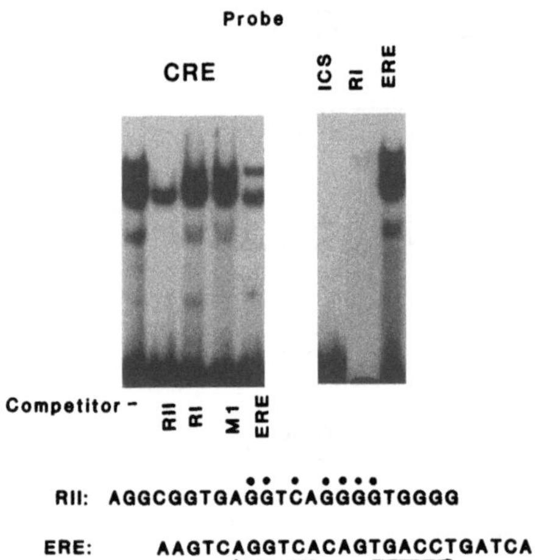

Fig. 4. Gel mobility shift assays for H-2RIIBP. H-2RIIBP expressed in the pBS vector was tested for mobility shift using ^{32}P-labeled CRE oligomer as a probe (left). Excess unlabeled oligomers corresponding to the native region II or the ERE compete for the specific retarded bands. Oligomers for region I and mutant region II fail to compete for the retarded bands. H-2RIIBP generates specific retarded bands by ^{32}P-labeled ERE but not by region I or ICS (right). Shown in the bottom of the figure is sequence similarity between the half palindrome of the ERE and the binding site of region II.

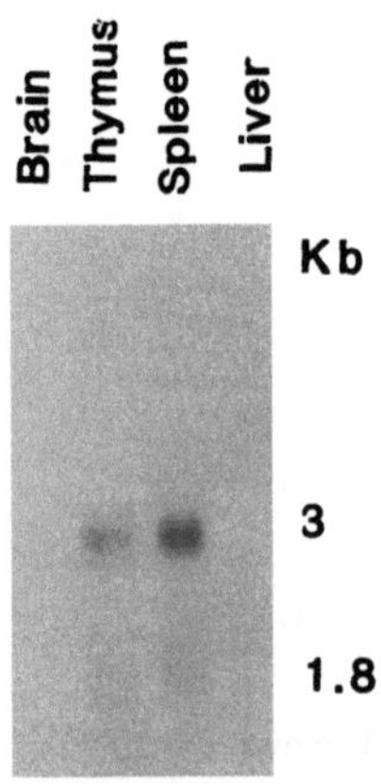

Fig. 5. Expression of ICSBPmRNA in various murine tissues. Poly A^{+} RNA (5 ug per lane) was separated in formaldehyde-gel electrophoresis, and probed with a ^{32}P-labeled ICSBP fragment.

The most remarkable finding emergerd from studies of cloned binding factors so far is that a single factor can bind to regulatory motifs present in many different genes. Thus, one can envisage that specific regulation of individual genes is controlled not so much by factors unique to individual genes, but rather by a combinatorial mechansism in which shared factors are used by different genes in different combinations.

ACKNOWLEDGEMENTS

We thank Ms. W.-H. Mak for expert technical work and Ms. D. Allor for sectretarial assistance.

REFERENCES

Baldwin AS, Sharp PA (1987) Binding of a nuclear factor to a regulatory sequence in the promoter of the mouse H-2K^b class I major histocompatibility gene. Mol Cell Biol 7:305-313

Baldwin AS, Sharp PA (1988) Two transcription factors, NF-kB and H-2TF1 interact with a single regulatory sequence in the class I major histocompativility complex promoter. Proc Natl Acad Sci USA 85:723-727

Beato M (1989) Gene regulation by steroid hormones. Cell 56:335-344

Blaner M, Baldwin AS, Flavell RA, Sharp PA (1989) A gamma-interferon-induced factor that binds the interferon response sequence of the MHC class I gene, H-2K^b. EMBO J 8:1139-1144

Burke PA, Hirschfeld S, Shirayoshi Y, Kasik J, Hamada K, Appella E, Ozato K (1989) Developmental and tissue-specific expression of nuclear proteins that bind the regulatory element of the major histocompatibility complex class I gene. J Exp Med 169:1309-1321

Burke PA, Ozato K (1989) Regulation of major histocompatibility complex class I genes. The year in immunology. Eds Cruse JM, Lewis RE. 4:23-40

Evans RM (1988) The steroid and thyroid hormone receptor superfamily. Science 240:889-895

Fan C-M, Maniatis (1989) Two different virus inducible elements are required for human β-interferon gene regulation. EMBO J 8:101-110

Handy DE, Burke PA, Ozato K, Coligan JE (1989) Site specific mutagenesis of the class I regulatory element of the Q10 gene allows expression in non-liver tissues. J Immunol 142:1015-1021

Hazel TG, Nathans D, Lau LF (1988) A gene inducible by serum growth factors encodes a member of the steroid and thyroid hormone receptor superfamily. Proc Natl Acad Sci USA 85:8444-8448

Israel A, Kimura A, Fournier A, Fellous M, Kourilsky P (1986) Interferon response sequence potentiates activity of an enahncer in the promoter region of a mouse H-2 gene. Nature 322:743-746

Israel A, Kimura A, Kieran M, Yano O, Kanellopoulos J, LeBail O, Kourilsky P (1987) A common positive trans-acting factor binds to enhancer sequences in the promoters of mouse H-2 and β-2 microglobulin genes. Proc Natl Acad Sci USA 84:2653-2657

Kasik J, Wan YJY, Ozato K (1987) A burst of c-fos gene expression occurs in the mouse at birth. Mol Cell Biol 9:3349-3352

Kimura A, Israel A, LeBail O, Kourilsky P (1986) Detailed analysis of the mouse H-2K^b promoter: Enhancer like sequences and their role in the regulation of class I gene expression. Cell 44:261-272

Korber B, Mermod N, Hood L, Stroynowski I (1988) Regulation of gene expression by interferons: Control of H-2 promoter responses. Science 239:1302-1306

Kumar V, Chambon P (1987) The estrogen receptor binds tightly to its responsive element as a ligand-induced homodimer. Cell 55:145-156

Miyamoto M, Fijita T, Kimura Y, Maruyama M, Harada H, Sudo Y, Miyata T, Taniguchi T (1988) Regulated expression of a gene encoding a nuclear factor, IRF-1, that specifically binds to IFN-β gene regulatory element. Cell 54:903-913

Miyazaki J, Appella E, Ozato K (1986) Negative regulation of the major histocompatibility class I gene in undifferentiated embryonal carcinoma cells. Proc Natl Acad Sci USA 83:9537-9541

Morrello D, Daniel, F, Baldacci P, Cayre Y, Gachelin G. Kourilsky P (1982) Absence of significant H-2 and β-2 microglobulin mRNA expression by mouse embryonal carcinoma cells. Nature (London) 296:260-262

Ozato K, Wan YJY, Orrison B (1985) Mouse major histocompatibility class I gene expression begins at midsomite stage and is inducible by interferon. Proc Natl Acad Sci USA 82:2427-2431

Reid LE, Brasnett AH, Gilbert CS, Porter ACG, Gewert DR, Stark GR, Kerr IM (1989) A sinlge DNA response element can confer inducibility by both β- and γ-interferons. Proc Natl Acad Sci USA 86:840-844

Silverman T, Rein A, Orrison B, Langloss J, Bratthauer G, Miyazaki J, Ozato K (1987) Establishment of cell lines from somite stage mouse embryos and expression of major histocompatibility class I genes in these cells. J Immunol 140:4378-4387

Singh H, LeBowitz JH, Baldwin AS, Sharp PA (1988) Molecular cloning of an enhancer binding protein: Isolation by screening of an expression library with a recognition site DNA. Cell 52:415-423

Shirayoshi Y, Miyazaki J, Burke PA, Hamada K, Appella E, Ozato K (1987) Binding of multiple nuclear factors to the 5' upstream regulatory element of the murine major histocomopatibility class I gene. Mol Cell Biol 7:4542-4548

Shirayoshi Y, Burke PA, Appella E, Ozato K (1988) Interferon-induced transcription of major histocompatibility class I gene accompanies binding of inducible nuclear factors to the interferon consensus sequence. Proc Natl Acd Sci USA 85:5884-5888

Sugita K, Miyazaki J, Appella E, Ozato K (1987) Interferons increase transcription of a major histocompatibility class I gene via a 5' interferon consensus sequence. Mol Cell Biol 7:2625-2630

Vinson CR, LaMarco KL, Johnson PF, Landschulz WH, McKnight SL (1988) In situ detection of sequence specific DNA binding activity specified by a recombinant bacteriphage. Genes Dev 2:801-806

Wan YJY, Orrison B, Lieberman R, Lazarovici P, Ozato K (1987) Induction of major histocompatibility complex class I antigens by interferons in undifferentiated F9 cells. J Cell Physiol 130:276-283

Transcriptional Regulation of MHC Class II Gene Expression: Are MHC Class II Genes Coordinately Regulated?

H.-C. Liou[1], L. B. Ivashkiv[1,2], C. J. Kara[1,3], and L. H. Glimcher[1,2]

[1] Department of Cancer Biology, Harvard School of Public Health, Boston, MA 02115, USA
[2] Department of Medicine, [3] Program in Immunology, Harvard Medical School, Boston, MA 02115, USA

INTRODUCTION

MHC (Major Histocompatibility Complex) Class II antigens are polymorphic cell surface glycoproteins which bind foreign antigens and form a complex which is recognized by T lymphocyte antigen receptors. Presentation of foreign antigens by self MHC molecules to T cells triggers antigen-specific responses in the immune system. Thus, increasing the expression of MHC class II molecules on the surface of antigen-presenting cells contributes to the initiation and amplification of immune responses.

Class II molecules consist of one heavy α chain (33-34 KD) and one light β chain (28-29KD), which associate noncovalently into a heterodimer which extends through the plasma membrane. In the mouse, there are four polymorphic α class II genes, Aα, Aβ, Eα and Eβ, each of which encodes one subunit of either I-A (AαAβ) or I-E (EαEβ) molecules.

Expression of the class II genes is subject to an elaborate program of regulatory controls which are of particular interest:

(1) Unlike the class I MHC gene products which are ubiquitously expressed, the expression of class II molecules is limited to certain cells of the immune system, e.g., macrophages, B lymphocytes, Langerhans cells, dendritic cells, thymic epithelial cells and other accessory cells.
(2) The level of class II gene expression can be modulated by various lymphokines and stimuli in different cell types. These stimuli include: interferon-gamma (γ-IFN), interleukin-4 (IL-4), LPS, Tetradecanoyl-Phorbol-Acetate (TPA), Prostaglandin E2 (PGE 2), and antibodies which crosslink B cell surface differentiation molecules.
(3) The class II genes are thought to be coordinately regulated. All class II genes are co-expressed during lymphokine stimulation and during stages of B cell differentiation and are extinguished when a class II positive cell is fused with a class II negative cell.

These properties of class II genes, tissue-specific, lymphokine inducible and coordinate expression, are controlled predominantly at the transcriptional level. Therefore, study of mechanisms involved in the transcriptional regulation of class II gene expression will be the main focus of this manuscript. In most of our work, we used the mouse class II Aα gene as a model system.

I.K. Egorov/C.S. David (Eds.)
Transgenic Mice and Mutants in MHC Research

TISSUE-SPECIFIC DNase I HYPERSENSITIVE SITE IN THE CLASS II PROMOTER REGION

Transcriptional regulation of eukaryotic genes relies on the interaction of cis-and trans-acting elements. Cis-acting elements are DNA sequences which serve as the targets for sequence-specific DNA binding factors (or trans-acting elements). The interaction of elements with the transcription machinery is a pre-requisite for transcription initiation to occur.

Changes in the chromatin configuration, as represented by increased sensitivity to DNase I treatment, have been shown to correlate with an active transcriptional state of a variety of genes. As a first step toward understanding the control of the class II MHC genes, we examined the DNase I hypersensitive sites in the murine class II Aα gene (Liou et al., 1988a). Nuclei isolated from various cell types were treated with increasing amounts of DNase I under conditions such that digestion occurs only at the most sensitive sites in the chromatin. The DNA was then purified from these nuclei, digested with restriction enzymes, and the resulting fragments were analyzed by Southern blotting using probes from within the body of the Aα gene. In summary, there are three DNase I hypersensitive sites in class II Aα gene. Two of these sites, termed D1 and D3, were located approximately at -1.4 kb and + 0.7 kb relative to the translation initiation site. These two sites are constitutively present in both class II positive and class II negative cells. In addition to these constitutive sites, however, there is another hypersensitive site which is tissue-specific. This site, termed D2, is located in the promoter region (approximately -200 bp to the translation start site) and is specific to cells which express class II genes at some stages of cell differentiation (e.g., pre-B cells, B cells, myeloma cells and macrophages), but not present in cells which do not normally express class II genes (e.g., T cells, fibroblasts and mastocytoma cells).

The presence of a tissue-specific D2 site led us to ask whether it serves as a binding site for positive or negative regulatory factors. Therefore, we examined two somatic cell hybrid cell lines produced by the fusion of the BW5147 T cell thymoma with normal splenic B cells for the presence of DNase I hypersensitive sites. These B-T hybrids did not express class II alleles from either T cell or B cell partners, although such molecules were expressed by the B cell partner before fusion. When DNase I hypersensitivity experiments were performed with two of these B-T hybrids, the tissue-specific D2 site of the B cell partner was extinguished from the chromatin of these hybrids. Thus, the D2 site appears to be under negative trans-regulatory control. We conclude that a target of tissue-specific trans-acting regulators is located at about -0.2 kb in the Aα promoter region (Fig. 1).

Sequence analysis of the promoter region revealed two DNA sequence motifs which are conserved among all human and mouse class II genes. They are designated as X and Y boxes, which are located between -50 to -150 bp upstream of the transcription initiation site (Fig. 1). The X and Y boxes are 14 and 8 bp in length respectively, and are separated by a 19-20 bp interspace region of conserved length but variable sequence. The Y box is, in fact, an inverted CCAAT box, which is a common element detected in many eukaryotic promoters (Chodosh et al., 1988; Dorn et al., 1987a).

The conservation of both X and Y boxes suggested that they may function in regulating gene experssion and may mediate tissue-specific and inducible regulation of class II MHC genes. Certain functional roles of both X and Y boxes have been demonstrated by several groups in both transient transfection assays and studies with transgenic mice (Boss et al., 1986; Dorn et al., 1987b; Basta et al., 1988). For both human and mouse class II genes several groups have demonstrated that:

(1) The X and Y boxes are essential for the constitutive and basal level of class II gene transcription.
(2) The X and Y boxes are the minimal sequences required for B cell specific expression of class II genes.
(3) Induction of expression of class II genes by γ-IFN requires at least X and Y boxes in addition to another indispensible motif, the W box, located just upstream of X box in β chain gene.
(4) The X box is thought to be the ultimate target of factors which are defective in several class II negative B cell mutants, since the X box constructs which function as transcription elements in normal cells could not direct expression of a heterologous promoter in these mutant cells.

Because both X and Y boxes are the only and/or the most conserved regions among different human and class II upstream regions, it has been proposed that the coordinate tissue-specific and inducible expression of the class II genes may be mediated through these two elements. A reasonable hypothesis is that all of these X boxes or Y boxes are recognized by a common trans-acting factor(s) and that the coordination of expression is achieved through interactions of the common factor(s) with these X or Y boxes. However, evidence obtained from our laboratory suggests that the regulation of class II genes is more complex.

NUCLEAR FACTORS BIND TO THE CONSERVED X AND Y BOXES OF THE Aα GENE

Gel retardation or mobility shift assays have been widely used to detect sequence-specific DNA binding activities which recognize sequence motifs in the promoter and enhancer regions. These sequence motifs usually serve as important functional cis-acting elements. In several cases, these DNA binding proteins are transcription factors which modulate transcription activity after binding to their cognate sites.

The DNase I hypersensitivity studies localized the putative regulatory region to the 200 bp within the upstream region of the Aα gene, which contains two conserved elements. Since the X and Y boxes are functionally important for both basal level and inducible transcription of class II genes, it is likely that they exert their function through an interaction with trans-acting nuclear factors. To detect any sequence motifs which may serve as targets of the trans-acting nuclear factors, the upstream 1.9 kb region of the Aα gene was analyzed by gel retardation assay (Boothby et al., 1989). A 220 bp fragment (Hf220) containing both X and Y boxes bound nuclear factors and generated two prominent retarded DNA-protein complexes. However, these binding activities were not unique to class II expressing cells, but present in both class II positive (e.g., pre-B, B cells, myeloma, macrophages) and class II negative cells (e.g., T cells, fibroblasts, Hela cells, etc.) analyzed. The intensity of the retarded complexes is also comparable in the R8-205 pre-B cell line either untreated or treated with IL-4 and in the macrophage-like cell line P388D1 untreated or treated with IFN-γ.

To test whether the retarded complexes from nuclear extracts of different cells represent the same or different factors, DNase I footprinting analyses were then performed. Nuclear extracts from a panel of cells, including class II positive and negative cells, pre-B cells, and macrophages treated with IL-4 or γ-IFN were analyzed using the labeled Hf220 fragment as probe. The footprinting patterns of all these cells were similar. The protected region spanned all but the first three residues of the X box and extended 10 residues into the X-Y interspace (Fig. 1). The Y box was also protected from DNase I cleavage (Fig. 1). Therefore, no difference in the footprint pattern could be discerned among these cells. We conclude that the binding activities to the X and Y boxes are constitutive and cannot solely determine the regulation of Aα expression. Other mechanisms such as changes in the chromatin configuration of the class II genes and/or post-transcriptional modification of these DNA binding factors in various cell types may possibly lead to differential expression of the class II genes.

The conservation among the X box sequences of different class II genes ranges from 50% to 95%. For instance, the X boxes of Eα and Eβ share only 57% and 67% identity to the X box of Aα gene (Table 1). The Y boxes are more highly conserved among different isotypes since they are reverse CCAAT boxes. To test models in which class II MHC gene transcription is coordinately controlled through a common set of DNA binding transcription factors, we assayed whether the X boxes of different isotypes recognize the same or different DNA binding factors. Oligonucleotides

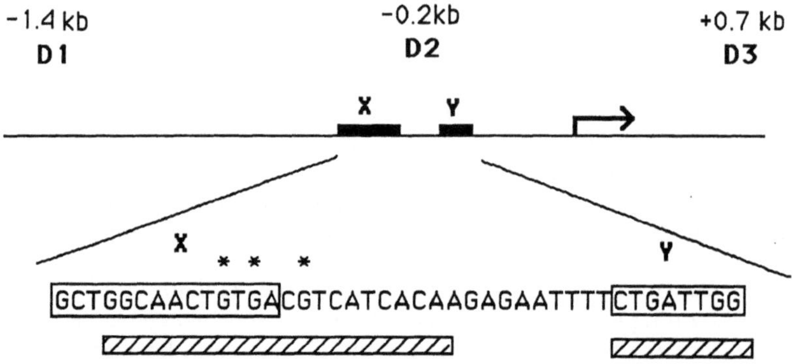

Figure 1. Schema of the Aα upstream region. D1, D2 and D3 represent three DNaseI hypersensitive sites. The X and Y elements are boxed. Shadowed boxes represent the footprinting regions. The contacted G residues in the methylation interference analysis are starred

Table 1 . Competition analysis of the Aα, Eα and Eβ X boxes

X box of Aα:	GCT GGC AAC TGTGA
X box of Eα:	CCT AGC AAC AGATG
X box of Eβ:	ACT AGC AAC TGATG

Competitor:	Probe X(Aα)	Probe X(Eα)	Probe X(Eβ)
X(Aα)	+++	-	-
X(Eα)	-	+++	+/-
X(Eβ)	-	+++	+++

X(Aα), X(Eα), and X(Eβ), corresponding to the X box and interspace sequences of the Aα, Eα, and **Eβ** genes were used in cross-competition assays. Although the retarded complexes generated with the X box oligonucleotide of Aα(X(Aα)) were readily competed by the unlabled X(Aα) oligonucleotide, they were not competed by either X(Eα) or X(Eβ) oligonucleotides (Table 1). By contrast, the retarded complexes of X(Eα) and X(Eβ) were not able to be competed by the X(Aα) oligonucleotide (Table 1). These competition experiments demonstrated that the nuclear proteins that specifically bind to the Aα X box differ from those that bind to Eα or Eβ X boxes. We conclude that there are multiple X box binding proteins, each of which recognizes X boxes of different isotypes. This observation challenges the concept that all class II genes are coordinately regulated through the conserved X boxes.

CLONING OF THE Aα X BOX BINDING PROTEINS

Although gel retardation assays help to localize the putative cis-acting elements involved in gene regulation, the information provided by this type of assay is limited since it provides no information about the composition of the retarded complexes or about their functions in transcription initiation. Therefore, in order to understand the mechanisms by which these DNA-protein complexes are involved in the class II gene regulation, cloning of these nuclear factors becomes an imporant step. A Southwestern method of library screening developed by Singh and Sharp (1988) was used to clone the nuclear factors which bind to the class II conserved X box. A murine spleen cDNA λgt11 library was screened with a DNA probe which contains both the X box and X-Y interspace region of the Aα gene. Positive clones obtained through three screenings were lysogenized and protein lysates were made from those lysogens and tested for binding activity.

Two cDNA clones, termed λmXBP-1 and λmXBP-2 (for murine-X-box-binding protein) were obtained which remained positive in the tertiary screening with XAα DNA probe but not with a non-specific DNA probe (Liou et al., 1988b). Lysogen lysates were prepared from these two clones and tested for their DNA binding specificities. Lysate proteins from these two clones bind to the X box probe of the Aα gene specifically since they do not bind probes from the irrelevant Y box or a more upstream element, BRE (Boothby 1989). DNase I footprinting analysis with these two proteins demonstrated that they protected the same X box and X-Y interspace regions as observed with native B cell nuclear extracts (Fig. 1). By methylation interference analysis, both mXBP clones and native nuclear extract were observed to contact two G residues in the X box and one G residue in the interspace region (Fig. 1).

Previous observations with gel retardation assays have suggested that Aα X box binding protein are different from the X box binding proteins of the Eα and Eβ genes. We then tested whether mXBP-1 and mXBP-2 can bind to the Eα and Eβ X boxes. As expected, these two clones do not recognize Eα or Eβ X boxes (Liou et al., 1988b). These observations confirmed our previous data and are consistent with the hypothesis that multiple X box binding proteins recognize X boxes of different class II genes.

The expression of the mXBP-1 RNA was analyzed by Northern blot hybridization. Two transcripts (approximately 6.2 and 3.0 kb) were observed with poly A^+ RNA from both class II positive and negative cells. Lymphokine treatment of B cells or macrophages did not affect mXBP-1 RNA levels (Liou et al., 1988b). This is consistent with previous data that the X box binding proteins are constitutively expressed.

The two mXBP-1 transcripts could be derived from two different genes with partial homology in the DNA binding domain or may represent alternative splicing of a single species of RNA. To differentiate between these two possibilities, we performed Southern blot hybridization with genomic DNA. Restriction enzyme digests of the genomic DNA generated multiple bands that hybridized to the mXBP-1 0.9 kb DNA as well as to two truncated mXBP-1 probes (150~200 bp in length) (Liou et al., 1988b). Thus, the mXBP-1 cDNA is probably represented in at least two genes.

Although mXBP-1 and mXBP-2 have similar DNA binding specificies, they are not identical (described below). Thus, at least two proteins recognize a single Aα X box motif. Reports on the Y box (or CCAAT box) binding proteins also identified several distinct binding activities which bind a single motif in native nuclear extracts. All of these data support a rather complicated regulatory pathway involved in class II gene regulation through these two conserved elements/sequences.

OTHER TARGETS FOR mXBP-1 AND mXBP-2

Footprinting analyses indicated that the protected region in the Aα gene for mXBP-1, mXBP-2 and the native B cell nuclear extract does not correspond exactly to the 14 bp consensus X box. The protected region overlaps the 3' part of the X box and the X-Y interspace (Fig. 1) and contains a consensus 8 bp cAMP-responsive element (CRE), TGACGTCA. CRE elements are found in many eukaryotic promoters which respond to cAMP induction, such as α-hCG, somatostin, etc. Others have reported that the target site for the activating transcription facotr (ATF), which is involved in EIA-mediated trans-activation of adenovirus early promoters, is identical to the CRE element (Lin et al., 1988). Furthermore, CREs are probably recognized by a common factor(s) indispensible for transcription initiation from CRE-containing promoters.

To test whether the cloned cDNAs encoding the mXBP proteins also recognize the CRE/ATF elements, oligonucleotides corresponding to the CRE/ATF sequences of several genes were used to compete with the X(Aα) probe in gel retardation assays (Table 2). For both mXBP-1 and mXBP-2, the XAα and AdE4 oligonucleotides were the most potent competitors. The CRE of the *fos* gene competed well with mXPB-2 binding to the XAα probe but not with mXBP-1 binding, indicating subtle differences in the DNA binding specificities of these two proteins. Although the CMV viral gene has a consensus CRE sequence, it competed poorly the mXPB-1 and mXBP-2 binding activities. In all cases, the retarded bands from B cell nuclear extracts were competed by all of the oligonucleotides since the nuclear extract contains pools of binding proteins which share similar binding activities. The competition data demonstrate that (1) mXBP-1 and mXBP-2 show similar but not identical DNA binding specificities, and (2) both mXBPs can bind to some of the CRE/ATF elements (in this sense, they are similar to CREB or ATF).

TABLE 2 . Binding specificities of mXBP-1 and mXBP-2.

Competitor	sequence	B cell nuclear hXBP-1 extract	mXBP-1	mXBP-2/CREB
XAa	TGACGTCA	++++	++++	++++
AdE4	TTACGTCA	+++	++++	+++
CMV	TGACGTCA	+/-	+/-	+
fos	TGACGTAG	++	-	+++
hCG	TGACGTCA	++	+++	++
Somat.	TGACGTCA	+++	++	+++

SEQUENCES OF mXBPs

Sequence analysis of the mXBP cDNA reveals interesting structural features characteristic of DNA binding transcription factors of the leucine-zipper class of proteins (Ivashkiv et al., and Kara et al., unpublished data). The open reading frame of mXBP-1 extends from nucleotides 1 to 939 and contains 3 ATG codons mXBP-1 contains a heptad repeat of five leucines and an adjacent highly basic region. This region of the mXBP-1 sequence shares striking similarities to the Fos, Jun, GCN4 and CREB proteins which also contain the leucine-zipper and basic regions. The basic regions and leucine-zipper domains of the FOS and JUN proteins have been reported to be required for DNA binding and FOS-JUN and JUN-JUN dimerization individually.

Partial sequence analysis of the mXBP-2 cDNA also revealed similar structures and indicated that mXBP-2 is probably the mouse homologue of the recent cloned CREB protein (Hoeffler et al., 1988 and Gonzalez et al., 1989).

This leucine-zipper class of proteins, in fact, recognize similar target sequences. The AP-1 site (TGACTCA) recognized by the FOS/JUN heterodimer has 7/8 bp identity to the CRE site. The optimal GCN4 site is identical to the AP-1 site and it has been reported that the JUN protein can direct HIS3 gene transcription through binding to the GCN4 site (Struhl 1988). The related sequence homology among the proteins in the leucine-zipper class as well as their striking target site similarity suggest that these proteins may share some overlapping functions in eukaryotic cells.

DISCUSSION

We observed a DNase I hypersensitive site D2 for the class II Aα gene which is uniquely present in cells that normally can express class II genes. This tissue-specific D2 site is located approximately 200 bp region upstream of the Aα gene transcription initiation site, a region which contains conserved X and Y boxes which play important roles in class II gene regulation. The DNase I site could correlate with the presence of active transcription complexes. However, gel retardation assays detected ubiquitous DNA binding activities which recognize both the X box plus the X-Y interspace region and the Y box. Therefore, the tissue-specific and/or inducible expression of the class II genes could either involve other as yet unidentified trans-acting factors or be controlled by the X and Y box-associated DNA binding transcription factors which are post-translationally modified.

Studies from our laboratory have identified several other DNA elements in the Aα gene which serve as target sequences for DNA binding factors. One DNA binding activity (NF-BRE) (Boothby, 1988) which binds two sites (-450 and -1100 bp) further upstream from the X box in the Aα promoter was detected. The binding activity of this protein correlates well with the status of class II expression in panels of cells and is responsible for IL-4 induction of class II genes in cells of the B lineage. Therefore, NF-BRE may mediate B cell specific and IL-4 induced expression of the Aα class II gene.

Evidence from cross-competiton studies showed that the DNA binding activities which recognize the X boxes of the Aα gene differ from those that bind Eα and Eβ X boxes. The fact that the two cloned X box binding proteins (mXBP-1 and mXBP-2) bind only the X box of Aα but not the X box of Eα or Eβ genes confirms this observation. Given the fact that mXBP-1 and mXBP-2 do not cross-hybridize and that mXBP-1 is represented in at least two genes, the number of proteins that recognize the MHC X box sequences may be quite large. This is also true for the Y box since several CCAAT box binding proteins with cross-reactivities to the CCAAT boxes of different genes have been purified or isolated.

Several interesing questions can be raised. What are the functions of these multiple DNA binding proteins? Could it be that some of these proteins serve as transcription activators and the others as repressors? Or they might be expressed in variable amounts in different tissues, and the relative levels of these factors in the same cell determines whether the class II gene is transcribed. Alternatively, these proteins might respond to different stimuli such as IL-4, γ-IFN, etc., which are known to modulate class II gene transcription.

The structural similarity of mXBPs to the Fos and Jun protein family may indicate functional relationship among these proteins. These proteins are known to recognize sequence motifs responsible for EIA, cAMP, and PMA inducibility and, therefore it will be interesting to see whether mXPBs respond to these stimuli. Moreover, since the mXBP proteins contain leucine zipper structure, it will also be interesting to test whether the mXBPs can form homodimers or heterodimers with other members of the leucine-zipper protein family.

One intriguing question still remains: Are class II genes coordinately regulated? Comparison of all class II upstream sequences obtained so far indicates that only the X and Y box regions are conserved in different class II isotypes. Since the Y box is a very common binding motif and different X boxes may be recognized by different DNA binding factors, it is unlikely that coordination of class II gene expression is controlled by the X or Y box binding factors. However, it is possible that these multiple X box proteins may interact with some other common transcription factor(s) yet unidentified whose activity is responsive to lymphokine or differentiation signals. Alternatively, coordination could be achieved through different motifs with equivalent functions (e.g., responding to the same stimuli) in different class II promoters. Thus, with arrays of sequence motifs which share overlapping but not identical functions for each class II isotype, all class II genes may coordinately respond to the same environmental signals, but through distinct signal transduction pathways. Alternatively, each class II gene may respond uniquely to particular stimuli through its own specific cis-acting motifs. Delineation of all class II upstream cis-acting motifs and characterization of their corresponding DNA binding trans-acting factors will help to unravel the molecular mechanisms of differential class II gene regulation.

REFERENCES

Basta, P.V., Sherman, P.A. and Ting, J.P.-Y. (1988). Detailed delineation of an interferon-γ-responsive element important in human HLA-DRα gene expression in a glioblastoma multiform line. Proc. Natl. Acad. Sci. USA 85, 8618-8622.

Boss, J.M. and Strominger, J.L. (1986). Regulation of a transfected human class II MHC gene in human fibroblasts. Proc. Natl. Acad. Sci. USA 83, 9139-9143.

Boothby, M., Gravallese, E., Liou, H.-C. and Glimcher, L.H. (1988). A DNA binding protein regulated by IL-4 and by differentiation in B cells. Science 242, 1559-1562.

Boothby, M., Liou, H.-C. and Glimcher, L.H. (1989). Differences in DNA sequence specificity among MHC class II X box binding proteins. J. Immunol. 142, 1005-1014.

Chodosh, L.A., Olesen, J., Hahn, S., Baldwin, A.S., Guarente, L. and Sharp, P.A. (1988). A yeast and a human CCAAT-binding protein have heterologous subunits that are functionally interchangeable. Cell 53, 25-35.

Dorn, A., Bollekens, J., Staub, A., Benoist, C. and Mathis, D. (1987a). A multiplicity of CCAAT box-binding proteins. Cell 50, 863-872.

Dorn, A., Durand, B., Marfing, C., LeMeur, M., Benoist, C. and Mathis, D. (1987b). Conserved MHC class II boxes-X and Y are transcriptional control elements and specifically bind nuclera proteins. Proc. Natl. Acad. Sci. USA 84, 6249-6255.

Gonzalez, G.A., Yamamato, K.K., Fisher, W.H., Karr, D., Menzel, P., Biggs, W., Vale, W.W. and Montminy, M.R. (1989). A cluster of phosphorylation sites on the cyclic AMP-regulated nuclear factor CREB predicted by its sequence. Nature 337, 749-752.

Hoeffler, J.P., Meyer, T.E., Yun, Y., Jameson, J.l. and Habener, J.F. (1988). Cyclic AMP-responsive DNA binding protein: Structure based on a cloned placental cDNA. Science 242, 1430-1433.

Lin, Y.-S. and Green, M.R. (1988). Interaction of a common cellular transcription factor, ATF, with regulatory elements in both Ela and cyclic-AMP inducible promoters. Proc. Natl. Acad. Sci. USA 85, 3396-3400.

Liou, H.-C., Polla, B.S., Aragnol, D., Leserman, L.D., Griffith, I.J. and Glimcher, L.H. (1988a). A tissue specific DNase I hypersensitive site in a class II Aα gene is under trans-regulatory control. Proc. Natl. Acad. Sci. USA 85, 2738-2742.

Liou, H.-C., Boothby, M.R. and Glimcher, L.H. (1988b). Distinct cloned class II MHC DNA binding proteins recognize the X box transcription element. Science 242, 69-71.

Singh, H., LeBowitz, J.H., Baldwin, A.S. and Sharp, P.A. (1988). Molecular cloning of an enhancer binding protein: Isolation by screening of an expression library with a recognition site DNA. Cell 52, 415-423.

Struhl, K. (1988). The JUN oncoprotein, a vertebrate transcription factor, activates transcription in yeast. Nature 332, 649-650.

The Effect of Copy Number on mRNA and Cell Surface Expression of an Aβ^k Transgene

S. GILFILLAN[1], S. AISO[1], S. A. MICHIE[3], and H. O. MCDEVITT[1,2]

[1]Departments of Microbiology and Immunology, [2]Medicine, and [3]Pathology, Stanford University, Stanford, CA 94303, USA

INTRODUCTION

Because allelic polymorphism of the class II antigens affects the immune response at several levels, we wanted to characterize the contribution of a particular chain or epitope in an *in vivo* system using transgenic mice. Initially, we introduced an Aβ^k genomic clone into [B10.S x SJL]F_2 (H-$2^{s/s}$) embryos and, from fourteen founders, have established twelve independent lines carrying from one to sixty-five copies of the transgene. The transgene was coexpressed with the endogenous allele in a tissue-specific manner, and Aβ^k mRNA expression correlated well with transgene copy number. Although the ratio of Aβ^k to Aβ^s expressed on the cell surface correlated well with the ratio of Aβ^k to Aβ^s mRNA, cell surface levels of the endogenous Aβ^s/Aα^s complex and total Ia were reduced in H-$2^{s/s}$ mice overexpressing Aβ^k mRNA. However, extremely high levels of Aα^k/Aβ^k cell surface expression were observed in Aβ^k x Aα^k double transgenic mice, which implies that the excess mRNA is translated in the high copy number mice and that pairing of Aβ^k with Aα^s is permissive but inefficient. Initial immune response experiments reflected the variation in cell surface levels of Aβ^k (and total Ia) and suggested that expression of the Aα^s/Aβ^k heterodimer has some effect on the secondary antibody response to the I-A^k-restricted synthetic antigen (H,G)-A--L [(histidine, glutamic acid)-alanine--lysine; McDevitt and Chinitz 1969]. Unexpectedly, high copy numbers (extreme overexpression) of the transgene were associated with a variety of defects including abnormal extramedullary hematopoiesis, depletion of B cells, an increased susceptibility to infection, and premature death of some animals. Collectively, these mice provide a well controlled system in which to study the effects of gene dosage on regulation, expression and function of the class II antigens *in vivo*.

RESULTS

An 11.6 kb XbaI-HindIII fragment of the pIAβ^k-1 clone (Choi 1983, Germain 1983) containing 2-3 kb of 5' sequence and over one kb of 3' sequence was injected into the pronuclei of [B10.S x SJL]F_2 (H-$2^{s/s}$) embryos. Founders were initially backcrossed to [B10.S x SJL]F_1 mice and then maintained as hemizygotes, or intercrossed to produce homozygous lines (summarized in Table 1). The transgene was inherited in

I.K. Egorov/C.S. David (Eds.)
Transgenic Mice and Mutants in MHC Research

a Mendelian manner consistent with integration on an autosome with the following exceptions: one founder carried approximately fifty copies of the transgene at each of two insertions sites (one on the X chromosome, the other on an autosome); at least two founders were mosaic (12 to 24% of their offspring carried the transgene); and two founders failed to transmit the transgene (one male did not breed and one female carrying approximately seventy copies of the transgene died prior to having offspring). The transgene apparently disrupted an essential gene in another lineage (21 of 37 offspring from six intercrosses were transgenic; no homozygotes have been identified).

Table 1. $A\beta^k$ Transgenic Lines

Founder mice	$A\beta^k$ copy number	mRNA Expression	Cell Surface Expression
$A\beta^k$-110 $A\beta^k$-130 $A\beta^k$-124 $A\beta^k$-131*	1-2	$A\beta^k < A\beta^s$ $A\beta \geq A\alpha$	$A\beta^k \lll A\beta^s$ Total Ia = F_1 control
$A\beta^k$-111 $A\beta^k$-115 $A\beta^k$-145	6-10	$A\beta^k \geq A\beta^s$ $A\beta > A\alpha$	$A\beta^k = A\beta^s$ Total Ia < F_1 Control
$A\beta^k$-70 $A\beta^k$-99 $A\beta^k$-92	20-30	$A\beta^k \gg A\beta^s$ $A\beta \gg A\alpha$	$A\beta^k > A\beta^s$ Total Ia << F_1 Control
$A\beta^k$-34# $A\beta^k$-40 $A\beta^k$-105 $A\beta^k$-90*	40-70	$A\beta^k \ggg A\beta^s$ $A\beta \ggg A\alpha$	$A\beta^k > A\beta^s$ Total Ia <<< F_1 Control

Founder had approximately 50 copies of $A\beta^k$ integrated in each of two insertion sites.
* Lines not established from these founders.

mRNA Expression

Northern blot analysis of total RNA isolated from the spleens of representative mice indicated that expression of $A\beta$ mRNA was elevated in mice carrying more than ten copies of the transgene. The ratio of $A\beta$ to $A\alpha$ mRNA progressively increased from one to more than sixty in mice carrying from one to sixty-five copies of the transgene (compared to nontransgenic controls, in which the ratio of $A\beta$ to $A\alpha$ hybridization intensities were normalized to one). RNase protection assays using a probe that distinguishes the $A\beta^k$ transcript from the endogenous $A\beta^s$ transcript demonstrated clearly that the excess $A\beta$ mRNA was encoded by the transgene. The ratio of $A\beta^k$ to $A\beta^s$ message was approximately one in [B10.A(4R) x nontransgenic]F_1 ([4R x NT]F_1) control mice. In mice carrying one copy of the transgene, the ratio of $A\beta^k$ to $A\beta^s$ mRNA was 0.1 and progressively increased in mice carrying higher copy numbers of the transgene. Surveys of tissues from mice of several lines indicated that the transgene was coexpressed with the endogenous allele. No $A\beta^k$ message was observed in the absence of $A\beta^s$ message, and the ratio of $A\beta^k$ to $A\beta^s$ message was similar in all tissues examined within each mouse, including the thymus.

Cell Surface Expression

Cell surface expression of the transgene and endogenous allele was assessed by Fluorescence activated Cell Sorter (FACS) and immunohistochemical analysis using the monoclonal antibodies MKS4 (anti-$A\beta^{s,f,u}$; Kappler 1981); 40.N (anti-$A\beta^{k,f,r,u}$; Pierres 1981); and 10-3.6 [anti-$A\beta^{s,k,f,r,u}$ (total Ia in these mice), Oi 1978]. B cells were stained with RA3 6B2 (anti-B220, Coffman and Weissman 1981) and macrophages/granulocytes with MI/70 (anti-Mac I, Springer 1979). Control mice included B10.A(4R) mice, nontransgenic littermates, and [4R x NT]F_1 mice (which express $A\alpha^k$, $A\beta^k$, $A\alpha^s$, and $A\beta^s$, hence should serve as the best controls with which to compare levels of transgene expression).

The lymphoid population expressing Ia appeared normal in mice carrying from one to thirty copies of the transgene; the numbers of B cells and Mac-1 positive cells (macrophages and granulocytes) were similar to those in controls. In mice carrying more than forty copies of $A\beta^k$, however, there was a consistent increase in the number of Mac-1 positive cells observed in spleen and peripheral blood. In addition, most individuals had reduced numbers of B cells, some having less than 20% the average number observed in controls. FACS analysis indicated that pre B cells ($B220^+$, IgM^-, Ia^-) were present in the bone marrow of high copy number $A\beta^k$ mice and often represented a relatively high percentage of the $B220^+$ population in the spleen. However, the $B220^+$ IgM^+ population was depleted in the bone marrow and represented the majority of mature ($B220^{++}$, IgM^+, Ia^+) cells in the spleen; very few IgM^+,IGD^+ cells were present. Tissue sections from several mice carrying from sixty to one hundred-thirty copies of the transgene confirmed the FACS data. All had an increased number of granulocytes in the spleen, lymph nodes, and thymus, and two had a severe depletion of B cells in the spleen and lymph nodes as well as extramedullary hematopoeisis in the spleen, lymph nodes, and liver.

FACS analysis of spleen and peripheral blood cells indicated that the ratio of $A\beta^k$ to $A\beta^s$ on the surface of B cells correlated well with transgene copy number and mRNA expression. Low cell surface levels of the transgene were detected in all mice in which the ratio of $A\beta^k$ to $A\beta^s$ mRNA was less than one (those carrying one to two copies of the transgene). The cell surface ratio of $A\beta^k$ to $A\beta^s$ in these mice was lower than that observed in [4R x NT]F1 controls, and the level of $A\beta^s$ on the cell surface was not affected by transgene expression. The cell surface ratio of $A\beta^k$ to $A\beta^s$ closely resembled [4R x NT]F1 controls in mice carrying ten copies of $A\beta^k$, in which the ratio of $A\beta^k$ to $A\beta^s$ mRNA was slightly greater than one. In mice carrying more than ten copies of the $A\beta^k$ transgene (those with $A\beta^k/A\beta^s$ mRNA ratios greater than one), the ratio of $A\beta^k$ to $A\beta^s$ on the cell surface varied among individuals, but was consistently higher than that observed in [4R x NT]F1 controls.

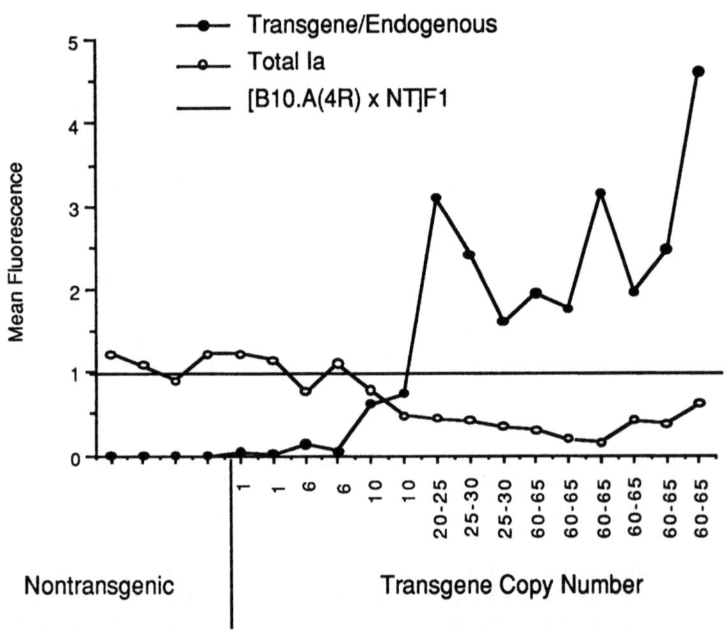

Figure 1. Relative Ia cell surface levels on mature splenic B cells from control and $A\beta^k$ transgenic mice. Cells were stained for Ia with 40.N ($A\beta^k$), MKS4 ($A\beta^s$) and 10-3.6 ($A\beta^k$ and $A\beta^s$). Mean fluorescence values obtained with each reagent were normalized to those observed on (B10.A(4R) x NT) F_1 B220++ spleen cells in the same experiment (assigned a value of one, represented by the solid line). The closed circles represent the ratio of 40.N to MKS4 (transgene/endogenous) and the open circles represent the relative levels of 10-3.6 (total Ia)

Although the ratio of $A\beta^k/A\alpha^s$ to $A\beta^s/A\alpha^s$ on the cell surface correlated well with the ratio of $A\beta^k$ to $A\beta^s$ mRNA, the amount of total Ia on the cell surface as well as the number of Ia positive cells progressively decreased in mice carrying more than ten copies of $A\beta^k$. This reduction in Ia expression was observed in every mouse in which the ratio of $A\beta^k$ to $A\beta^s$ on the cell surface was greater than or equal to that observed in [4R x NT]F1 control mice (Figure 1). Expression of the transgene in thymic tissue paralleled that observed on B cells. A mouse carrying one copy of $A\beta^k$ expressed low levels of the transgene in both the cortex and medulla but control levels of $A\beta^s$. Again, approximately equal levels of the transgene and endogenous allele were observed in both the thymic medulla and cortex from a mouse carrying ten copies of $A\beta^k$, but the levels of both $A\beta^k$ and $A\beta^s$ were lower than those observed in [4R x NT]F_1 controls. As the ratio of $A\beta^k$ to $A\beta^s$ progressively increased in mice carrying more than ten copies of the transgene, the level of $A\beta^s$ (and total Ia) decreased. This decrease in Ia expression was more marked in the cortex; mice carrying from fifty to one hundred-thirty copies of $A\beta^k$ had a severe reduction in cell surface expression of Ia in the thymic cortex and slightly decreased expression in the medulla. The numbers and CD4/CD8 ratios of T cells were similar to controls in mice from all of the $A\beta^k$ transgenic lines, including those with markedly reduced Ia expression in the thymic cortex.

I-A^k Expression in $A\beta^k$ x $A\alpha^k$ Double Transgenic Mice

Providing the appropriate pairing partner did dramatically increase cell surface expression of $A\beta^k$ (demonstrating that the excess mRNA is being translated), but did not correct the defects observed in the high copy number $A\beta^k$ mice. When a mouse carrying sixty-five copies of the $A\beta^k$ was mated to a mouse carrying approximately twenty copies of $A\alpha^k$ (from the $A\alpha^k$-26 line, Lemeur 1989), $A\beta^k$ x $A\alpha^k$ double transgenic offspring expressed extremely high levels of $A\alpha^k/A\beta^k$ on the surface of spleen cells as well as on cortical and medullary cells in the thymus. However, B cell numbers were reduced, and pre-B cells (B220 dull, Ia^-) represented a relatively high percentage of $B220^+$ cells in the spleen. In addition, the numbers of Mac-1 positive cells were significantly increased, and abnormal extramedullary hematopoiesis occurred in the spleen, lymph nodes, and liver. Similar results were obtained by mating a mouse carrying sixty-five copies of the transgene to a B10.A(4R) mouse. Offspring carrying one copy of $A\alpha^k$ and approximately sixty-six copies of $A\beta^k$ expressed [4R x NT]F_1 control cell surface levels of $A\alpha^k/A\beta^k$, very low levels of $A\alpha^s/A\beta^s$, and shared the defects observed in all other high copy number mice. Cell surface expression of the transgene was also enhanced in double transgenic mice carrying six or ten copies of $A\beta^k$ and twenty copies of $A\alpha^k$, but these mice did not share the defects observed in $A\beta^k$ high copy number mice.

Immune Response

Mice from five $A\beta^k$ lines carrying from six to sixty-five copies of the transgene, $A\alpha^k$-26 mice carrying twenty copies of $A\alpha^k$, and three combinations of $A\beta^k$ x $A\alpha^k$ double transgenic mice were immunized with (H,G)-A--L, an I-A^k-restricted synthetic polymer. Five mice from the $A\beta^k$-40 high copy number lineage were also immunized with Bovine Serum Albumin (BSA), a nonspecific antigen. A fairly rigorous immunization protocol was followed to ensure maximum responses. Control groups for each experiment consisted of non-transgenic littermates, [B10.SxSJL]F_1

mice, and [4R x NT]F_1 mice. The antibody responses to BSA of mice from the $A\beta^k$-40 lineage were indistinguishable from those of controls, indicating that (at least using this protocol) these high copy number mice are capable of generating a secondary antibody response to BSA.

Table 2. Secondary antibody response to (H,G)-A--L. Mice were immunized with 50µg of (H,G)-A--L (Miles Scientific) emulsified in complete Freunds adjuvant injected subcutaneously in the hind footpads, and were boosted 3 weeks later with 10 µg of the antigen emulsified in incomplete Freunds adjuvant (again subcutaneously in the hind footpads). Mice were bled 8 days later and serum titers to (H,G)-A--L assessed in an ELISA asay. Antibody titers to (H,G)-A--L were determined as the last two or five fold serum dilution with an absorbance at 405 nm four standard deviations above the mean of normal mouse serum control wells. NT, non transgenic

Mice	$A\beta^k : A\alpha^k$ copy number	# mice responding/ # mice immunized	(H,G)-A--L antibody titers in responding animals
[B10.S x SJL]F_1	0:0	0/23	-
NT littermates	0:0	0/30	-
[B10.A(4R) x NT]F_1	1:1	32/32	1/1,000 - 1/10,000
$A\alpha^k$-26	0:20	1/10	1/20
$A\beta^k$-111	6:0	4/8	1/40 - 1/800
$A\beta^k$-145	10:0	6/22	1/20 - 1/400
$A\beta^k$-99	20:0	0/8	-
$A\beta^k$-34	50:0	1/6	1/160
$A\beta^k$-40	65:0	0/15	-
$A\beta^k$-111 x $A\alpha^k$-26	6:20	6/6	1/800 -1/5,000
$A\beta^k$-145 x $A\alpha^k$-26	10:20	6/6	1/400 - 1/10,000
$A\beta^k$-40 x $A\alpha^k$-26	65:20	8/8	1/80 - 1/1,000

All $A\alpha^k$ x $A\beta^k$ double transgenic mice and a few mice from several $A\beta^k$ transgenic lines responded signigicantly better to (H,G)-A--L than did non-transgenic littermates or [B10.SxSJL]F_1 negative controls (Table 2). Antibody titers to (H,G)-A--L in the double transgenic mice were similar to those of [4R x NT]F_1 positive controls, though the responses were slightly lower in several $A\beta^k$-40 x $A\alpha^k$-26 mice (probably due to reduced numbers of B cells). Although antibody titers to (H,G)-A--L were well above negative controls in some of the $A\beta^k$ transgenic mice, all were lower than those observed in the positive controls. The highest titer from the $A\beta^k$ transgenic group was approximately 1/10 that of the mean of positive controls.

Illness and Death in High Copy Number Lineages.

Initial breeding, particularly attempts to establish homozygous lines, indicated that the premature death rate among mice carrying high copy numbers of the transgene was extremely high (Table 3). Many of the females in these lines were poor mothers and lost part or all of most litters. Because crosses between as well as within the $A\beta^k$ lines produced the same effect, integration site(s) did not appear to be responsible for this phenomenon. In addition, three high copy number founder females died before reaching six month of age, one prior to having offspring. Six of the ill/dead mice were autopsied at three to four months of age. All had extramedullary granulopoiesis in the liver, five had congestion and atelectasis in the lung, and three had extramedullary hematopoiesis and granulatomous inflammation within the myocardium. Other findings were consistent with viral or blood borne bacterial infections. Mice carrying more than one hundred-fifty copies of the transgene became ill and generally died at three to four weeks of age; three autopsied had a severe myeloproliferative disease involving granulocytes.

Table 3. Survival of $A\beta^k$ Transgenic Mice. Numbers reflect outcrossing of all lines as well as intercrossing between and within lines; all mice were monitored until at least 6 months of age. T, Transgenic/Number screened

$A\beta^k$ Copy Number	Number Born	T	Dead < 1 week of age	T	Dead < 6 Months of age	T
0-20	233	(96/220)	13 (6%)	(?)	1 (4%)	(0/1)
20-50	315	(141/302)	13 (4%)	(?)	4 (1%)	(4/4)
40-170	334	(169/238)	110 (33%)	(9/14)	29 (9%)	(29/29)

Because the autopsy reports implicated a variety of infectious agents and high copy number mice were difficult to maintain in our conventional facility, two litters from a line carrying appproximately sixty-five copies of $A\beta^k$ were derived into Specific Pathogen Free (SPF) conditions. Survival did improve considerably, suggesting that an increased susceptibility to infection is one aspect (result) of the defect in these mice. Though several homozygotes have been identified and bred, they have been difficult to maintain even in SPF conditions. One mouse autopsied after dying in the SPF facility shared many clinical findings with those dying in the conventional facility (liver congested with multifocal neutrophilic infililtrates, heart extensively infiltrated with neutrophils and granulocytic precursors, increased granulopoietic activity in the bone marrow). Thus, mice carrying approximately one hundred-thirty copies of the transgene may be susceptible to normally non-pathogenic bacteria present in the SPF colony or have more severe (but variable) intrinsic defects caused by the transgene that result in early death.

DISCUSSION

The fact that $A\beta^k$ mRNA levels correlated with transgene copy number was unexpected in context of early reports in the transgenic literature (for a review see Palmiter and Brinster 1986). Such correlation has since been noted in mice carrying the human HLA-B7 gene (Chamberlain 1988) and the human CD2 gene (Lang 1988), and DNA sequences conferring this property have been identified in mice carrying a human β-globin gene flanked by distal 5' and 3' sequences containing erythroid-specific DNAase 1-hypersensitive sites (Grosveld 1987), and the murine Thy-1.1 gene linked to T cell-specific DNAase 1-hypersensitive sites from the 3' flanking region of the human CD2 gene (Greaves 1989). Chromosomal position may have had some effect on transgene expression in our system, in that $A\beta^k$ mRNA expression in mice carrying one copy of the transgene was less than that observed in [4R x NT]F_1 controls.

Because $A\beta^k$ mRNA expression did correlate with transgene copy number, these mice allowed us to assess competition of $A\beta^k$ with $A\beta^s$ for pairing with $A\alpha^s$ *in vivo*. *In vitro* studies have shown that haplotype-matched α and β chains pair more efficiently than do chains from different haplotypes; expression of some mixed heterodimers is fairly permissive ($A\beta^k/A\alpha^b$) whereas others ($A\beta^k/A\alpha^d$) require high levels of mRNA in order to achieve relatively low cell surface levels in transfected cells (Germain 1985). In our system relative levels of mRNA clearly affect expression of the $A\beta^k/A\alpha^s$ heterodimer *in vivo*. Mice with an $A\beta^k/A\beta^s$ mRNA ratio less than or equal to one (those carrying one to two copies of $A\beta^k$) expressed low but detectable levels of the transgene on all Ia^+ cell types examined. Cell surface levels of the endogenous $A\beta^s/A\alpha^s$ heterodimers were not affected by transgene expression in these mice, and were similar to those of nontrangenic litermates. In the majority of mice in which the ratio of $A\beta^k$ to $A\beta^s$ mRNA was greater than one (those carrying from ten to one hundred-thirty copies of $A\beta^k$), the ratio of $A\beta^k$ to $A\beta^s$ on the cell surface was greater than or equal to that observed in [4R x NT]F_1 control mice. However, the levels of $A\beta^s/A\alpha^s$ were reduced in these mice and none expressed control levels of Ia on the cell surface of any Ia^+ cell examined. This implies that $A\beta^k/A\alpha^s$ pairing is possible but not efficient. In the high copy number $A\beta^k$ mice, excess $A\beta^k$ protein apparently pairs with most of the available $A\alpha^s$, but only a fraction of the $A\beta^k/A\alpha^s$ heterodimers reaches the cell surface. Improperly paired and unpaired proteins are presumably degraded. Preliminary pulse-chase experiments suggest that Ia composed primarily of $A\beta^k/A\alpha^s$ heterodimers may be turning over more rapidly in B cells than does Ia in control mice.

Le Meur et al. (1989) have described two lines of mice carrying ten and fifteen copies of $A\beta^k$ on both $H\text{-}2^{s/s}$ and $H\text{-}2^{b/b}$ backgrounds which resemble our low copy number transgenic lines (one-six copies of $A\beta^k$) in terms of cell surface expression. The reason for this discrepancy is not obvious, but may reflect differences in Aβ^k mRNA expression. Clearly, in both systems, when the natural pairing partner is provided by mating Aβ^k mice to either B10.A(4R) or Aα^k transgenic mice, Aα^k/Aβ^k heterodimers form efficiently and are expressed on the cell surface of all Ia$^+$ cells. The level of expression seems to be determined by the amount of protein being synthesized, correctly paired and transported to the cell surface. The maximum level may be determined simply by how much protein the cell can tolerate.

The immune response experiments indicate that the Aβ^k transgene encodes a functional protein. Secondary antibody responses to the I-A^k-restricted antigen (H,G)-A--L were elicited in all Aα^k/Aβ^k double transgenic mice and were similar to those observed in [4R x NT]F_1 controls. The serum titers of anti (H,G)-A--L antibodies reflected B cell numbers and were reduced in several high copy number double transgenic mice. Only one of the Aα^k transgenic mice responded slightly better to (H,G)-A--L than did negative control mice (which is probably due to the extremely low levels of Aα^k cell surface expression in these mice, Le Meur 1989). Expression of the Aβ^k chain in association with the Aα^s chain did have an effect on the response to (H,G)-A--L. That the Aβ^k transgenic responses were lower than those of [4R x NT]F_1 controls may be due to the fact that the Aα^s/Aβ^k heterodimer binds (H,G)-A--L (peptides) less efficiently than does Aα^k/Aβ^k or that T cells capable of being activated by (H,G)-A--L in context of Aα^s/Aβ^k are relatively scarce in these mice. Only a few of the Aβ^k transgenic mice responded, and the majority were from the Aβ^k-111 and 145 lines. The density of Ia expressed on antigen presenting cells does affect the efficiency of T cell activation (McNicholas 1982, Matis 1982; Matis 1983; Lechler 1985). The minumum density of Aα^s/Aβ^k heterdimers required to generate a response to (H,G)-A--L may be more likely to occur on antigen presenting cells in mice carrying less than ten copies of Aβ^k (in which the cell suface levels of Ia are not reduced more than 50%).

The most unexpected result was the toxicity of high copy numbers of the transgene. Mice carrying high copy numbers of the Aβ^k transgene shared several defects, including abnormal extramedullary hematopoiesis (primarily granulopoiesis) in spleen, lymph node and liver; increased numbers of MI/70$^+$ cells (primarily granulocytes); a variable but often severe reduction in B cell number; and a relatively high premature death rate, partially due to an increased susceptibility to infection. This phenomenon does not appear to be an artifact unique to this particular Aβ^k transgene, as we have produced a similar phenotype by introducing an Aβ^d genomic clone into NOD mice (H. Acha-Orbea, unpublished data). The Aβ^k transgene appears to be expressed in appropriate tissues but at extremely high levels. The fact that modulating cell surface expression of the transgene by mating the high copy number Aβ^k mice to Aα^k or B10.A(4R) mice did not correct the defects (or save the mice) suggests that the initiating factor is overexpression of a normal protein which at some point functionally impairs at least some of those cells it is expressed in.

The toxic effect on B cells is variable and dependent on the degree of overexpression. Murine pre-B cells are Ia negative whereas mature B cells constitutively express Ia (Mond 1980; Kearney 1977). Ia is expressed at low levels on immature B cells and increases thereafter. The B cell depletion observed in the high copy number $A\beta^k$ mice correlates well with the onset of Ia expression; overexpression of Ia at this time may directly block the development of the majority of B cells. Whether the other Ia expressing cells are functionally intact is not known; Ia expression and regulation in macrophages has not yet been examined. Thymuses from several high copy number mice appeared normal (except for class II expression) in histological sections.

This situation is reminiscent of that observed in transgenic mice in which the insulin-secreting β cells are functionally impaired and variably killed by insulin promoter-driven expression of class II and class I proteins (Sarvetnick 1988, Lo 1988, Allison 1988). High level expression of MHC protein can apparently disrupt insulin synthesis and/or secretion without necessarily killing the β cells (for a discussion see Parham, 1988). Whereas any damage to β cells that affects insulin secretion will result in characteristic "diabetes", similar impairment of Ia expressing cells would be phenotypically much more complex. The defects we have observed in the high copy number $A\beta^k$ transgenic mice are not inconsistent with such a phenomenon.

Despite the high copy number defects, these $A\beta^k$ mice demonstrate that an exogenous $A\beta$ chain can be introduced into the germline of mice and coexpressed with the endogenous allele in the context of a normal immune system. Copy number dictates the level of transgene expression and thus is an important (albeit difficult to control) parameter in such mice. By crossing the $A\beta^k$ transgene onto other haplotypes, pairing, expression, and the effect of various $A\beta^k/A\alpha$ heterodimers on the response to defined antigens can be examined. Because relative cell surface levels of $A\beta^k/A\alpha$ heterodimers can be modulated by mating appropriate pairs, these mice provide a well controlled system in which to assess the effect of extreme variation in Ia cell surface expression in the thymus on the selection of the T cell repertoire. They may also allow the identifcation of T cell receptors biased for recognition of $A\alpha^k/A\beta^k$, which is expressed at extremely high levels in $A\alpha^k$ x $A\beta^k$ transgenic mice.

ACKNOWLEDGEMENTS

We thank Mary Vadeboncoeur, Donna Louie and Peggy Sullivan for excellent technical assistance; Antonio Ferraz for derivation and care of the SPF mice; Alan Stall and members of the Herzenberg FACS facility for reagents and advice; Drs. Diane Mathis and Christophe Benoist for the $A_\alpha{}^k$-26 mice, and Drs. Hans Acha Orbea, Chaim Jacob and David Wraith for helpful discussions. This work was supported by NIH grant AI19512; S.Gilfillan was supported by a predoctoral tranining grant (NIH 5T32 CA 09302) in the Cancer Biology Program.

REFERENCES

Allison J, Campbell IL, Morahan G, Mandel TE, Harrison LC, Miller JFAP (1988) Diabetes in transgenic mice resulting from overexpression of class I histocompatibility molecules in pancreatic β cells. Nature 333: 529-533

Chamberlain JW, Nolan JA, Conrad PJ, Vasavada HA, Vasavada HH, Ganguly S, Janeway CA Jr, Weissman SM (1988) Tissue-specific and cell surface expression of human major histocompatibility complex class I heavy (HLA-B7) and light (β2-microglobulin) chain genes in transgenic mice. Proc Natl Acad Sci USA 85: 7690-7694

Choi E, McIntyre K, Germain RN, Seidman JG (1983) Murine I-Aβ chain polymorphism: nucleotide sequences of three allelic I-Aβ genes. Science 221: 283-286

Coffman RL, Weissman IL (1981) B220- A B-cell specific member of the T-200 glycoprotein family. Nature 289: 681-683

Germain RN, Norcross MA (1983) Functional expression of a transfected murine class II MHC gene. Nature 306: 190-194

Germain RN, Bentley DM, Quill H (1985) Influence of allelic polymorphism on the assembly and surface expression of class II MHC (Ia) molecules. Cell 43: 233-242

Greaves DR, Wilson FD, Lang G, Kioussis D, (1989) Human CD2 3'-flanking sequences confer high-level T cell-specific, position-independent gene expression in transgenic mice. Cell 56: 979-986

Grosveld F, van Assendelft GB, Greaves DR, Kollias G (1987) Position-independent, high-level epression of the human β-globin gene in transgenic mice. Cell 51: 975-985

Kearney JF, Cooper MD, Klein J, Abney ER, Parkhouse RME, Lawton AR (1977) Ontogeny of Ia and IgD on IgM-bearing B lymphocytes in mice. J Exp Med 146: 297-301

Lang G, Wotton D, Owen MJ, Sewell WA, Brown MH, Mason DY, Crumpton MJ, Kioussis D (1988) The structure of the human CD2 gene and its expression in transgenic mice. EMBO J 7: 1675-1682

Lechler RI, Norcross MA, Germain RN (1985) Qualitative and quantitative studies of antigen-presenting cell function by using I-A-expressing L cells. J Immun 135: 2914-2922

Le Meur M, Waltzinger C, Gerlinger P, Benoist C,Mathis M (1989) Restricted assembly of MHC Class II molecules in transgenic mice. J Immun 142: 323-327

Lo D, Burkly LC, Widera G, Cowing C, Flavell RA, Palmiter RD, Brinster RL (1988) Diabetes and tolerance in transgenic mice expressing class II MHC molecules in pancreatic beta cells. Cell 53: 159-168

Matis LA, Jones PP, Murphy DB, Hedrick SM, Lerner EA, Janeway CA Jr, McNicholas JM, Schwartz R H (1982) Immune response gene function correlates with the expression of an Ia antigen. II. A quantitative deficiency in A_e:E_α complex expression causes a corresponding defect in antigen presenting cell function. J Exp Med 155: 508-523

Matis LA, Glimcher LH, Paul WE, Schwartz RH (1983) Magnitude of response of histocompatibility-restricted T-cell clones is a function of the product of the concentrations of antigen and Ia molecules. Proc Natl Acad Sci USA 80: 6019-6023

McDevitt HO, Chinitz, A (1969) Genetic contol of the antibody reponse: relationship between immune response and histocompatibility (H-2) type. Science 163, 1207-1208

McNicholas JM, Murphy DB, Matis LA, Schwartz RH, Lerner EA, Janeway, CA Jr, Jones PP (1982) Immune response gene function correlates with the expression of an Ia antigen. I. Preferential association of certain A_e and E_a chains results in a quantitative deficiency in expression of an A_e: E_a complex. J Exp Med 155: 490-507

Mond JJ, Kessler S, Finkelman FD, Paul WE, Scher I (1980) Heterogeneity of Ia expression on normal B cells, neonatal B cells, and on cells from B cell-defective CBA/N mice. J Immun 124: 1675-1682

Oi VT, Jones PP, Goding JW, Herzenberg LA, Herzenberg LA (1978) Properties of monoclonal antibodies to mouse Ig allotypes, H-2 and Ia antigens. Curr Topics in Micro and Immun 81: 115-129

Palmiter RD, Brinster RL (1986) Germline transformation of mice. Ann Rev Genetics 20: 465-501

Parham P, (1988) Intolerable secretion in tolerant transgenic mice. (News and Views) Nature 333: 500-503

Pierres M, Devaux C, Dosseto M, Marchetto S, (1981) Clonal analysis of B- and T- cell responses to Ia antigens. Immunogenetics 14: 481-495

Sarvetnick N, Liggitt D, Pitts SL, Hansen S, Stewart TA (1988) Modelling autoimmunity in transgenic mice: Class II MHC and interferon-gamma induce insulin-dependent diabetes mellitus. Cell 52: 773-782

Springer, T, Galfre, G, Secher, DS, Milstein, C (1979) Mac-1: A macrophage differentiation antigen identified by a monoclonal antibody. Eur J Immun 9: 301-306

Construction of Retroviral Vectors for the Study of Tissue-Specific Regulation of Human MHC Class II Genes

G. DEL POZZO, A. MAFFEI, C. PERFETTO, N. OMBRA, and J. GUARDIOLA[1]

[1] International Institute of Genetics and Biophysics, Via Guglielmo Marconi 10, 80125 Naples, Italy

INTRODUCTION

Retroviral vectors are now used for the transfer of exogenous genetic material into cultured cells and in vivo to the end of correcting genetic defects. Studies concerning the activity of tissue specific promoters can also be performed. We have constructed a vector which can be useful for the analysis of the expression of human MHC class II genes in various wild type and mutant cell lines and in bone marrow cells.

CONSTRUCTION OF VECTORS

The regulatory region upstream to the transcriptional start site of the HLA-DQα gene has been dissected by subcloning and has been characterized for the ability to direct correctly regulated gene expression in transient CAT assays (N. Ombra et al., unpublished). A DNA fragment, namely dq-7, carrying the ubiquitary X and Y regulatory boxes and the start of transcription of HLA-DQα has been subcloned and sequenced (Auffray et al., 1987; Del Pozzo and Guardiola, 1989); this fragment was capable of promoting MHC class II gene transcription and to allow expression of a CAT reporter gene in Ia-positive B-cells but not in a previously characterized Ia-negative B-cell mutant (Accolla, 1983; Guardiola et al., 1986). A transcriptional unit encompassing the dq-7 promoter region and a bacterial hph gene conferring hygromycin resistance was thus cloned into the XhoI site of the retroviral vector pLJ, hoping that the construct could confer hygromycin resistance to Ia-positive cells only.

Characterization of retroviral vectors

The structure of two retroviral vectors (respectively R1 and R2) was analyzed by restriction enzyme mapping and Southern hybridization experiments. The orientation of the dq-7/hph transcriptional unit relative to the direction of transcription of the 5' retroviral LTR and the internal SV40 promoter was determined (see Fig. 1).

The vectors were then introduced into PA317, a cell line carrying a helper provirus; the virus particles thus generated were used to infect various cell lines (see Table 1). Preliminary experiments on bulk populations of infected cells indicated that the dq-7 promoter region cloned in the R1 vector, although active in B-cells in transient transfection assays, is not sufficient to direct gene expression in retrovirus infected cells. The evidence available so far indicates that lack of expected promoter activity in some Ia positive cell lines is not due to an artifact of the retroviral construct , but is indicative of a regulatory property of the recipient cells and of the integrated state of the DNA carrying the promoter.

I.K. Egorov/C.S. David (Eds.)
Transgenic Mice and Mutants in MHC Research

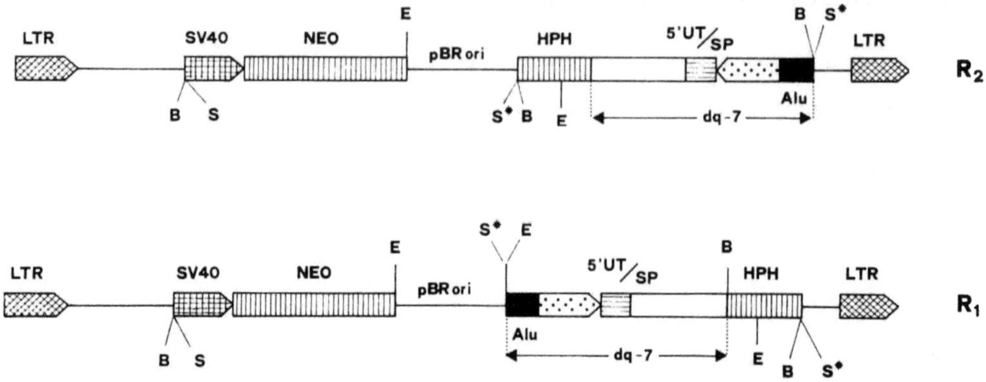

Fig. 1. R1 and R2 retroviral vectors carrying a hybrid DQα-promoter/hygromycin-resistance-gene transcriptional unit. The two vectors differ for the orientation of this transcriptional unit relative to the direction of transcription of the 5' LTR and the SV40 promoters. LTR = long terminal repeat; SV40 = SV40 promoter; neo = neomycin (G-418) resistance gene; pBRori = origin of replication of pBR322 plasmid; hph = hygromycin resistance gene; 5'UT/SP = 5' untranslated and signal peptide sequence-containing exon; Alu = Alu repeat; stippled area = DQα regulatory boxes; B = Bam HI; S = Sal I; S* = inactivated Sal I site; E = Eco RI.

Table 1. Characteristics of cell lines carrying DQα promoter-containing retroviral vectors[a]

Vector	Phenotype	Cell Lines				
		NIH 3T3	PA317	Raji	RJ2.2.5	HUT-78
	neo^r	+	+	+	+	+
R1	hyg^r	-	-	-	-	+
	Φ	-	+	-	-	-
	neo^r	+	+	nt	nt	nt
R2	hyg^r	-	-	nt	nt	nt
	Φ	-	-	nt	nt	nt

[a] PA317 was transfected with vector DNA by standard calcium phosphate/DNA precipitation procedures or infected with virus supernatants derived from transfected clones; NIH 3T3 was obtained by supernatant infection; Raji and RJ2.2.5 (a B-lymphoma Ia-positive cell line and its Ia-negative mutant, respectively) and HUT-78 (an Ia-positive T-cell line) were infected by co-cultivation with populations of vector-transfected PA317 clones. Resistance to G-418 (400-800 μg/ml for PA317 and NIH 3T3; 1200-1500 μg/ml for lymphoid cell lines), resistance to hygromycin (50-250 μg/ml) and virus production (Φ) were measured.

CONCLUSIONS

Construction of retroviruses carrying tissue-specific promoter sequences fused to reporter genes may be a useful tool for the analysis of gene regulation in differentiating cells. The construct illustrated here, although exhibiting properties that might help to elucidate some aspects of the regulatory circuits controlling MHC class II gene expression, has a limited range of activity. A more detailed examination of the cis-acting signals required to direct gene expression in different cells may indicate other regulatory region to be incorporated in the retroviral construct R1 in order to allow its use in the study of MHC class II gene activity in bone marrow cells. The ability of human MHC class II cis-acting signals to promote gene expression in response to mouse trans-acting regulatory factors (Accolla et al., 1985a, 1985b and 1986) suggests that the R1 construct and its future improved versions might prove useful to study human gene expression in an in vivo heterologous system.

ACKNOWLEDGMENTS

This work was supported by grants from AIRC (Associazione Italiana per la Ricerca sul Cancro) and CNR PF Biotecnologie e Biostrumentazione. GDP gratefully acknowledges a fellowship from AIRC

REFERENCES

Accolla RS (1983) J Exp Med 157:1053-1058

Accolla RS, Jotterand-Bellomo M, Scarpellino L, Maffei A, Carra G, Guardiola J (1986) J Exp Med 164:369-374

Accolla RS, Scarpellino L, Carra G, Guardiola J (1985b) J Exp Med 162:1117-1123

Accolla RS, Carra G, Guardiola J (1985a) Proc Natn Acad Sci USA 82:5145-5149

Guardiola J, Scarpellino L, Carra G, Accolla RS (1986) Proc Natn Acad Sci USA 83:7415-7419

Del Pozzo G, Guardiola J (1989) Nature, in press

Auffray C, Lillie JN, Korman AJ, Boss JM, Frèchin N, Guillemot F, Cooper J, Mulligan RC, Strominger JL (1987) Immunogenetics 26:63-73

IV. Expression and Function of Class I Genes in Transgenic Mice

Male Sterility in HLA-B27-Transgenic Mice

P. IVANYI and W. LOKHORST[1]

[1] Central Laboratory of the Netherlands Red Cross Blood Transfusion Service and Laboratory for Experimental and Clinical Immunology, University of Amsterdam, P.O.Box, 9406, 1006 AK Amsterdam, The Netherlands

The major interest when producing MHC-transgenic mice (TGM) is in studying the immunocompetence of the transgenic animal. In the case of HLA, the TGM provide the unique possibility to examine *-in vivo-* the function of the individual HLA molecule(s).

What happens with the mouse genome due to the introduction of the (human) transgene is comparable to the "dark side of the moon". The question is very complex and the answer might depend on the nature of the DNA fragment ("the gene") which was used for injection and, mainly, on the site of insertion into the mouse genome. The site of insertion is random, hence reserach of the dark side of the moon is "intentional serendipity".

Really, we have not seen a written resolution for plans to systematically explore the consequence of insertional mutations; reported observations of pathology as a consequence of insertion are the result of "random" situations which rather interfere with the original plans of the researcher.

In transgenic mice, embryonal or postnatal mortality as well as other deviations have been recorded. Moreover, decreased reproductive performance (for various reasons, of various degree and in one or both sexes) is also frequently observed in affected TGM's. Some observations are not studied in details and represent only a minor aspect of the main (TGM) story, and thus escape attention.

Apparently, the most "precise" hit was described by Krulewski *et al.* (1989) in TGM for a dihydrofolate reductase gene with a Simian virus 40 early promotor (a 4400-bp plasmid, named pFR400). In search for insertional mutations, Krulewski *et al.* bred all 17 transgenic mouse lines generated from independent founders to homozygosity. In one line (P432), offspring with abnormal neurologic (and reproductive) function was found. In short, the integration of about 400 copies of the plasmid at a single site in the mouse genome resulted in a recessive insertional mutation which is allelic to an already known ("natural") mutation named pcd (Purkinje cell degeneration). Thus, the transgene was inserted in a gene critical for (1) normal development of cerebellar cells (and some other cell types) and (2) for normal spermatogenesis (sperm cells are also abnormal).

In our Institute, various lines of HLA-TGM mice were collected (Table 1); at present, they occur in various breeding programmes with the aim to produce series of HLA-TGM's with various H-2 haplotypes on the C57Bl/B10 background (Kievits *et al.*, 1989a). Some of them were bred to homozygosity (Table 1).

As shown in Table 1, from the total of five TGM in which homozygous males were obtained, male sterility was observed in one line. Thus, insertional mutations (for gross phenotypical changes) are not frequent among MHC-TGM lines.

I.K. Egorov/C.S. David (Eds.)
Transgenic Mice and Mutants in MHC Research

Table 1
Inventory of TGM lines

Designation of TGM line	First description, origin of the TGM	Homozygous[1]	Observations
B27.2 (F22)	Krimpenfort *et al.* 1987	yes	normal
B27.2 (F42)	Krimpenfort *et al.* 1987	probably	male sterility
B27.5 (875.5)	Weiss *et al.* 1989	yes	normal
Hu-β_2m (F55)	Krimpenfort *et al.* 1987	yes	normal
Cw3 (22)	Dill *et al.* 1986, 1988	n.d.	normal[2]
SLA-PD1 (1-4)	Frels *et al.* 1985	yes	normal

[1]The TGM lines are not homozygous, but individual mice were bred to homozygosity.
[2]Heterozygous individuals appear normal.

In our laboratory, most work was done with the B27.2 TGM produced by Krimpenfort *et al.* (1987). Immunobiology and cell membrane expression of B27.2 TGM was studied by Kievits *et al.* (1989a,b).

HLA-B27.2-heterozygous TGM males and males homozygous for B27 from founder No. 22 were normal. The presence of the hu-β_2m gene was also not related to male sterility, because hu-β_2m-homozygous males were also normal. Further, B27 "double" heterozygous males, i.e. heterozygous for B27 from Fo No. 22 and for B27 from Fo No. 42 (and for the human β_2m gene) were also normal. Thus, mice harboring three human genes were healthy and fertile.

Sterile males were observed among the progeny of TGM mice obtained from intercrosses between progeny of two founders (Fo No. 22 and Fo No. 42). However, sterile males were observed only in situations where individuals homozygous for B27 descendent from Fo No. 42 could occur. Among the progeny of B27-heterozygous parents, the percentage of sterile male corresponded to the expected number of homozygotes. Males, potentially homozygous for B27.2 (from Fo No. 42) but without hu-β_2m were also sterile. So far, the association of male sterility with B27.2 was absolute, i.e. all of the 16 sterile males were B27.2-positive.

Taken together, these data strongly indicate that males, homozygous for B27.2 derived from Fo No. 42, are sterile. Experiments are in progress to ascertain the origin of the sterility gene. This must be achieved in the absence of a line derived directly from the progeny of Fo No. 42, because this line was lost.

Females from the same families, in which sterile males were found, were normally fertile. The male sterility was absolute. Males were found either normally fertile or completely sterile. The sterile males appeared and behaved as normal healthy individuals. By autopsy, we could observe no abnormalities. Their vesicular glands were of normal size and weight. Thus, target organ sensitivity to testosterone was not impaired.

The direct cause of the male sterility was simple to establish. Testis weight was reduced to about one-third as compared to normal sib males (50-60 mg paired testis weight in sterile males), and histology showed complete absence of spermatogenesis. Spermatogonia do not differentiate further than to the stage of spermatogonia grade B. Sertoli and Leydig cells are well developed (Fig. 1a, b, c, d).

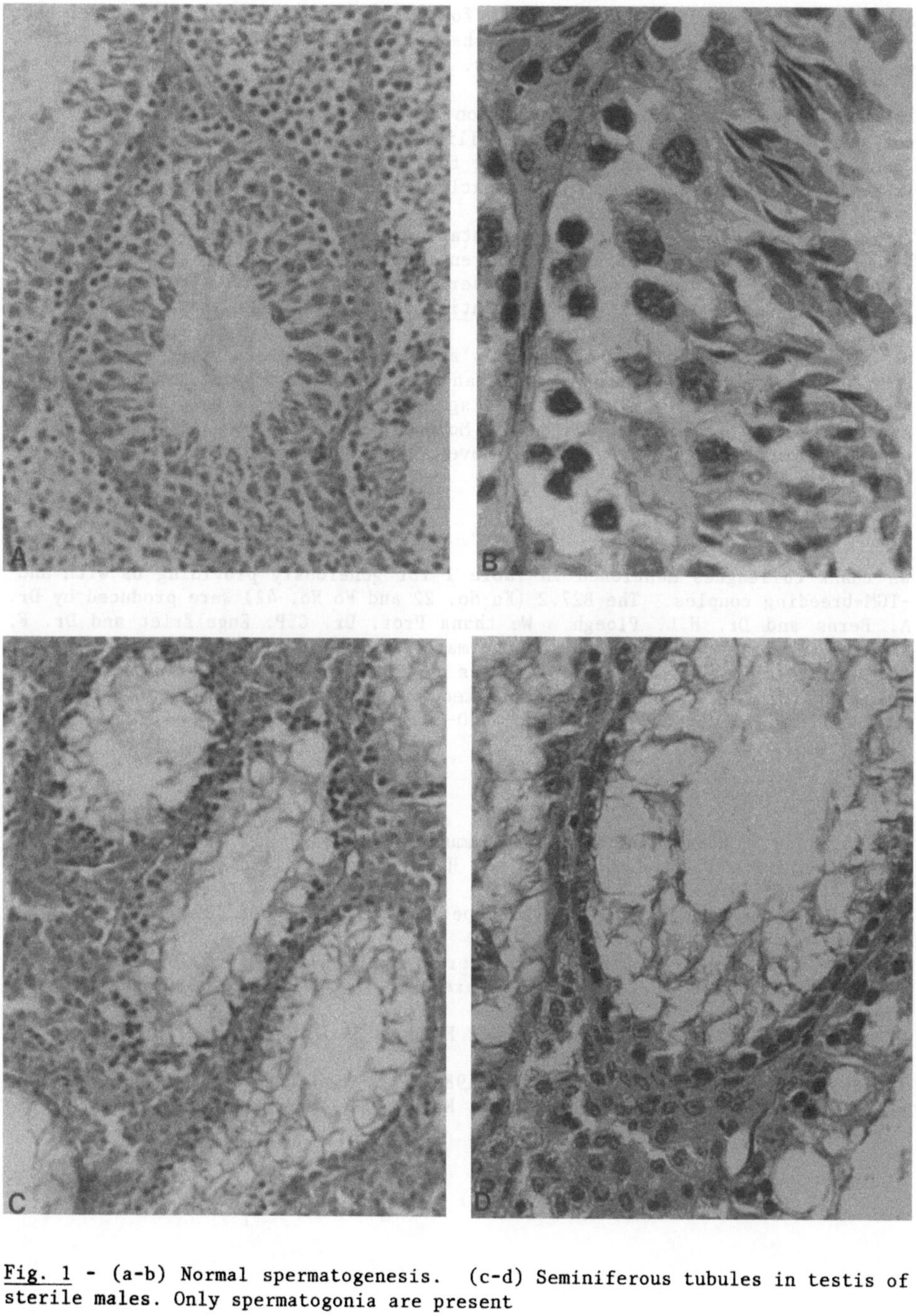

Fig. 1 - (a-b) Normal spermatogenesis. (c-d) Seminiferous tubules in testis of sterile males. Only spermatogonia are present

Although formal proof for homozygosity for B27.2 from Fo No. 42 has not yet been obtained, all data strongly indicate that this is the case. With this premise, we summarize the observation as follows.

The B27.2 gene, used for the production of TGM (Krimpenfort *et al.*, 1987), is not the direct cause of the male sterility, as shown by normally fertile B27.2 (homozygous) males derived from another founder TGM (No. 22). A mutation in the inserted B27.2 gene itself cannot be excluded but is not probable.

In founder mouse No. 42, a recessive mutation occurred by insertional disruption of a gene which is critical for differentiation of spermatogonia. To our knowledge, this is the first case of an insertional mutation in TGM which results in an isolated complete arrest of differentiation of early spermatogonia.

We designate the mutation provisionally as spermatogonia arrest (sga); the location of the gene in the mouse genome and its function are not known. However, the TGM should make the cloning of the sga gene possible, as well as learning to understand its product and function. Moreover, it will be possible to examine whether human (and other species) do have a homologous gene.

ACKNOWLEDGEMENTS

We thank colleagues mentioned in Table 1 for generously providing us with MHC--TGM-breeding couples. The B27.2 (Fo No. 22 and Fo No. 42) were produced by Dr. A. Berns and Dr. H.L. Ploegh. We thank Prof. Dr. C.P. Engelfriet and Dr. F. Kievits for reading and commenting the manuscript. We thank Mrs. Jetty Gerritsen for typing the manuscript. This work was supported by the Foundation for Medical Research (MEDIGON), which is subsidized by the Netherlands Organization for Scientific Research (NWO) (grant no. 900-509-110).

REFERENCES

Dill O, Koch S, Hämmerling GJ (1986) Immunobiology 173: 264 (Abstract)

Dill O, Kievits F, Koch S, Ivanyi P, Hämmerling GJ (1988) Proc Natl Acad Sci USA 85: 5664-5668

Frels RI, Bluestone JA, Hodes RJ, Capecchi MR, Singer DS (1985) Science 228: 577-580

Kievits F, Wijffels J, Lokhorst W, Boerenkamp WJ, Ivanyi P (1989) This volume

Kievits F, Wijffels J, Lokhorst W, Boerenkamp WJ, Ivanyi P (1989) Tissue Antigens, in press

Krimpenfort P, Rudenko G, Hochstenbach F, Guessow D, Berns A, Ploegh HL (1987) EMBO J 6: 1673-1676

Krulewski TF, Neumann PE, Gordon JW (1989) Proc Natl Acad Sci USA 86: 3709-3712

Weiss EH, Schliesser G, Riethmüller G, Kievits, F, Ivanyi P, Brem G (1989) This volume

The Lyt-2 Accessory Molecule is Responsible for the Weak Mouse Anti-HLA Xeno-Response

G. J. HÄMMERLING, U. KALINKE, and B. ARNOLD[1]

[1] Institute for Immunology and Genetics, German Cancer Research Center, Im Neuenheimer Feld 280, 6900 Heidelberg, FRG

SUMMARY

For the last two decades it remained a puzzle why CTL responses to xeno MHC antigens (e.g. mouse anti-human HLA) were weak in comparison to the very strong allo-responses (H-2 against another H-2 allele). One possibility was that in mice the T cell repertoire was not positively selected towards the recognition of HLA molecules, due to the absence of HLA in mice. However, HLA-Cw3 transgenic mice mounted the same weak xeno-response against another HLA molecule (HLA-A2) as did normal mice. In further studies it was observed that the murine Lyt-2 molecule failed to interact with the $\alpha 3$ domain of the HLA class I molecule. Following this observation transgenic mice were constructed with an HLA/H-2 hybrid gene containing the $\alpha 1$ and $\alpha 2$ domains of HLA-B27 and the $\alpha 3$ domain of H-2K^d. Splenocytes from normal mice made a very strong xeno-response against the cells from the transgenic mice expressing the HLA/H-2 hybrid molecule. These findings show that the weak xeno-responses are due to the failure of the Lyt-2 molecules interact with HLA class I, and not to the selection of the mouse T cell repertoire.

INTRODUCTION

It is well established that a large fraction of T lymphocytes can recognize allogeneic major histocompatibility antigens (MHC). Estimates of the respective precursor frequencies of cytotoxic T lymphocytes (CTL) with specificity for one particular MHC alloantigen range between 1/300 to 1/2000. These are about 100 to 1000 times higher than the precursor frequencies of CTL recognizing foreign antigens or peptides thereof in the context of syngeneic MHC (reviewed in Dutton et al 1978, Klein 1986). The peptide model of MHC restricted antigen recognition offers an explanation for this high alloreactivity (Bjorkman et al 1987, Kourilsky et al 1986, Marrack et al 1988). According to this hypothesis the peptide binding grooves of the MHC molecules are always occuppied by a large number of different self-peptides, thus creating a multitude of epitopes which can be recognized by allogeneic T lymphocytes.

In contrast to the high alloreactivity stands the relatively weak response to xenogeneic MHC antigens which has been documented in many experimental systems (Engelhard et al 1978, Hines et al 1976, Widmer et al 1972, Wilson et al 1971). In recent studies, when mice were immunized with murine cells transfected with HLA genes, again only a relatively weak response against the respective HLA antigen was observed. In addition, most CTL clones

I.K. Egorov/C.S. David (Eds.)
Transgenic Mice and Mutants in MHC Research

generated in this manner recognized the HLA molecules in an H-2 restricted way (Achour et al 1985, Holterman et al 1986, Maryanski et al 1986). In further studies it has been found that peptides of the HLA molecules were recognized together with the murine H-2 class I molecules (Maryanski et al 1986). These observations are in striking contrast to murine CTL responses to allogeneic cells or mouse cell lines transfected with allogeneic H-2 class I antigens which always resulted in the exclusive or predominant generation of CTL which were not restricted by the H-2 class I molecules of the responding mouse.

It can be assumed that the overall structures of HLA and H-2 class I antigens are very similar (Bjorkman et al 1987). It is not clear why mice see the HLA antigens predominantly like a nominal antigen and therefore so drastically different from murine H-2 antigens. On the other hand, recent work with HLA class I transgenic mice established that these mice can use the transgenic HLA antigens as restriction elements for nominal antigens similarly to their own H-2 molecules although not as efficiently. In HLA-Cw3 transgenic mice generated in this laboratory the Cw3 antigen was not only expressed in the same tissues as the endogeneous H-2 class I molecules, including the thymus, but functioned also as an MHC restriction molecule in CTL responses to influenza virus and to Sendai virus (Dill et al 1988). Similar observations have been reported for HLA-B27 transgenic mice (Kievits et al 1987).

A possible explanation for these apparent discrepancies should be that murine and human MHC antigens carry species-specific determinants which might be involved in the selection of the T cell repertoire. Following this hypothesis murine T cells could be predominantly selected in an H-2 positive thymus to see H-2 antigens but not HLA, and vice versa. This raised the question of whether in HLA class I transgenic mice the T cell repertoire was altered towards a stronger and predominantly unrestricted recognition of allogeneic HLA class I antigens. Alternatively, it is possible that the xeno-response is due to a species-specific barrier in interaction between the murine Lyt-2 molecule and the human HLA molecule. The subsequent studies will demonstrate that the weak xeno-response of mice against HLA molecules is not caused by an "undeveloped" mouse T cell repertoire but rather by the inability of the Lyt-2 molecule to interact efficiently with HLA.

RESULTS AND DISCUSSION

H-2 restriction and frequency of HLA specific CTL generated in normal and in HLA transgenic mice

The present study was designed to investigate whether HLA-Cw3 transgenic C57Bl/6 mice would generate a stronger and less H-2 restrictable response against HLA/H-2 antigens than normal C57Bl/6 mice (Hämmerling et al, 1989). For this purpose CTL were generated against the C57Bl/6 derived fibrosarcoma Bc20 (H-2^b) transfected with HLA-A2 (Bc20.A2). The resulting CTL were tested against H-2 unrelated L cell targets (H-2^k) transfected with either HLA-A2 alone or in addition with the H-$2K^b$ and H-$2D^b$ genes (L.A2.K^b.D^b). The L.A2 target should be lysed only by the H-2 unrestricted CTL with specificity for HLA-A2,

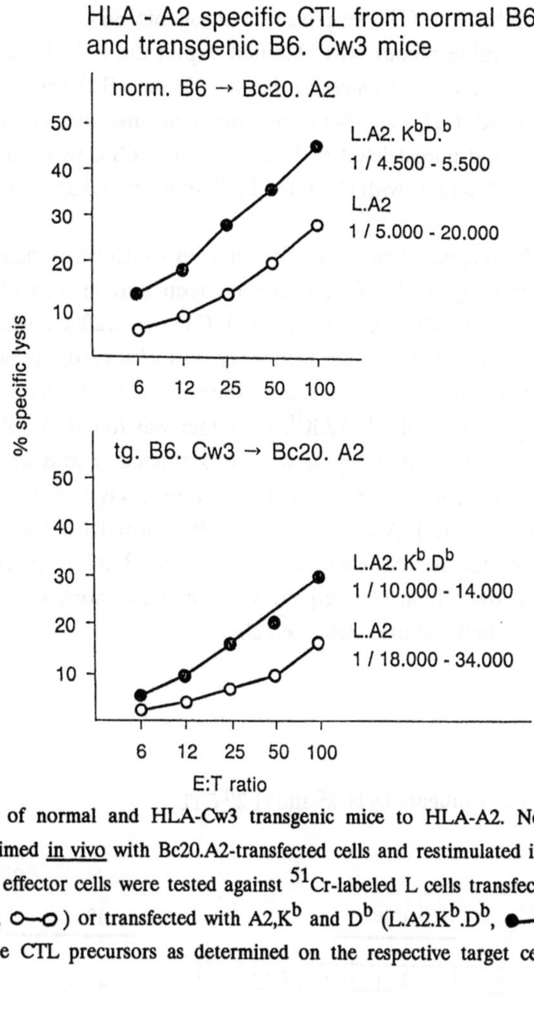

Fig.1. CTL response of normal and HLA-Cw3 transgenic mice to HLA-A2. Normal B6 and transgenic B6.Cw3 mice were primed in vivo with Bc20.A2-transfected cells and restimulated in vitro with the BC20.A2 cells. After 5days the effector cells were tested against ^{51}Cr-labeled L cells transfected with either the HLA-A2 gene alone (L.A2, o—o) or transfected with A2,K^b and D^b (L.A2.K^b.D^b, ●—●). The figure gives also the frequencies of the CTL precursors as determined on the respective target cells in a limiting dilution system

whereas the L.A2.K^b.D^b target should identify the sum of unrestricted and H-2K^b or H-2D^b restricted CTL. Transfectants expressing comparable amounts of the HLA-A2 antigen were used.

No in vitro primary response against the Bc20.A2 transfectant was observed. Therefore the mice were first immunized in vivo with Bc20.A2 and 9-14 days later their splenocytes were restimulated in vitro against Bc20.A2. A representative bulk culture experiment is shown in Figure 1. It can be seen that CTL generated in normal B6 mice against Bc20.A2 lysed the L.A2.K^b.D^b target more efficiently than the L.A2 target suggesting that the CTL population consisted of a mixture of H-2^b restricted and unrestricted A2-specific CTL. Only marginal lysis of L cells transfected with K^b and D^b alone or with the neomycin resistance gene alone or against the Bc20 cell line was observed. These results demonstrated the specificity of the

CTL for A2 epitopes. When the CTL response of transgenic B6.Cw3 mice against BC20.A2 was investigated a comparable picture was obtained. Again the L.A2.K^b.D^b target was lysed more strongly than the L.A2 target suggesting that also in the HLA transgenic mouse a large fraction of H-2^b restricted CTL had been generated against the HLA-A2 antigen. The existence of A2 specific and unrestricted or H-2^b restricted CTL was verified by respective blocking studies of the CTL targets with HLA and H-2^b monoclonal antibodies.

Figure 1. depicts also the respective precursor frequencies which were measured in a limiting dilution system (Hämmerling et al 1989). It can be seen that in normal mice the primed CTL_p frequency for A2-specific and unrestricted CTL is 1/20000 whereas against the L.A2.K^b.D^b target a considerably higher frequency was observed, namely 1/4500. In the same experiment in transgenic mice the CTL_p frequency for unrestricted CTL was 1/32000 and again a higher frequency for the L.A2.K^b. D^b target was found (1/10500). Although the precursor frequencies varied considerably between experiments a consistent picture emerges when the individual experiments are analysed. There were always more cells positive on the L.A2.K^b.D^b target than on the L.A2 target in both the normal B6 and transgenic B6.Cw3 mice. In addition, the precursor frequencies obtained in the B6.Cw3 transgenic mice appear to be in general slightly lower than the frequencies obtained in normal B6 mice. These data are in agreement with the bulk cultures described above.

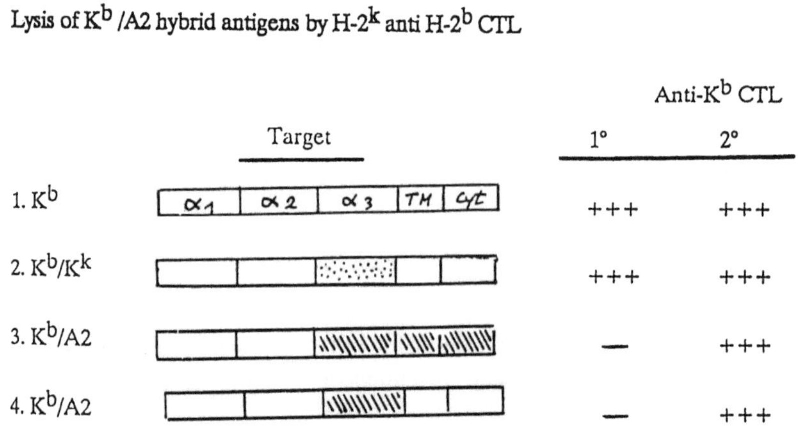

Fig.2. Influence of the a3 domain on the recognition of K^b/A2 hybrid antigen by anti-K^b CTL. Anti-K^b specific CTL were generated in a primary and in a secondary in vitro culture using splenocytes from H-2^k mice as responders. The respective CTL bulk cultures were tested against L cells transfected with 1.) the complete K^b gene; 2.) a K^b gene in which the a 3 domain was replaced by the a3 exon of the K^k gene; 3.) a Kb/A2 hybrid gene in which a1 and a2 exons were of Kb origin and the rest of A2 origin. 4.) a Kb/A2 hybrid gene in which the a3 exon was of A2 origin. Transfectants expressing comparable amounts of K^b epitopes were used as determined by FACScan analysis

The mouse Lyt-2 molecule interacts with the $\alpha 3$ domain of mouse class I but not human class I molecules

Recent evidence suggests that the interaction between the Lyt-2 and class I molecules increases the avidity of the interaction between CTL and stimulator or target cells (Moretta et al 1980, Shimonkevitz et al 1985, Dembic et al 1987 and Gabert et al 1987). The question of a species-specific interaction between the murine Lyt-2 and class I molecules from other species was investigated as follows: K^b specific CTL were generated in either a primary or, after restimulation, in a secondary in vitro culture. The respective CTL bulk cultures were tested on L cell targets which had been transfected with either the K^b gene or a modified K^b in which the $\alpha 3$ domain was exchanged against the $\alpha 3$ domain of the HLA-A2 gene. Comparable cell surface expression was verified with monoclonal K^b antibodies. The respective data are summarized in Figure 2. It can be seen that primary CTL do not lyse the target cells when a K^b molecule contains the human $\alpha 3$ domain (line 3). In contrast, secondary CTL which are Lyt-2 independent lyse efficiently this target. As a control the $\alpha 3$ domain of another H-2 gene (K^k) was also introduced in a K^b gene. This did not effect lysis of the respectively transfected target cells by primary CTL (line 2). Similar studies have been performed by Maziarz et al (1985).

Since the primary CTL are Lyt-2 dependent these data suggest that 1.) mouse Lyt-2 molecule can not efficiently interact with the human HLA molecule and 2.) that the $\alpha 3$ domain of the class I molecule is responsible for this interaction. These conclusions are illustrated in Figure 3. They are in agreement with other reports suggesting that the $\alpha 3$ domain (e.g. amino acids 227 and other neighbouring residues) is involved in the interaction between CD8 and class I molecules (Potter et al 1989, Salter et al 1989). In contrast to the present study Salter et al did not observe a species barrier in the interaction of CD8 with H-2. This discrepancy could be due to the different techniques employed. Whereas Salter et al used a rosetting assay we employed a more physiological functional system.

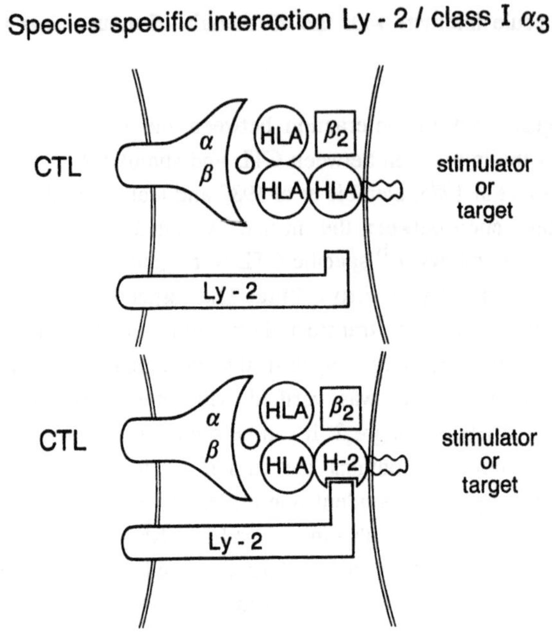

Fig.3. Species-specific interaction of the Lyt-2 molecule with the a3 domain of class I molecules. This schematic diagram shows that the mouse Lyt-2 molecule cannot interact with the a3 domain of human HLA class I molecules due to a species barrier. However, efficient interaction can be obtained when the a3 domain of the target class I molecule is also of mouse origin

Transgenic splenocytes expressing HLA/A-2 hybrid antigens containing the murine *a*3 domain stimulate a strong xeno-response

Based on the observations described above a hybrid gene containing the *a*1 and *a*2 domains of the HLA-B27O5 gene and the *a*3 domain of the mouse H-2K^d gene was constructed and transgenic mice were generated. A hybrid transgene was expressed on the surface of the lymphocytes, albeit to a lower extend than the endogenous H-2 antigens. Splenocytes from this transgenic mouse (designated HLA-B27/*a*3) were used to stimulate splenocytes from B6D2F1 mice in a primary in vitro culture. The CTL were tested on Bc20 derived fibroblast transfected with either the B27/*a*3 hybrid gene or with the complete HLA-B2705 gene. It can be seen from Figure 4 that a good primary CTL response against the B27 epitopes was obtained. However, for efficient lysis, the target B27 molecule had to contain the murine *a*3 domain in order to allow interaction of the mouse Lyt-2 molecule with the target HLA molecule. No lysis was observed when the target cells expressed the unmodified HLA-B27 molecule (lower panel of Figure 4). When the precursor frequencies against the B27/*a*3 hybrid molecule were measured they were found to be of the same magnitude as the CTL^p frequencies for H-2 molecules, e.g. for H-2^b anti-H-2^k CTL$_p$. The upper panel of Figure 4 depicts that splenocytes from transgenic mice expressing the unmodified HLA-B27 molecule were not able to induce a primary B27 specific response.

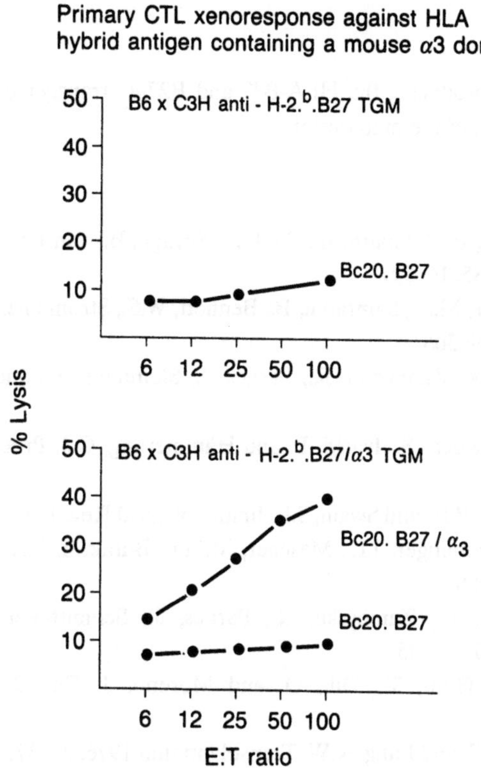

Fig.4. The primary CTL xeno-response against HLA antigens is determined by the origin of the a3 domain. A hybrid gene containing the a1 and a2 domains of HLA-B27 and the a3 domain of H-2K^d was used to generate transgenic mice (designated H-2^b.B27/a3 TGM). The lower panel shows the primary in vitro CTL response of B6D2 splenocytes against irradiated spleen cells expressing the B27/H-2 hybrid antigen. The CTL were measured on day 5 against Bc20(H-2^b) fibroblast transfected with either the B27/a3 hybrid gene (Bc20.B27/a3) or against Bc20 cells transfected with the native B27 gene (Bc20.B27). The upper panel shows the failure to obtain a primary anti-HLA-B27 response when the stimulator cells are irradiated splenocytes from transgenic mice expressing the unmodified B27 gene

CONCLUDING REMARKS

The experiments demonstrate that the weak xeno-response is not due to a lack of positive selection of HLA reactive T lymphocytes. The data show that normal mice contain a large number of CTL precursors which are able to recognize and be stimulated by the HLA-B27 molecule. However, there exists clearly a species-specific barrier in the interaction between the mouse Lyt-2 molecule and the target class I antigens. The site of interaction of Lyt-2 with class I resides in the $a3$ domain of the class I molecule. If the $a3$ domain is of mouse origin, then an efficient stimulation of CTL with specificity for B27 determinants is obtained. Thus, the interaction between the accessory molecule Lyt-2 with the $a3$ domain of class I determines the magnitude of the xeno-MHC response.

ACKNOWLEDGEMENTS

We thank Ms. S. Beck for producing the HLA-B27 and B27/a_3 transgenic mice and Ms. B. Ermshaus for the preparation of the manuscript.

REFERENCES

1. Achour, A., Beguue, B., Gomard, E., Paul, P., Sayagh, B., van Pel, A. and Levy, J.P. Eur. J. Immunol. 1985. 16: 597
2. Bjorkman, P.J., Saper, M.A., Samraoui, B., Bennett, W.S., Strominger, J.L. and Wiley, D.C. Nature 1987. 329: 506
3. Dembic, Z., Haas, W., Zamoyska, R., Parnes, J., Steinmetz, M. and v. Boehmer, H. Nature 1987. 326:510
4. Dill, O., Kievits, F., Koch, S., Ivanyi, P. and Hämmerling, G.J. Proc. Natl. Acad. Sci. USA 1988. 85: 5664
5. Dutton, R.W., Panfili, P.R. and Swain, S.L. Immunological Rev. 1978. 42: 20
6. Engelhard, V.H., Strominger, J.L., Mescher, M. and Burakoff, S.J. Proc. Natl. Acad. Sci. USA 1978. 75: 5688
7. Gabert, J., Langlet, C., Zamoyska, R., Parnes, J., Schmitt-Verhulst, A.-M. and Malissen, B. Cell 1987. 40:545
8. Hämmerling, G.J., Hunt, T., Dill, O. and Moreno, J. Eur. J. Immunol. 1989. 19: 599-604
9. Hines, D.L., Cowall, C. and Lang, R.W. Transplantation 1976. 21: 375
10. Holterman, M.J. and Engelhard, V.H. Proc. Natl. Acad. Sci. USA 1986. 83: 9699
11 Kievits, F., Ivanyi, P., Kripenfort, P., Berns, A. and Ploegh, H.L. Nature 1987: 329: 447
12. Klein, J. 1986. "Natural History of the Major Histocompatibility Complex". Nature of Alloreactivity: Frequency of Responding Cells. John Wiley and Sons, p. 417
13. Kourilsky, P. and Claverie, J.-M. Annal. Inst. Pasteur 1986. 137: D3
14. Marrack, P. and Kappler, Immunol. Today 1988. 9: 285
15. Maryanski, J.L., Moretta, A., Jordan, B., De Plaen, E., Van Pel, A., Boon, T. and Cerottini, J.C. Eur. J. Immunol. 1985. 15: 1111
16. Maryanski, J.L., Accolla, R.S. and Bertran, J. J. Immunol. 1986. 136: 4340
17. Maryanski, J.L., Pala, P., Corradin, G., Jordan, B. and Cerottini, J.C. Nature 1986. 578: 581
18. Maziarz, R., Allen, H., Strominger, J.L., Flavell, R., Biro, P.A. and Burakoff, S.J. Proc. Natl. Acad. Sci. USA 1985. 82: 6276-6280
19. Moretta, A., Pantaleo, G., Mingari, M.C., Moretta, I. and Cerottini, J.C. J. Exp. Med. 1980. 159:921
20. Potter, T.A., Rajan, T.V., Dick II, R.F. and Bluestone, J.A. Nature 1989. 337: 73-75
21. Salter, R.D., Norment, A.M., Chen, B.P. Clayberger, C., Krensky, A.M., Littman, D.R. and Parham, P. Nature 1989. 338: 345-347
22. Shimonkevitz, R., Luescher, B., Cerottini, J.-C. and MacDonald, H.R. J. Immunol. 1985. 135:892
23. Widmer, M.B. and Bach, F.H. J. Exp. Med. 1972. 135: 1204
24. Wilson, D.B. and Fox, D.H. J. Exp. Med. 1971. 134: 857

Human HLA-B27 Antigen in Transgenic Mice can Function as a Major Histocompatibility Antigen

M. PLA, A. SAMAAN, D. GILLET, M. REBOUL, B. FRANGOULIS, A. OPOLSKI, M. CHOPIN, and L. DEGOS[1]

[1]Mouse Immunogenetics, INSERM U93, Hôpital Saint-Louis, 2, place du Dr. A. Fournier, 75010 Paris, France

INTRODUCTION

The revolution of recombinant DNA technology and advances in the manipulation of mouse embryo have made possible experiments only dreamt of before - the ability to isolate and sequence genes, engineer them in specific ways, modulate their expression and even introduce the foreign DNA sequences into the germ line of mice. These DNA sequences, once inserted into germ line are stably transmitted from generation to generation.

Because of the outstanding complexity of the human MHC, it is important for functional studies to use cells expressing only one HLA product. Nature does not provide us with such cells, but they can be produced by transfecting selected individual HLA genes in cells from other species. In the past years, different human class I and class II genes have been successfully transfected into mouse cells. The recent development of the transgenic technology has allowed the introduction of cloned HLA-B27 (Krimpenfort et al. 1987), B7 (Chamberlain et al. 1988) and Cw3 (Dill et al. 1988) genes into the fertilized eggs of mice. The HLA molecules are expressed on the cell surface of these transgenic mice (Krimpenfort et al. 1987, Chamberlain et al. 1988, Dill et al. 1988), induce a rapid skin graft rejection (Dill et al. 1988) and can function as a restriction element for virus-specific cytotoxic T cells (Kievits et al. 1987, Dill et al. 1988).

The transgenic mice used in this work were obtained from the crosses of 27M TGM (transgenic mice with HLA-B27.2 and human ß2-microglobulin genes) and mice from the following inbred strains : C57BL/10, C3H.B10, A.BY (H-2^b) ; B10.D2 (H-2^d) ; B10.BR (H-2^k) ; B10.P, C3H.NB (H-2^p) ; B10.G, C3H.Q (H-2^q). The original 27 TGM (transgenic mice with HLA-B27.2 genes) and M TGM (transgenic mice with human ß2-microglobulin genes) were produced in The Netherlands Cancer Institute (Krimpenfort et al. 1987). The genes encoding HLA-B27 heavy and light chains have been described elsewhere (Seeman et al. 1986, Rein et al. 1987, Krimpenfort et al. 1987).

CELL SURFACE AND SERUM EXPRESSION OF TRANSGENIC HLA-B27 PROTEIN AND HUMAN ß2-MICROGLOBULIN

We have tested the ability of the HLA-B27 genes to be expressed in the transgenic mice by cytofluorometric analysis (Fig. 1) or microlymphocytotoxic test using peripheral and spleen lymphocytes as target cells. The level of cell membrane expression of HLA-B27 molecules in transgenic mice is dependent on the presence of human ß2-microglobulin. However, a monoclonal antibody isolated from nontransgenic mice immunized with cells from mice carrying only HLA-B27 transgenes was able to detect on the cell surface HLA-B27 heavy chains associated with mouse ß2-microglobulin (Fig. 1). The level of expression of mouse class I molecules is not influenced by the incorporation of HLA-B27 and/or human ß2-microglobulin genes into the mouse genome. There is no effect of H-2 alleles (b, d, k, p and q) on the level of expression of HLA-B27 molecules on transgenic lymph node cells. The human ß2-microglobulin transgenic mice have a detectable level of human ß2-microglobulin in their sera and express to various degrees on their cell surface the mouse class I heavy chains in association with human ß2-microglobulin.

I.K. Egorov/C.S. David (Eds.)
Transgenic Mice and Mutants in MHC Research

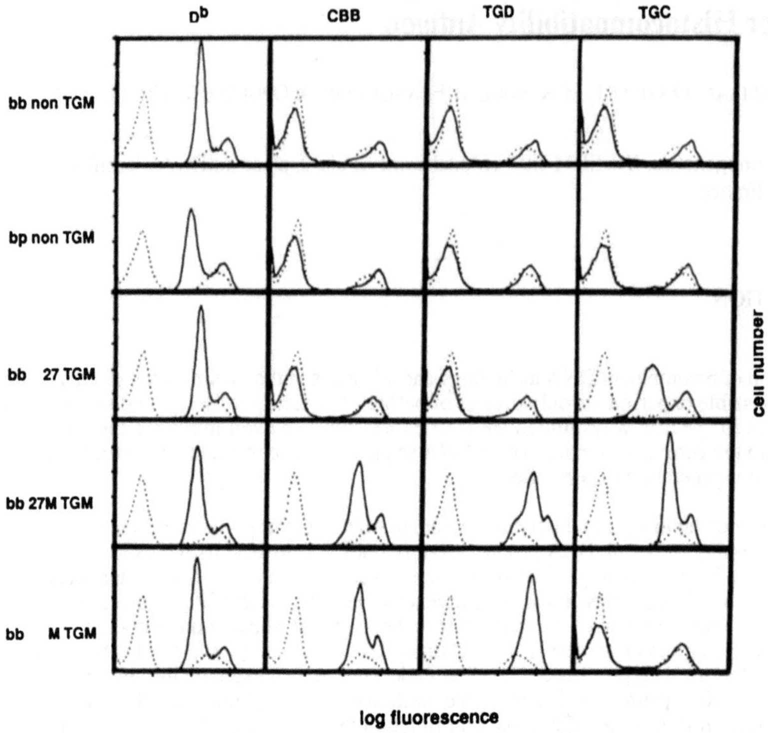

Fig. 1. Cytofluorometric analysis of the expression of H-$2D^b$ and transgenes products (HLA-B27 and human ß2-microglobulin) on lymph node cells from non transgenic (H-2^b and H-$2^{b/p}$), 27, 27M and M transgenic mice.The following monoclonal antibodies were used : 28.14.8 (H-$2D^b$-specific), CBB (recognizing a determinant resulting from the interaction between the H-$2D^b$ heavy chain and human ß2-microglobulin), TGD (human ß2-microglobulin-specific) and TGC (recognizing B27 heavy chains associated with either human or mouse ß2-microglobulin)

ALLOANTIGENIC FUNCTION OF HLA-B27 ON TRANSGENIC CELLS

Skin grafts from 27 and 27M transgenic mice were found to be rejected by normal H-2 matched mice (graft survival about 18 days), suggesting that the foreign HLA antigen is recognized as a functional transplantation antigen.

Antibodies could be induced by both single (27) and double (27M) TGM lymphocytes in nontransgenic H-2 matched recipients and anti-TGM monoclonal antibodies (around 50) were isolated. All of these monoclonal antibodies were HLA class I-specific with various reaction patterns on human cells. 50 % of them displayed a monomorphic pattern of reactivity, the others a polymorphic one which varied from very broad (e.g. HLA-Bw4-specific) to rather narrow. Interestingly, some of these monoclonal antibodies cross-react with mouse cells, on which they recognize H-2 class I molecules in a polymorphic manner. HLA-transgenic mice offer completely new ways of producing the HLA-specific monoclonal antibodies. Reliable HLA typing antibodies have so far been obtained mostly from multiparous women or polytransfused individuals, and monoclonal antibodies have been produced in mice immunized with human cells or isolated HLA gene products. Most antibodies obtained after such immunizations recognize human proteins, among which HLA antigens, in a monomorphic manner. HLA-transgenic mice thus represent a very interesting system, because they allow the production of HLA-specific antibodies in a quasi allogeneic situation. HLA-transgenic mice can be immunized with lymphocytes from other transgenic mice expressing different HLA genes or with mouse cells transfected with HLA genes (intact or modified by genetic engineering).

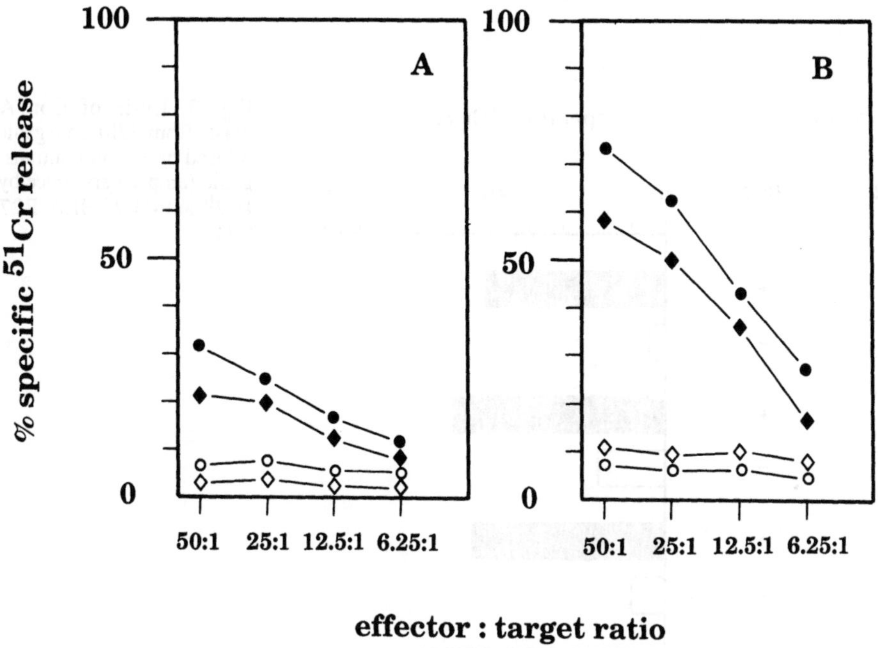

Fig. 2. Lysis of Con A blasts from 27M transgenic (closed circles) and nontransgenic (open circles) H-2^b mice, and of HLA-B27-positive (closed squares) and HLA-B27-negative (open squares) EBV cell lines by the H-2^b anti-H-2^b HLA-B27 CTL raised from unprimed (A) and in vivo primed (B) mice.

Cell-mediated cytotoxicity assays were performed to determine whether normal mice could generate a cytotoxic T cell (CTL) response against HLA-B27-expressing transgenic spleen cells (Samaan et al. 1989). CTL were generated in vitro in unidirectional mixed lymphocyte cultures, using spleen cells from B10 mice as responder cells and irradiated spleen cells from H-2 matched 27M transgenic mice as stimulator cells. After 5 days, the CTL generated in these primary bulk cultures were tested in the ^{51}Cr-release assay on human EBV cell lines and Con A blasts from nontransgenic or 27M transgenic mice. The CTL exerted a significant killing of Con A blasts from 27M transgenic mice and were also capable of killing HLA-B27-expressing human EBV cell lines. The HLA-B27-negative cell lines (Fig. 2A) as well as Daudi cells (a human cell line which fails to express cell surface HLA molecules) were not killed. When B10 mice intraperitoneally immunized with one injection of 10^7 splenocytes 4-6 weeks before the in vitro assays were used for generation of HLA-B27 cytotoxic T cells (Fig. 2B), the percent specific ^{51}Cr release was about three times higher. The availability of 27M transgenic mice carrying different H-2 haplotypes, recently developed in our laboratory, allows us to test the H-2^b anti-H-2^b B27 CTL on the panel of Con A blasts from 27M transgenic and nontransgenic H-2 matched and mismatched mice (Fig. 3). Only target cells expressing HLA-B27 molecules were significantly lysed, whether they were H-2-matched (H-2^b) or mismatched (H-2^d, H-2^k, H-2^p, H-2^q). Interestingly, a low level of lysis was observed on nontransgenic allogeneic H-2^d and H-2^p target cells. The results indicate that HLA-B27 expressed on spleen cells from 27M transgenic mice can serve as an immunogenic element and that a significant portion of anti-HLA CTL response involved in recognition of HLA-B27 is unrestricted by H-2. Lysis of human targets and the H-2-mismatched HLA-B27-positive mouse cells indicate that HLA-B27 molecule on the immunizing transgenic spleen cells can be recognized as a major histocompatibility antigen.

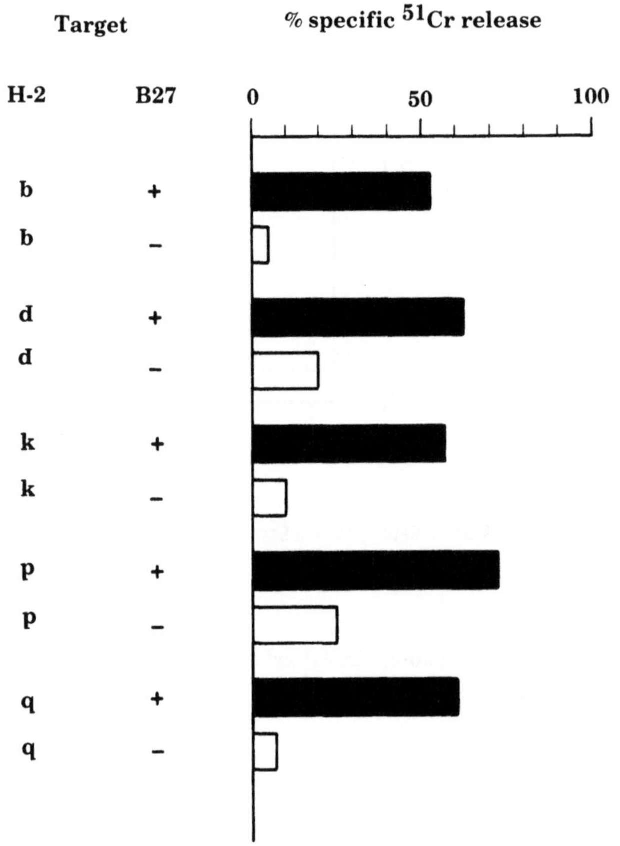

Fig. 3. Lysis of Con A blasts from 27M transgenic (closed bars) and nontransgenic (empty bars) mice by H-2^b anti-H-2^b HLA-B27 CTL.

Long term HLA-B27-specific CTL lines and clones have been generated by stimulation with spleen cells from either 27M transgenic mice (Fig. 4A) or from 27 transgenic mice expressing HLA-B27 heavy chains associated with mouse β2-microglobulin (Fig. 4B). Both T cell lines were found to be able to lyse H-2 mismatched HLA-B27 expressing target cells. These CTL also recognized allogeneic H-2^k (HLA-B27 negative) target cells.

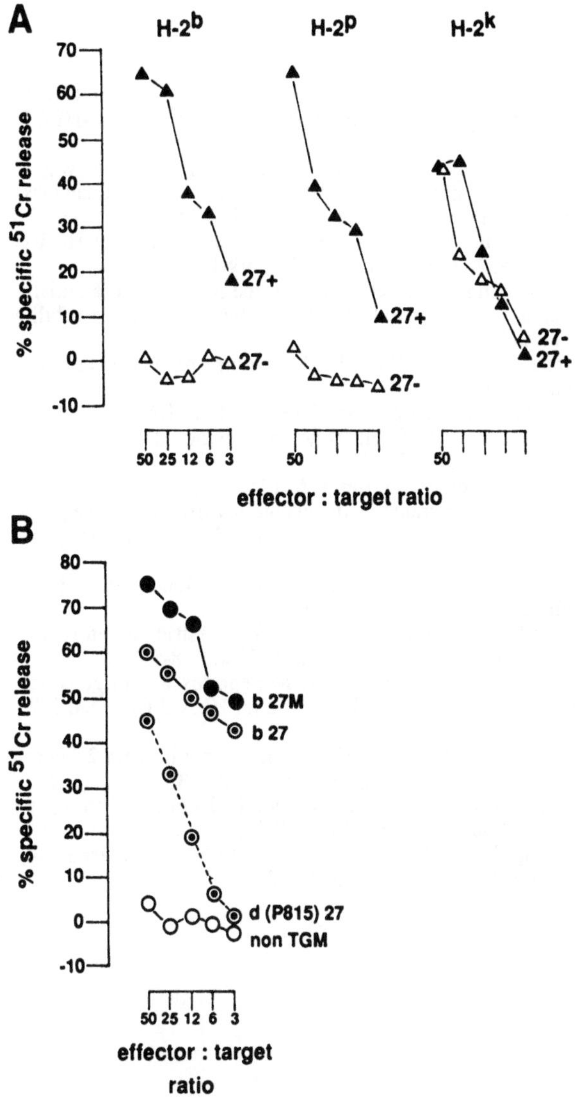

Fig. 4. Reactivity of mouse H-2^b anti-H-2^b HLA-B27-specific CTL lines induced by stimulation with spleen cells from 27M (A) and 27 (B) transgenic mice.

The availability of mouse cells expressing individual HLA products provides an interesting possibility to immunize syngeneic animals against human MHC antigens on mouse cell background. Using selected CTL clones or lines and P 815- or L-HLA class I transfectants, it has been shown that CTL recognition of xenogeneic class I histocompatibility gene products is restricted by mouse class I gene products (Achour et al. 1986, Holterman and Engelhard 1986, Maryanski et al. 1986a, b, 1988). We used the HLA-transgenic mice model to study the uncloned CTL response against HLA-B27 antigens. Maryanski and her colleagues (Maryanski et al. 1986b) suggested that a membrane bound form of HLA is recognized by alloreactive and by HLA-restricted anti-viral human CTL, whereas a processed fragment of HLA is recognized as a nominal antigen by class I-restricted mouse CTL. However, our results (Samaan et al. 1989) similarly to those of others (Chamberlain et al. 1988, Kievits et al. 1989), indicate that the mouse T cell repertoire does contain T cells able to recognize HLA molecules in an H-2 unrestricted manner. At present there are insufficient experimental data to allow us to decide whether H-2 unrestricted anti-HLA CTL recognize a membrane bound form of HLA molecules or HLA peptide(s) presented by these molecules.

REFERENCES

Achour A, Begue B, Gomard E, Paum P, Sayagh B, Van Pel A, Lévy JP (1986) Specific lysis of murine cells expressing HLA molecules by allospecific human and H-2-restricted anti-HLA T killer lymphocytes. Eur J Immunol 16 : 597-604

Chamberlain JW, Nolan JA, Gromkowski SH, Kelley KA, Eisenstadt JM, Herrup K, Janeway Jr CA, Weissman SM (1988) Cell surface expression and alloantigenic function of a human class I MHC heavy gene (HLA-B7) in transgenic mice. J Immunol 140 : 1285-1292

Dill O, Kievits F, Koch S, Ivanyi P, Hämmerling GJ (1988) Immunological function of HLA-C antigens in HLA-Cw3-transgenic mice. Proc Natl Acad Sci USA 85 : 5664-5668

Holterman MJ, Engelhard V H (1986) HLA antigens expressed on murine cells are preferentially recognized by murine cytotoxic T cells in the context of the H-2 major histocompatibility complex. Proc Natl Acad Sci USA 83 : 9699-9703

Kievits F, Ivanyi P, Krimpenfort P, Berns A, Ploegh H (1987) HLA-restricted recognition of viral antigens in HLA transgenic mice. Nature 329 : 447-449

Kievits F, Wijffels J, Lokhorst W, Ivanyi P (1989) Recognition of xeno-(HLA,SLA) major histocompatibility complex antigens by mouse cytotoxic T cells is not H-2 restricted : A study with transgenic mice. Proc Natl Acad Sci USA 86 : 617-620

Krimpenfort P, Rudenko G, Hochstenbach F, Guessow D, Berns A, Ploegh H (1987) Crosses of two independently derived transgenic mice demonstrate functional complementation of the genes encoding heavy (HLA-B27) and light (ß2-microglobulin) chains of HLA class I antigens. EMBO J 6 : 1673-1676

Maryanski J A, Accolla R S, Jordan B (1986a) H2-restricted recognition of cloned HLA class I gene products expressed in mouse cells. J Immunol 136 : 4340-4347

Maryanski JL, Pala P, Corradin G, Jordan B R, Cerottini J C (1986b) H-2-restricted cytolytic T cells specific for HLA can recognize a synthetic HLA peptide. Nature 324 : 578-579

Maryanski JL, Pala P, Cerottini JC, Corradin G (1988) Synthetic peptides as antigens and competitors in recognition by H-2-restricted cytolytic T cells specific for HLA. J Exp Med 167 : 1391-1405

Rein RS, Seemann HA, Stam N, Neefjes JJ, Ploegh H L (1987) Association with beta 2-microglobulin controls expression of transfected human class I genes. J Immunol 138 : 1178-1183

Samaan A, Gillet D, Chopin M, Degos L, Pla M (1989) Mouse cytotoxic T cells can recognize HLA-B27 antigen without H-2 restriction. Immunogenetics 29 : 210-212

Seemann GHA, Rein RS, Brown CS, Ploegh HL (1986) Gene conversion like mechanisms may generate polymorphism in human class I genes. EMBO J 5 : 547-552

Cytotoxic T Cell Responses Against Human Class I Molecules in Normal and HLA-A2.1 Transgenic Mice[1]

V. H. Engelhard[2], E. J. Bernhard[2], M. J. Holterman[2], A.-X. T. Le[3], R. Henderson[2], J. P. Ridge[2], S. Strub[2], J. Barbosa[4], and E. Lacy[5]

[1]This work was supported by National Institutes of Health Grants AI20963, AI21393, AI24815, and American Cancer Society Grant ACS-IM-457. E. Lacy is the recipient of an American Heart Association Established Investigatorship.

Departments of [2]Microbiology and [3]Surgery, University of Virginia Medical Center, Charlottesville, VA 22908, USA
[4]Molecular Diagnostics, Inc., West Haven, CT 06516, USA
[5]Dewitt Wallace Research Laboratory, Program in Molecular Biology, Memorial Sloan-Kettering Cancer Center, New York, NY 10021, USA

INTRODUCTION

A considerable amount of recent work in a number of laboratories has led to significant increases in our understanding of the recognition of antigens by T cells. T cells utilize an antigen receptor that is structurally similar to the antigen receptor present on B cells. Despite this apparent similarity, the B cell repertoire enables recognition of a virtually limitless array of foreign antigenic structures, while T cells recognize antigens only when presented in the context of class I or II molecules of the Major Histocompatibility Complex (MHC). One issue which has not been satisfactorily resolved is why and how T cell recognition is focused on MHC antigens. One possibility is that the structures that make up the T cell recognition apparatus are inherently biased toward recognition of MHC molecules. Alternatively, it is possible that TCR molecules that are expressed early during T cell ontogeny in the thymus engender a repertoire similar to that of B cells, but that these receptors are focussed through positive selection mediated by self-MHC antigens. A second issue that also remains unresolved is to what extent the alloantigenic determinants recognized by T cells on foreign class I and class II MHC molecules are dependent upon the presentation of processed host cell proteins.

We have approached both of theses issues through an analysis of the murine cytotoxic T cell response to human class I MHC molecules. Despite the multitude of foreign proteins expressed on human stimulator cells, productive T cell recognition in this xenogeneic system is focused only on human MHC antigens (Engelhard et al.1980). However, the magnitude of this response is extremely low (Holterman and Engelhard, 1986; Engelhard et al.1988). In this paper, we have explored the reasons for this weak response through the use of DNA mediated gene transfer, exon shuffling to create chimeric mouse/human class I molecules, and finally through the construction of transgenic mice that express the human class I molecule, HLA-A2.1 (Le et al.1989). These same approaches have also been used to provide evidence for the presentation of endogenous, species-specific molecules by human class I molecules to T cells (Bernhard et al.1987,1988).

CELL LINES

Human B lymphoblastoid cell lines included: JY (HLA-A2.1,2.1;-B7,7;-DR4,6); MST (A3,3; B7,7; DR2,2); WT49 (HLA-A2.2,2.2;-B17,17;-DR 3,3); C1R.LON.HYM2.C1R, abbreviated C1R (HLA-A negative;-B negative;-CW4;-DR8), and C1R-A2.2, a transfectant of C1R expressing HLA-A2.2 (Le et al.1989). Human T lymphoblastoid cell lines included: HSB (HLA-A1,2.1;-B12, 17;-DR negative) and JM (HLA-A3,25;-B7,37;-DR negative).

MHC transfectants were produced in the murine thymoma EL4 (H-2^b) (Engelhard et al.1988; Le et al.1989) and included: EL4-A2.1, (HLA-A2.1); EL4-B7, (HLA-B7); EL4-L^d, (H-$2L^d$); EL4-

I.K. Egorov/C.S. David (Eds.)
Transgenic Mice and Mutants in MHC Research

L^d/B7, (H-2L^d/HLA-B7 chimeric molecule); EL4-B7/L^d, (HLA-B7/H-2L^d chimeric molecule); and EL4-A2.2, (HLA-A2.2). All of the transfectants expressed the introduced molecules at 20-60% of levels seen on the human cell lines JY or WT49, or the murine cell line P815.

CYTOTOXIC T LYMPHOCYTE PRECURSOR (CTL) FREQUENCY ANALYSIS

Mice were immunized where indicated by intraperitoneal injection of $2x10^7$ allogeneic or xenogeneic cells or allantoic fluid containing 300 HAU of influenza virus (A/JAP/57 strain), provided by Dr. Bill Biddison, NIH) and used one to two months later. Measurement of CTLp by limiting dilution culture has been described (Holterman and Engelhard 1986). For responses involving human MHC antigens, 5 serial responder dilutions ranging from $3x10^5$ to $1.9x10^4$ cells were added, while for responses involving H-2, responder cells ranged between 10^5 to $5x10^3$ cells were added. Ninety-six replicate wells were set up for each dilution. Viral infection of splenic stimulator cells was carried out by incubating cells at a concentration of 10^7/ml with 300 HAU/ml for two hours. Cells were washed twice and irradiated (2000R). Viral infection of target cells was carried out in a similar fashion except that target cells were labelled with ^{51}Cr during exposure to the virus. Cells were washed twice and then incubated for 5 hours to facilitate viral protein expression and processing. Cells were washed one time prior to initiation of ^{51}Cr release assay. Calculation of CTLp frequency was carried out as previously described (Holterman and Engelhard, 1986).

CTL RESPONSES TO HUMAN CLASS I MOLECULES IN NORMAL MICE

Comparison of Allogeneic and Xenogeneic CTL Responses

Limiting dilution analyses were done to compare the frequency of cytotoxic precursors (CTLp) capable of responding to allogeneic or xenogeneic class I antigens in C57BL/6 (H-2^b) mice (Holterman and Engelhard, 1986; Engelhard et al.1988). Using a transfectant of the syngeneic EL4 cell line that expresses H-2L^d (EL4-Ld), the frequency of CTLp in naive B6 spleen cells capable of responding to the H-2L^d antigen was found to be 45 per 10^6 cells (Table 1). In contrast, the frequency of CTLp capable of responding to HLA-A2.1 or -B7 on a human lymphoblastoid cell line was less than 1 in 10^6 spleen cells. Spleen cells from animals primed with the HLA-A2.1 positive human line, JY, and restimulated in vitro with human cells sharing only HLA-A2.1 or HLA-B7 with JY, gave specific CTLp frequencies of about 2.6 per 10^6 (Table 2).

Table 1. Frequency of Allospecfic and Xenospecific CTL Precursors In Naive C57BL/6 Spleens

Stimulator	Target	Antigen	CTLp per 10^6 Cells
EL4-L^d	EL4-L^d	H-2L^d	45
JY	MST	HLA-B7	0.9
JY	HSB	HLA-A2.1	0.6
JM-L^d	JM-L^d	H-2L^d	20
	EL4-L^d	H-2L^d	36
EL4-A2.1	EL4-A2.1	HLA-A2.1	3.3
	HSB	HLA-A2.1	1.9
EL4-B7	EL4-B7	HLA-B7	2.1
	MST/JM	HLA-B7	2.6

The cells are described in the text. All values represent the means of at least two experiments, calculated as described.

Table 2. Frequency of CTL Precursors in Primed C57BL/6 Spleens Capable of Responding to HLA Class I Antigens

stimulator cell used:			
in vivo	in vitro	target	CTL precursors/106 cells
JY	MST	JM	2.9
JY	JY/HSB	HSB	2.3
EL4-A2.1	EL4-A2.1	EL4-A2.1	56
		HSB	3.0
EL4-B7	EL4-B7	EL4-B7	45
		JM	3.0

The cells are described in the text. All values represent the means of at least two experiments, calculated as described.

These results indicate that the frequency of murine CTLp specific for human class I molecules can be increased significantly by in vivo priming, and argues against suppression as an explanation for the low frequency in naive animals. Despite this increase, the secondary response to xenogeneic HLA antigens was still one to two orders of magnitude weaker than a primary allogeneic response.

Influence of Cellular Background on Recognition of Xenoantigens

One possible explanation for these results is that stimulation requires a species specific interaction between accessory molecules on the T cells and ligands on the stimulator cells. Therefore, the response to H-2L^d and HLA-A2.1 expressed on either human (JM) or syngeneic murine (EL4) cells was compared (Engelhard et al.1988). H-2L^d positive JM or EL4 transfectants stimulated approximately equal responses in naive B6 animals (Table 1), indicating that the cellular background plays little, if any, role in the ability of CTLp to respond to H-2 alloantigens. Syngeneic murine transfectants expressing HLA-A2.1 (EL4-A2.1) or HLA-B7 (EL4-B7) stimulated CTLp frequencies in naive animals of approximately 2-3 per 10^6 cells, which is somewhat higher than the values seen using human stimulator cells (Table 1). Most of these CTL were also capable of recognizing human cells expressing these antigens. Although the cellular background appears to augment the stimulation of CTLp, the frequencies seen are still at least an order of magnitude lower than those achieved in response to H-2 alloantigens. This indicates that there is still a strong disparity in the response of murine CTL to murine and human class I molecules.

Surprisingly, when the frequencies of CTLp specific for EL4-A2.1 or EL4-B7 were evaluated in mice primed with these cells, values of about 50 per 10^6 cells were obtained (Table 2). This is one order of magnitude higher than the response elicited by HLA-A2.1 expressing human cells in primed animals. However, in contrast to the results with naive animals the frequency of these CTL that could lyse human targets expressing the same antigen was only 3 per 10^6 cells, virtually the same value as obtained in naive mice. This result indicates that in vivo priming with human class I expressing murine transfectants selectively primes a population of CTLp capable of recognizing HLA-A2.1 expressing murine cells, and suggests a strong influence of the murine cellular background on CTL stimulation. However, we have shown elsewhere (Holterman and Engelhard, 1986) that when expanded, all clones obtained with this type of specificity recognize only syngeneic, and not allogeneic murine transfectants. Taken together with antibody blocking data (Holterman and Engelhard, 1986) and transfection of H-2K^b molecules into allogeneic murine cells (unpublished data), this indicates that these clones recognize human class I molecules in the context of syngeneic H-2^b restricting elements. Similar results have also been obtained by two other groups (Achour et al.1986; Maryanski et al.1986b).

The above results indicate that expression of HLA antigens on a murine cellular background does increase CTLp recognition of these molecules. On the other hand, priming of these animals results in the selective expansion of cells that recognize these antigens as nominal antigens, rather than as independent "transplantation" antigens. In either case, the response to these molecules in this independent mode is one to two orders of magnitude below that to a murine class I alloantigen. The participation of the cellular background, presumably due to accessory molecule interactions, may be more important during CTLp stimulation by human class I molecules than during stimulation by murine alloantigens, but it is clearly insufficient to account for the large disparity in these responses. These results are in accord with the hypothesis that there have been evolutionary changes between HLA and H-2 antigens that render xenogeneic antigens less recognizable as transplantation antigens, and potentially less efficient as restricting elements in mice.

Structures on Class I Molecules that Determine the Magnitude of the Response

The regions of the HLA molecule determining the strength of this response were mapped using chimeric H-2/HLA antigens (Engelhard et al.1985; Engelhard et al.1988). These molecules consist of the $\alpha1+\alpha2$ domains derived from one species and the $\alpha3$, TM, and CT domains from the other. CTLp frequencies in naive mice stimulated with EL4 transfectants expressing either H-2L^d or the L^d/B7 chimeric antigen were similar (Table 3). The majority of clones stimulated with either antigen exhibited cross reactive recognition of the other. CTLp frequencies in naive mice stimulated with EL4 transfectants expressing either HLA-B7 or the B7/L^d chimeric antigen were also similar. These results indicate that the frequency of CTLp is controlled by the $\alpha1+\alpha2$ domain while the species of origin of the $\alpha3$, TM, and CT domains is largely irrelevant. However, these latter domains do influence the fine specificity of HLA-B7 specific clones (Engelhard et al.1988). A possible explanation for this is that this region directly alters epitopes on the HLA-B7

Table 3. Frequency of CTL Precursors in Naive C57BL/6 Mice Responding to Murine Transfectants Expressing H-2L^d or L^d/B7

Stimulator	Target	CTL Precursors/10^6 Cells
EL4-L^d	EL4-L^d	23
	EL4-L^d/B7	33
EL4-L^d/B7	EL4-L^d/B7	71
	EL4-L^d	25
EL4-B7	EL4-B7	3.1
	EL4-B7/L^d	2.0
	MST	2.3
EL4-B7/L^d	EL4-B7/L^d	1.9
	EL4-B7	0.9
	MST	1.0

The cells are described in the text. All values represent the means of at least two experiments, calculated as described.

molecule. Alternatively, this region may contain epitopes that are processed and become associated with H-2 molecules for recognition by H-2 restricted CTL.
The finding that the species of origin of the $\alpha3$ domain does not influence the magnitude of the CTL response is particularly striking, since the $\alpha3$ domain is highly conserved within a species, but

exhibits significantly lower homology between species. In addition, class I antigens act as specific ligands for CD8 (Norment et al.1988; Blue et al.1988; Rosenstein et al.1988), and recent results suggest that the $\alpha3$ domain contains at least part of the CD8 interaction site. Thus, the data presented here suggest either that the low response to xenogeneic class I molecules results from poor recognition of these molecules by parts of the TCR other than CD8, or that CD8 may bind to $\alpha1+\alpha2$ as well as $\alpha3$.

CHARACTERISTICS OF HLA-A2.1 TRANSGENIC MICE

Expression of HLA-A2.1 in lymphoid tissues of HLA-A2.1 transgenic mice

Transgenic mice were constructed by microinjection of the 7 kb EcoRI fragment containing the full length HLA-A2.1 gene into fertilized eggs from C57BL/6 mice as described (Hogan et al.1986). The expression of the HLA-A2.1 and H-2D^b genes in C57BL/6 and HLA-A2.1 transgenic mice was assessed by flow cytometry analysis using the HLA-A2 specific mAbs MA2.1 and PA2.1, and the H-2D^b specific monoclonal antibody 28.14.8. Spleen, bone marrow, and thymus cells all expressed significant quantities of the HLA-A2.1 molecule (Le et al.1989). The level of expression varied, being highest in the spleen and lowest in the thymus. However, the level of expression was similar to that of the endogenous H-2D^b molecule in each of these tissues. In addition, staining of frozen thymus sections indicated that the human class I molecules were expressed strongly on epithelial cells as well as medullary thymocytes, while cortical regions stained much more weakly (unpublished data). These results suggest that the expression of HLA-A2.1 in these immunologically relevant tissues was appropriate.

Frequency of CTLp specific for the human class I alloantigens HLA-B7 and HLA-A2.2.

If the bias of the murine CTL response toward murine, as opposed to human, class I alloantigens were due to positive selection of developing T cells by host MHC antigens, then one would expect that the response of HLA-A2.1 transgenic mice to other human class I MHC alloantigens would be enhanced. This prediction was tested by comparing the CTL response of HLA-A2.1 transgenic mice to the human HLA-B7 alloantigen and normal C57BL/6 by limiting dilution analysis (Le et al.1989). Mice were primed in vivo and restimulated in vitro with the HLA-B7 positive human cell line JY. Using the HLA-B7 positive human cell line JM as a target, it was

Table 4. Frequency of CTL Precursors in C57BL/6 or HLA-A2.1 Transgenic Spleen Cells Responding to HLA-B7 or HLA-A2.2 Expressed on Human Cells

Responder	*In vivo* Stimulus	*In vitro* Stimulus		CTLp/10^6 cells	
Set I			*Target:*	JM	EL4-B7
Transgenic	JY	JY		1.5	0.8
C57BL/6	JY	JY		3.8	2.6
Set II			*Target:*	C1R-A2.2	EL4-A2.2
Transgenic	WT49	WT49		4.6	0.2
C57BL/6	WT49	WT49		4.5	0.7
Set III			*Target:*	P815	
Transgenic	-	EL4-Ld		64	
C57BL/6	-	EL4-Ld		45	

The cells are described in the text. All values represent the means of at least two experiments, calculated as described.

found that the transgenic mice exhibited a frequency of specific CTLp that was not increased relative to that of normal C57BL/6 mice (Table 4). However, the frequency of CTLp for the H-$2L^d$ alloantigen in the transgenic mice was comparable to that of normal C57BL/6. Thus, the lack of an enhanced response to HLA-B7 in the transgenic mice was not due to a generalized hyporesponsive state.

We next considered the possibility that enhancement of the response to human class I alloantigens in the transgenic mice might be limited only to alloantigens that were structurally more related to HLA-A2.1. Therefore, the response to HLA-A2.2, which differs from HLA-A2.1 at only 4 amino acid positions (Holmes et al.1987), was evaluated. Mice were primed in vivo and restimulated in vitro with the HLA-A2.2 positive human cell line WT49. Using the human target cell C1R-A2.2, which expresses the HLA-A2.2 gene, it was found that the frequencies of specific CTLp in transgenic and C57BL/6 mice were equivalent (Table 4). When these same CTL were assayed using the murine HLA-A2.2 positive transfectant EL4-A2.2 as a target, frequencies of less than 1 in 10^6 were obtained for both responder populations. Thus, there was no enhancement of the response to the human class I MHC alloantigen HLA-A2.2 in HLA-A2.1 transgenic mice. This result stands in strong contrast to the observations made using H-$2K^{bm}$ mutants, whose amino acid sequence differences with the H-$2K^b$ molecule are similar to the differences between HLA-A2.1 and -A2.2, but which are comparable to other murine alloantigens in the frequency of CTLp that they stimulate in H-$2K^b$ mice (von Boehmer et al.1982).

It was considered possible that selection of HLA class I specific T cells in the HLA-A2.1 transgenic mice might have resulted in CTL that were specific for HLA alloantigens only when presented on murine cells. However, the frequencies of CTLp in naive C57BL/6 and transgenic mice stimulated with either EL4-B7 or HLA-A2.2, and assayed on these targets were similar, or in some cases actually lower (Le et al.1989). Taken together, these results indicate that there has been no quantitative expansion of CTLp for HLA-B7 or HLA-A2.2 in HLA-A2.1 transgenic mice regardless of the species of the cell on which these antigens are presented.

Frequency of H-2 and HLA-restricted, influenza specific CTLp in transgenic mice

We next investigated the ability of the transgenic mice to use HLA-A2.1 as a restricting element. There was no difference in the frequency of CTLp from transgenic and C57BL/6 mice capable of lysing virus infected EL4 targets (Table 5). In contrast, the frequency of transgenic and C57BL/6 CTLp capable of recognizing virus-infected JY targets was much lower. A similar poor ability of HLA-A2 to present influenza B virus, or of HLA-B7 to present either influenza viral strain to HLA-B7 transgenic T cells has been observed (unpublished data), although it has been possible to derive HLA-B7 restricted CTL clones by selective restimulation. Nonetheless, it is clear that the HLA transgenic products function poorly as restricting elements for influenza viruses as compared to the H-2^b antigens.

Table 5. Frequency of CTL Precursors in C57BL/6 or HLA-A2.1 Transgenic Spleen Cells Responding to Influenza Infected Target Cells Expressing H-2^b or HLA-A2.1 as Restricting Elements.

Responder	Target	Antigen recognized	CTLp/10^6 cells
Transgenic	EL4+virus	H-2^b+virus	41
C57BL/6	EL4+virus	H-2^b+virus	39
Transgenic	JY+virus	HLA-A2.1+virus	1.2
C57BL/6	JY+virus	HLA-A2.1+virus	0

Mice were primed by in vivo immunization with intraperitoneal injection of 300 HAU of virus in 0.5 ml PBS. Spleen cells from both types of animals were restimulated in vitro with irradiated transgenic spleen cells that had been infected with influenza virus as described in the text. All values represent the means of at least two experiments, calculated as described.

As discussed above, the low response to human class I molecules in normal mice probably reflects evolutionary changes in one or more structures on the class I molecule that are important for their recognition by T cells. However, the expression of human class I molecules in transgenic mice is not sufficient to positively select for an expanded population of CTLp that can recognize other human class I molecules, nor utilize the transgenic molecule as a restricting element. This failure to overcome the low xenogeneic response by positive selection indicates that the part of the T cell recognition apparatus involved is not clonally variable, but instead highly conserved. Our overall view is that the poor xenogeneic response results from a poor interaction between one or more species-specific structures on class I molecules and complementary species-specific structures on some part of the T cell recognition apparatus.

It is possible that poor recognition of conserved determinants on human MHC molecules is due to species specific differences in TCR V regions. Specific V regions have recently been suggested to have germ-line determined specificity for particular MHC molecules (Kappler et al.1987). These results, if generalized, are consistent with the idea that TCR V regions have evolved to recognize the MHC antigens expressed within the species. However, the results of Kappler et al (Kappler et al.1987) did not demonstrate that this specificity was independent of CD4 or CD8.

An alternative explanation for our results involves CD8. In TCR transgenic mice, selection of $CD8^+$ cells was dependent upon MHC haplotype (Teh et al.1988; Sha et al.1988). Based on this evidence, it seems likely that CD8 is directly involved in determining the specificity of the TCR for class I molecules. Thus, the failure of human class I molecules to be efficiently recognized by murine T cells could be due to poor interactions between these antigens and CD8. In this regard, human and murine CD8 share only 40% amino acid sequence homology in their extracellular domains (Littman, 1987). This predicts that T cells from transgenic mice expressing the human CD8 molecule should have an enhanced ability to recognize human MHC antigens. However, as mentioned above, if CD8 interactions are the basis for the poor xenogeneic response, then this must be mediated through an interaction with structures in the $\alpha1+\alpha2$, rather than the $\alpha3$, domains of the class I molecule.

<u>HLA-A2.1 transgenic mice are tolerant to HLA-A2.1 expressing murine cells, but still recognize HLA-A2.1 expressed on human cells.</u>

In order to determine whether expression of the HLA-A2.1 molecule in the transgenic mice had resulted in the induction of tolerance, responses of transgenic and C57BL/6 mice to HLA-A2.1 expressing EL4 cells (EL4-A2.1) were compared. C57BL/6 mice that had been primed once in vivo and then restimulated in vitro showed a CTLp frequency that was about 100-fold higher than that from similarly immunized HLA-A2.1 transgenic mice (Bernhard et al.1988). The response of the primed C57Bl/6 animals was elevated by about one order of magnitude, while the response of in vivo primed transgenic mice was considerably lower, than the frequency of CTLp in naive C57BL/6 mice. In contrast, after in vivo priming and in vitro restimulation with the HLA-2.1 expressing cell line JY, the frequency of CTLp that recognized the HLA-2.1 positive human cell line HSB was almost as high in the transgenic mice as in the C57BL/6 (Bernhard et al.1988). However, it was found that while most of these CTL elicited in the C57BL/6 mice also recognized the murine transfectant EL4-A2.1, virtually none of the CTL from the HLA-A2.1 transgenic did so. This is consistent with the idea that these animals are tolerant to HLA-A2.1, but only when the antigen is expressed on a murine cell. These results suggest that many of the HLA-A2.1 epitopes expressed on the human cells are distinct from those expressed on murine transfectants, in agreement with previous observations (Bernhard et al.1987).

The nature of the epitopes recognized was examined using 23 independent clones derived from limiting dilution culture. All clones lysed a wide variety of human cells expressing HLA-A2.1, including several different EBV transformed B-lymphoblastoid cell lines, normal human PBL, normal human fibroblasts, and HLA-A2.1 expressing transfectants of the human rhabdomyosarcoma cell line RD, and the xeroderma pigmentosum fibroblast M1 (Bernhard et al.1988). However, only 3 of 23 lysed monkey transfectants expressing HLA-A2.1. None of these clones lysed a variety of other HLA-A2.1 negative cell lines, including those expressing HLA-A2.2 and A2.3. Finally, none of the clones recognized HLA-A2.1 expressing murine transfectants derived from EL4, the C127 mammary carcinoma, or the P815 mastocytoma. These results demonstrate the existence of species-specific epitopes on the HLA-A2.1 molecule, and suggest that these epitopes are formed by the association of class I MHC products with one or more endogenous species-specific molecules.

In order to investigate the possibility that these epitopes depended upon the association of HLA-A2.1 with human β2-microglobulin (β2m), recognition of HLA-A2.1 transfected murine cells was examined after exchange of human for the endogenous β2m (Bernhard et al.1988), or using a murine cell also expressing a transfected human β2m gene (Table 6). It was found that the coexpression and association of HLA-A2.1 with human β2m on murine cells did not reconstitute the epitope recognized by any of these CTL clones. Similarly, the dependence of these epitopes upon the carbohydrate moiety of the HLA-A2.1 molecule was examined using a carbohydrate-negative mutant of HLA-A2.1 (Barbosa et al.1987). The recognition of this mutant molecule in human transfectants was somewhat lower than recognition of the wild-type molecule, reflecting the lower level of surface expression. However, all clones exhibited some degree of recognition of this cell (Table 7). In contrast, recognition of the HLA-A2 molecule was not restored when the carbohydrate negative molecule was expressed in monkey cells (Table 8). Taken together, these results indicate that the epitopes recognized by these CTL clones are not dependent upon specific human carbohydrate modifications, nor are such epitopes blocked by carbohydrate modification on nonhuman cells. Thus, none of the known posttranslational modifications to the extracellular domains of the HLA-A2.1 molecule can account for the difference in recognition of this molecule on murine and human cells.

These findings demonstrate the existence of epitopes on HLA-A2.1 that exist when the molecule is expressed on human cells but are largely absent when HLA-A2.1 is expressed on cells of other species. The recognition of human HLA-A2.1 expressing cells from different individuals and tissues suggests that these epitopes do not arise from the association of the HLA-A2.1 molecule

Table 6. Co-expression of HLA-A2.1 and Human β_2m on Murine Transfectants Does Not Reconstitute the Epitopes Recognized by CTL Clones From HLA-A2 Transgenic Mice

		Molecules	% Specific Lysis using CTL Clone:				
Target	Species	Expressed	AT1-8	AT1-21	AT2-1	AT2-3	AX23-1
HSB	Human	A2,Huβ_2M	71	67	71	45	29
EL4-A2.1	Murine	A2,Muβ_2M	1	1	0	1	36
LJ13M52	Murine	A2,Huβ_2M	0	0	0	2	38
LJ1M52	Murine	B7,Huβ_2M	-1	-1	-2	0	4
A9	Murine	Muβ_2M	1	-1	2	0	0

All CTL clones with the prefix AT were derived from transgenic mice. Clone AX23-1 was derived from a normal mouse, and recognizes HLA-A2.1 on both murine and human cells. The target cells LJ13M52 and LJ1M52 are murine L cells that have been transfected with human genomic DNA containing the human β_2m gene, and secondarily with genomic clones encoding HLA-A2.1 or -B7. A9 is an untransfected L cell derivative. Specific lysis values were obtained at an E:T of 10:1.

with human minor histocompatibility or tissue specific antigens. Instead, it appears that these CTL clones recognize epitopes formed by one or more endogenous, highly conserved but species specific molecules in the context of HLA-A2.1.

Table 7. Lysis of Human RD Cells Expressing HLA-A2.1 is Unaffected by Lack of Carbohydrate at Residue 86

	% Specific ^{51}Cr Release from:		
CTL clone	RD-A2.1	RD-A2.1(88)	RD
AT1-7	77	48	5
AT1-19	70	18	-2
AT2-1	77	46	-4
AT2-2	79	38	0
AH10-4[a]	57	24	-4
AX21-12[a]	68	36	-1

All CTL clones with the prefix AT were derived from transgenic mice. Clone AX21-12 was derived from a normal mouse, and recognizes HLA-A2.1 on both murine and human cells. The target cells are transfectants of the RD human rhabdomyosarcoma cell line. RD-A2.1 expresses HLA-A2.1, while RD-A2.1(88) expresses an HLA-A2.1 mutant Ser_{88}->Gly that results in loss of the carbohydrate side chain at Asn_{86}. Specific lysis values were obtained at an E:T of 1.25:1, except as indicated below.

[a]E:T= 10:1.

Table 8. Lysis of Monkey CV1O1 Cells Expressing HLA-A2.1 is Unaffected by Lack of Carbohydrate at Residue 86

	% Specific ^{51}Cr Release from:				
CTL clone	HSB	CV-A2.1	CV-A2.1(88)	CV-mock	EL4-A2.1
AT1-8	78	0	0	1	2
AT1-17	81	-1	-1	1	1
AT2-1	83	0	0	0	-1
AT2-2	70	44	48	0	0
AX1H11	68	52	1	23	23

All CTL clones with the prefix AT were derived from transgenic mice. Clone AX1H11 was derived from a normal mouse, and recognizes HLA-A2.1 expressed on both mouse and human cells. The target cells are transfectants of the simian CV101 cell line. CV-A2.1 expresses wild type HLA-A2.1, CV-A2.1(88) expresses an HLA-A2.1 mutant Ser_{88}->Gly that results in loss of the carbohydrate side chain at Asn_{86}, and CV-mock expresses a neomycin resistance gene only. Specific lysis values were obtained at an effector to target ratio of 10:1.

Given the demonstration that peptides may associate with class I molecules to form epitopes recognized by CTL (Townsend et al.1986; Maryanski et al.1986a; Clayberger et al.1987; Song et al.1988), the most likely explanation for our results is that the CTL produced here detect a complex of an MHC class I molecule with one or more peptides that are highly conserved among humans, but divergent in other species. An alternative is that the epitopes are formed by peptides that are conserved on both mouse and human cells, but altered by differences in mouse and human processing systems. This was addressed through the use of somatic cell hybrids made between murine cells and the cells of an HLA-A2.1 positive human. It can be seen that the lysis of such cells was dependent upon the expression of HLA-A2.1 (Table 9). However, it is also clear that only some CTL are able to recognize these somatic cell hybrids, while others are not. Since these cells contain virtually all of the murine genome, but only a fraction of the human genome, these results are compatible with the idea that the epitopes recognized by at least some clones are formed by species-specific human molecules associated with HLA-A2.1. However, for those

Table 9. Lysis of Human/murine Somatic Cell Hybrids is Specific for HLA-A2.1

Target	HLA Expression (% Positive)		% Specific ^{51}Cr Release using CTL Clone:				
Cell	A2	Class I	AH9-2	AH10-4	AT2-1	AT1-3	AX21-12
HSB	>99	ND	63	54	76	83	64
L201	>99	ND	0	-2	-1	7	69
IIIMA9-14a	97	ND	0	0	*29*	*23*	*30*
IIIMA9-14b	<0.5	7	2	0	2	3	0
IIIMA9-14c	<1	77	1	0	1	6	1
IIIMA9-14d	<0.5	<1	1	1	1	5	0

HSB and L201 are human and murine HLA-A2.1 positive cells, while the IIIMA9-14 cells are human (A2.1)$^+$ x murine somatic cell hybrids sorted for expression of A2.1 and other human class I antigens by flow cytometry. Lysis was assessed after 24 hr gamma interferon induction at an E:T of 12.5:1. Lysis of targets was assessed simultaneously to analysis of surface expression.

clones whose epitopes were not present on the hybrids, it is as yet uncertain whether this is due to the absence of the appropriate human gene, or to the involvement of the murine processing system.

The data presented here raise the possibility that epitopes recognized by alloreactive T cells may also result from the presentation of endogenously derived molecule in association with a class I MHC molecule on the surface of the stimulator cell. These results provide an explanation for the frequently observed failure of HLA class I specific CTL to recognize these antigens on murine transfectants. The failure of certain human CTL specific for HLA class I molecules to recognize these molecules when expressed on murine cells has been repeatedly observed (van de Rijn et al.1984; Bernabeu et al.1984; Barbosa et al.1984; Mentzer et al.1986; Koller et al.1987). It has been suggested that this is due to an inability of accessory molecules on human CTL to bind to their ligands on murine target cells. However it has also been observed that certain murine CTL raised against HLA antigens expressed on human cells fail to recognize these antigens on murine transfectants (Herman et al.1983). This result cannot readily be explained by a lack of participation of accessory molecules, since the CTL and the target are of the same species. It was instead suggested based on fine specificity patterns that this phenomenon is due to differences in the HLA class I molecule expressed on human and murine cells (Yannelli et al.1985; Bernhard et al.1987). The results of this study provide definitive evidence for the existence of one or more CTL-defined epitopes on the HLA-A2.1 molecule expressed on human cells which are entirely absent from this molecule when expressed on murine cells.

REFERENCES

Achour A, Begue B, Gomard E, Paul P, Sayagh B, Van Pel A, Levy JP (1986) Specific lysis of murine cells expressing HLA molecules by allospecific human and murine H-2-restricted anti-HLA T killer lymphocytes. Eur J Immunol 16:597-604

Barbosa JA, Mentzer SJ, Minowada G, Strominger JL, Burakoff SJ, Biro PA (1984) Recognition of HLA-A2 and -B7 antigens by cloned cytotoxic T lymphocytes after gene transfer into human and monkey, but not mouse, cells. Proc Natl Acad Sci USA 81:7549-7553

Barbosa JA, Santos-Aguado J, Mentzer SJ, Strominger JL, Burakoff SJ, Biro PA (1987) Site directed mutagenesis of class I HLA genes. Role of glycosylation in surface expression and functional recognition. J Exp Med 166:1329-1350

Bernabeu C, Maziarz RT, Spits H, de Vries JE, Burakoff SJ, Terhorst CP (1984) Coexpression of the human HLA-A2 or HLA-B7 heavy chain gene and human beta 2-microglobulin gene in L cells. J Immunol 133:3188-3194

Bernhard EJ, Le A-XT, Yannelli JR, Holterman MJ, Hogan KT, Parham P, Engelhard VH (1987) The ability of cytotoxic T cells to recognize HLA-A2.1 or HLA-B7 antigens expressed on murine cells correlates with their epitope specificity. J Immunol 139:3614-3621

Bernhard EJ, Le A-XT, Barbosa JA, Lacy E, Engelhard VH (1988) Cytotoxic T cells from HLA-A2 transgenic mice specific for HLA-A2 expressed on human cells. J Exp Med 168:1157-1162

Blue M-L, Craig K, Anderson P, Schlossman SF (1988) Evidence for specific association between class I major histocompatibility antigens and the CD8 molecules of human suppressor/cytotoxic cells. Cell 413-421.

Clayberger C, Parham P, Rothbard JB, Ludwig DS, Schoolnik GK, Krensky AM (1987) HLA-A2 peptides can regulate cytolysis by human allogeneic T lymphocytes. Nature 330:763-765

Engelhard VH, Kaufman JF, Strominger JL, Burakoff SJ (1980) Specificity of mouse cytotoxic T lymphocytes stimulated with either HLA-A and -B or HLA-DR antigens reconstituted into phospholipid vesicles. J Exp Med 152:54s-64s

Engelhard VH, Yannelli JR, Evans GA, Walk SF, Holterman MJ (1985) Construction of novel class I histocompatibility antigens by interspecies exon shuffling. J Immunol 134:4218-4225

Engelhard VH, Le A-XT, Holterman MJ (1988) Species-specific structural differences in the α_1 and α_2 domains determine the frequency of murine cytotoxic T cell precursors stimulated by human and murine class I molecules. J Immunol 141:1835-1839

Herman A, Parham P, Weissman SM, Engelhard VH (1983) Recognition by xenogeneic cytotoxic T lymphocytes of cells expressing HLA-A2 or HLA-B7 after DNA-mediated gene transfer. Proc Natl Acad Sci USA 80:5056-5060

Hogan, B., Constantini, F. and Lacy, E. *Manipulating the mouse embryo - laboratory manual*, Cold Spring Harbor, New York:Cold Spring Harbor Laboratory, 1986.

Holmes NJ, Ennis P, Wan AM, Denney DW, Parham P (1987) Multiple genetic mechanisms have contributed to the generation of the HLA-A2/A28 family of class I MHC molecules. J Immunol 139:936-941

Holterman MJ, Engelhard VH (1986) HLA antigens expressed on murine cells are preferentially recognized by murine cytotoxic T cells in the context of the H-2 major histocompatibility complex. Proc Natl Acad Sci USA 83:9699-9703

Kappler JW, Wade T, White J, Kushnir E, Blackman M, Bill J, Roehm N, Marrack PC (1987) A T cell receptor V_β segment that imparts reactivity to a class II major histocompatibility complex product. Cell 49:263-271

Koller TD, Clayberger C, Maryanski JL, Krensky AM (1987) Human allospecific cytolytic T lymphocyte lysis of a murine cell transfected with HLA-A2. J Immunol 138:2044-2049

Le A-XT, Bernhard EJ, Holterman MJ, Parham P, Lacy E, Engelhard VH (1989) Cytotoxic T cell responses in HLA-A2.1 transgenic mice: recognition of HLA alloantigens and utilization of HLA-A2.1 as a restriction element. J Immunol 142:1366-1371

Littman DR (1987) The structure of the CD4 and CD8 genes. Ann Rev Immunol 5:561-584

Maryanski JL, Pala P, Corradin G, Jordan BR, Cerottini J-C (1986a) H-2-restricted cytolytic T cells specific for HLA can recognize a synthetic HLA peptide. Nature 324:578-579

Maryanski JL, Accolla RS, Jordan BR (1986b) H2-restricted recognition of cloned HLA class I gene products expressed in mouse cells. J Immunol 136:4340-4347

Mentzer SJ, Barbosa JA, Strominger JL, Biro PA, Burakoff SJ (1986) Species-restricted recognition of transfected HLA-A2 and HLA-B7 by human CTL clones. J Immunol 137:408-413

Norment AM, Salter RD, Parham P, Engelhard VH, Littman DR (1988) Cell-cell adhesion mediated by CD8 and MHC class I molecules. Nature 336:79-81

Rosenstein Y, Ratnofsky SE, Burakoff SJ, Herrmann SH (1989) Direct evidence for binding of CD8 to HLA class I antigens. J Exp Med 169:149

Sha WC, Nelson CA, Newberry RD, Kranz DM, Russell JH, Loh DY (1988) Selective expression of an antigen receptor on CD8-bearing T lymphocytes in transgenic mice. Nature 335:271-274

Song EH, Linsk R, Olson CA, McMillan M, Goodenow RS (1988) Allospecific cytotoxic T lymphocytes recognize an H-2 peptide in the context of a murine major histocompatibility complex class I molecule. Proc Natl Acad Sci USA 85:1927-1931

Teh HS, Kisielow P, Scott B, et al (1988) Thymic major histocompatibility complex antigens and the ‡§ T-cell receptor determine the CD4/CD8 phenotype of T cells. Nature 335:229-233

Townsend AR, Rothbard JB, Gotch FM, Bahadur G, Wraith DC, McMichael AJ (1986) The epitopes of influenza nucleoprotein recognized by cytotoxic T lymphocytes can be defined with short synthetic peptides. Cell 44:959-968

van de Rijn M, Weiss J, Seidman JG, de Vries JE, Spits H, Terhorst CP (1984) Recognition of HLA-A2 by cytotoxic T lymphocytes after DNA transfer into human and murine cells. Science 226:1083-1085

von Boehmer H, Bennink JR, Langhorne J (1982) The frequency of mutant-specific killer precursors in related and unrelated mouse strains. Eur J Immunol 12:892-895

Yannelli JR, Moore LC, Engelhard VH (1985) Multiple epitopes on human and murine cells expressing HLA-B7 as defined by specific murine cytotoxic T cell clones. J Immunol 135:900-905

Analysis of the HLA-Cw3 Specific CTL Response of HLA-B7 x Human $\beta 2$ Microglobulin Transgenic Mice

C. BARRA[1], B. PERARNAU[1], P. GERLINGER[2], M. LEMEUR[2], A. GILLET[1], P. GIBIER[1], and F. A. LEMONNIER[1]

[1] Centre d'Immunologie INSERM-CNRS de Marseille-Luminy, Case 906, 13288 Marseille Cedex 9, France
[2] Laboratoire de Génétique Moléculaire des Eucaryotes du CNRS, Université de Strasbourg 1, 11, rue Human, 67085 Strasbourg Cedex, France

INTRODUCTION

The T cell receptor repertoire of peripheral cytolytic T lymphocytes (CTL) schematically results from the positive selection of precursor cells, followed by the deletion of potentially autoaggressive cells (Kappler and Marrack, 1988). Assuming both phenomena to be driven by MHC class I and class II molecules in the thymus, one might expect HLA-B7 x human β2-microglobulin (hβ2m) double transgenic mice to have positively selected peripheral T lymphocytes utilizing the transgenic molecules as restricting elements, as long as the tissue distribution of the transgenic molecules is correct. In fact, previous data from HLA-B27 and -Cw3 transgenic mice support this assumption (Kievits 1987, Dill 1988). If positive selection by contact with major histocompatibility molecules in the thymus is the major mechanism responsible for the generation of the T cell repertoire and if allo-reactivity is a by product of the self-MHC restricted antigenic stimulation of T lymphocytes (as suggested by the high frequency of self-restricted, allo-cross-reactive T lymphocytes), one might expect HLA-B7 x hβ2m mice to mount a potent allogeneic response against other HLA molecules. Therefore, we investigated the CTL response of HLA-B7 x hβ2m double transgenic mice against cells expressing HLA-Cw3 molecules. Normal (non transgenic) DBA/2 ($H\text{-}2^d$) mice immunized with HLA-Cw3 transfected P815 ($H\text{-}2^d$) mastocytoma cells develop a potent, $H\text{-}2K^d$-restricted response against a peptide derived from the N-terminal part of the second domain of the HLA-Cw3 heavy chain (Maryanski *et al*, 1986). We evaluated whether HLA-B7 x hβ2m double transgenic mice would be able to mount potent, non-H-2 restricted HLA-Cw3-specific CTL responses.

EXPRESSION OF HLA-B7 x hβ2m TRANSGENIC MOLECULES AT THE SURFACE OF LYMPHOID AND THYMIC CELLS

Comparative indirect immunofluorescence analysis of resting, Con A-activated or LPS-activated spleen cells from non-transgenic, HLA-B7 single transgenic and HLA-B7 x hβ2m double transgenic DBA/2 mice indicated that simultaneous expression of human β2m and HLA-B7 heavy chains enhanced cell surface expression of HLA-B7 molecules (Fig. 1, upper part), as previously observed with transfected cells (Pérarnau et al. 1988). However, even after activation, the expression of HLA-B7 x human β2m transgenic molecules remained 3 times lower than that of endogeneous $H\text{-}2K^d$ class I molecules.

Immunostaining of thymic cryosections from double transgenic mice not only indicated expression of HLA-B7 class I molecules by the thymic cells, but further suggested that the $H\text{-}2K^d$ and HLA class I molecules in the thymus had similar pattern of expression (Fig. 1, lower part).

I.K. Egorov/C.S. David (Eds.)
Transgenic Mice and Mutants in MHC Research

Mouse strain	Activation state	mAb and mean fluorescence intensity			
		B1.23.2	B2.62.2	20-8.4S	H-84.17.2
DBA/2	non-activated	1.5	0.7	143	0.7
	+ Con A	1.6	1.6	390	1.3
	+ LPS	7.7	9.7	556	8.5
HLA-B7 DBA/2	non-activated	1.2	0.9	161	0.6
	+ Con A	10	2.8	607	3.1
	+ LPS	14.8	7.6	862	1.4
HLA-B7 x h β 2m DBA/2	non-activated	11	74	135	0.7
	+ Con A	45	189	452	2.3
	+ LPS	79	300	600	7

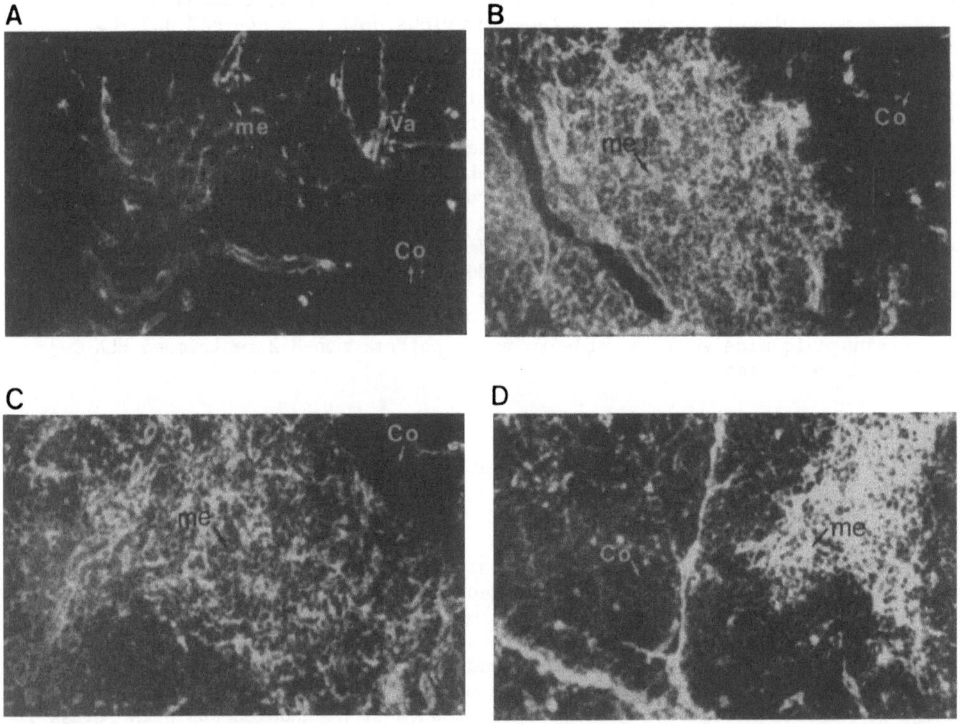

Fig. 1. Expression of HLA-B7 molecules on splenocytes and thymic cells in transgenic mice.
Upper part : Ficoll hypaque purified DBA/2 splenocytes either non activated, Con A- or LPS-activated were incubated (30 min at 4°C) with saturating concentrations of either HLA-B/C specific (B1.23.2) hβ2m specific (B2.62.2), H-2K^d specific (20-8.4S), H-2K^k specific (H-84.17.2) biotinylated monoclonal antibodies (mAb). Following 3 washes in DME-5% FCS medium, cells were stained with an excess of FITC-streptavidine for 30 min at 4°C, washed, 1% fixed with formaldehyde and analyzed by FACS. The mean

fluorescence of each population (reported in the figure) was calculated using each fluorescence channel as an arbitrary unit. Corrections were introduced corresponding to the lower gains used for highly positive cells.
Lower part: Following incubation of thymic cryosections prepared from HLA-B7xhβm double transgenic mice with H-84.17.2 (picture A), 20-8.4S (picture B), B1.23.2 (picture C), B2.62.2 (picture D) mAbs, mAb fixation was revealed with rhodamine-conjugated, mouse Ig-specific goat antibodies. The following thymic ultrastructural elements could be identified : cortex (Co), medulla (me), external capsula (ec), perivascular spaces (Va)

DERIVATION OF HLA-Cw3-SPECIFIC, NON-H-2 RESTRICTED CTL CLONES FROM HYPERIMMUNE DOUBLE TRANSGENIC MICE

Non transgenic, HLA-B7 single and HLA-B7 x hβ2m double transgenic DBA/2 male mice were injected intraperitonealy with P815-Cw3 or P815-Cw3xhβ2m transfected cells. Three hundred CTL clones were isolated which specifically lysed HLA-Cw3 expressing, P815 transfected cells. Lysis was inhibited in most cases (292 out of 300) by a $H\text{-}2K^d$-specific mAb (20-8.4S). In only 8 cases did we observe specific and complete inhibition by HLA-B/C-specific B1.23.2 mAb, indicating non-H-2 restricted recognition of cell surface expressed HLA-Cw3 molecules. These 8 HLA-Cw3-specific non-H-2 restricted clones were all derived from P815-Cw3xhβ2m immunized double transgenic mice, suggesting that coexpression of HLA class I heavy chain and hβ2m by both the responder mice (HLA-B7xhβ2m) and the stimulating cells (P815-HLA-Cw3xhβ2m) favors the development of an HLA-Cw3-specific non-H-2 restricted CTL response. Note however that even in such a situation, the $H\text{-}2K^d$ restricted response remained dominant.

The fine specificity of the 8 CTL clones isolated was next determined, using as target cells either P815 transfected cells expressing various HLA-A, -B, or -C specificities (HLA-AI.3, -A2, -A3, -B7, -B8, -B27, -B44, -B51, -Bw60) or human EBV transformed cells expressing or not the HLA-Cw3 specificity. As illustrated in Table 1 (for two of the ten transfectants tested), only P815 Cw3xhβ2m transfected cells were lysed. Similarly, HLA-Cw3 positive human target cells were significantly lysed. However, cross-reactivity was also observed with some (LG38) but not all (HHK) HLA-Cw3 negative cells. These results indicate recognition of HLA-Cw3 molecules expressed at the surface of both human and mouse cells by the 8 HLA-Cw3-specific, non-H-2 restricted CTLs, and also cross-reactivity with other as yet undefined HLA-class I specificities.

INHIBITION BY Ly-2/3 AND LFA.1-SPECIFIC mAbs

Experiments were devised to evaluate whether the mouse Ly-2/3 and LFA.1 adhesion molecules were involved in the non-H-2 restricted recognition of the HLA-Cw3 molecules at the surface of mouse (P815) or human (GGR) target cells.

Profound inhibition of lysis of the human (GGR) cells was observed in the presence of Ly-2/3-specific mAb (H35-17.2) with the eight clones tested, whereas lysis of mouse cells P815 Cw3 x hβ2m transfected target cells by the same CTLs was not reduced (Table 2). In contrast, addition of LFA.1-specific mAb resulted in the inhibition of lysis of both mouse and human cells (data not shown).

These results indicate primarily that heterospecific interaction can occur between mouse Ly-2/3 and human HLA-Cw3 molecules as had already been demonstrated by Swain *et al.* (1983). They further suggest that HLA-Cw3 molecules associate with slightly different peptides in mouse (P815) or human (GGR) cells, assuming cell surface density and accessibility of HLA-Cw3 molecules to be equivalent on P815 and GGR cells).

Table 1. Characterization of anti HLA-Cw3 non-H-2 restricted CTLs isolated from HLA-B7 x hβ2m double transgenic mice

CTLs[a]	P815 transfectants and antibodies					Human EBV-transformed cells		
	B7xhβ2m	A3xhβ2m	Cw3xhβ2m[b]			GGR[c]	HHK RO	LG38
			0	20-8.4S	B1.23.2	3 ; 7;w3	3 ; 7;w7	25 ; 18;w6
35B1	0	2	71.5	71	0	69.5	0	25
1A4	2.5	0	74	89	8.5	71	1.5	35.5
1B1	4	2	70	72	6	72	0	34
2A2	0	15	55.5	71.5	0.3	70	1	36
2B5	13	0	95	95	5	68	0	47
3A4	0.8	0	90	80	0	63	0	23
3C2	7	7	78	78	2	73	1.4	41
4A2	1.5	21.5	81.5	83	3	79	1.7	56
1B5*	ND	ND	67.5	0.5	94	ND	ND	ND

[a]. HLA-Cw3-specific non-H-2 restricted CTL clones derived from hyperimmunized HLA-B7 x hβ2m DBA/2 double transgenic mice were tested at effector to target ratio of 10:1 and 20:1 against P815 transfectants and EBV-transformed cells respectively. An HLA-Cw3-specific H-2K[d] restricted CTL clone (1B5*)was tested in parallel. Spontaneous release was below 10% (P815) and 25% (EBV) of total release; [b]. Lysis of P815 Cw3 x hβ2m transfectants in the presence of either H-2K[d]-specific (20-8.4S) or HLA-B/C-specific (B1.23.2) mAb ; [c]. Expressed HLA-A, -B and -Cw specificities.

ANALYSIS OF THE T CELL RECEPTOR α AND β GENE REARRANGEMENTS

Southern blot analysis of the DNA isolated from the eight non-H-2 restricted clones was performed following Hind III, Bam HI, or Eco RI digestion to evaluate more precisely the clonal diversity of the H-2 unrestricted HLA-Cw3 specific CTL response of HLA-B7 x hβ2m transgenic mice. Probes corresponding to the Dβ1, Cβ1, Jβ2 and Vβ14 (Kronenberg 1985 ; Malissen, 1986) subregions of the T cell receptor β locus and 4 single-copy DNA probes encompassing the 3' part of the T cell receptor Jα gene cluster (Malissen, 1988) were used for this study. As illustrated in Figure 2, similar patterns of hybridization were observed whichever probe was used for 4 of the 8 clones tested. Moreover, further analysis led to the conclusion that TcR β genes rearrangements had occured on both chromosomes, whereas as far as the 3' part of the Jα region is concerned, rearrangements could only be documented for one chromosome. As only one TcR β gene rearrangement should be implicated in the expression of the T cell receptor responsible for the H-2 unrestricted recognition of HLA-Cw3 molecules, the fact that the eight clones underwent similar TcR β gene rearrangements on both chromosomes strongly suggests that they all derived from a single parental cell.

Table 2. Inhibition of the lysis of HLA-Cw3 expressing human cells by Ly-2/3 specific mAb

CTLs[a]	Target cells and antibodies			
	P815 Cw3-hβ2m		GGR	
	0	H35.17.2	0	H35.17.2
35B1	27.5	30	42	4
1A4	44.8	44.8	49.2	19.5
1B1	30	37.5	75	26
2A2	24	29	44	17.2
2B5	35.5	29.5	56	23
3A4	36	34.5	49	7.5
3C2	16	15	22	4
4A2	60	55	88	35

[a] CTL clones from hyperimmunized HLA-B7x human β2m DBA/2 double transgenic mice were tested at a 20:1 ratio against either P815-HLA-Cw3 x hβ2m double transfected cells or GGR human EBV-transformed cells. The cytolytic assay was performed in the absence or presence of 2 μg of protein A-purified Ly-2/3 specific (H35.17.2) mAb. Results are expresssed in % of lysis, the spontaneous release was 10 % and 20 % of total release for P815 and GGR cells respectively.

CONCLUSION

HLA-Cw3-specific, non-H-2-restricted CTL clones could only be isolated from immune HLA-B7 x hβ2m double transgenic mice (a result confirmed by limiting dilution analysis - C. Barra, data not shown). However, the striking observation was the persisting low frequency of such clones compared to the frequency of alloreactive CTLs in the same mice. Several mutually non exclusive hypotheses can be put forward to explain such results. Note first that these low frequencies cannot be attributed to the lack of expression of human β2m by responder animals, as might have been proposed to explain earlier reports (Hämmerling 1989; Le 1989). Quantitative parameters should next be considered. Assuming the interaction of T lymphocytes with target cells to be of low avidity (Schreck 1989), a suboptimal cell surface expression of HLA class I molecules and a reduced affinity of heterospecific mouse Ly-2/3 for HLA class I molecule might result in inefficient usage of HLA-B7 molecules during the positive selection process in transgenic mice. Finally, one should consider the possibility of a genetically controlled relative inability of the mouse T cell receptors to recognize heterospecific HLA molecules.

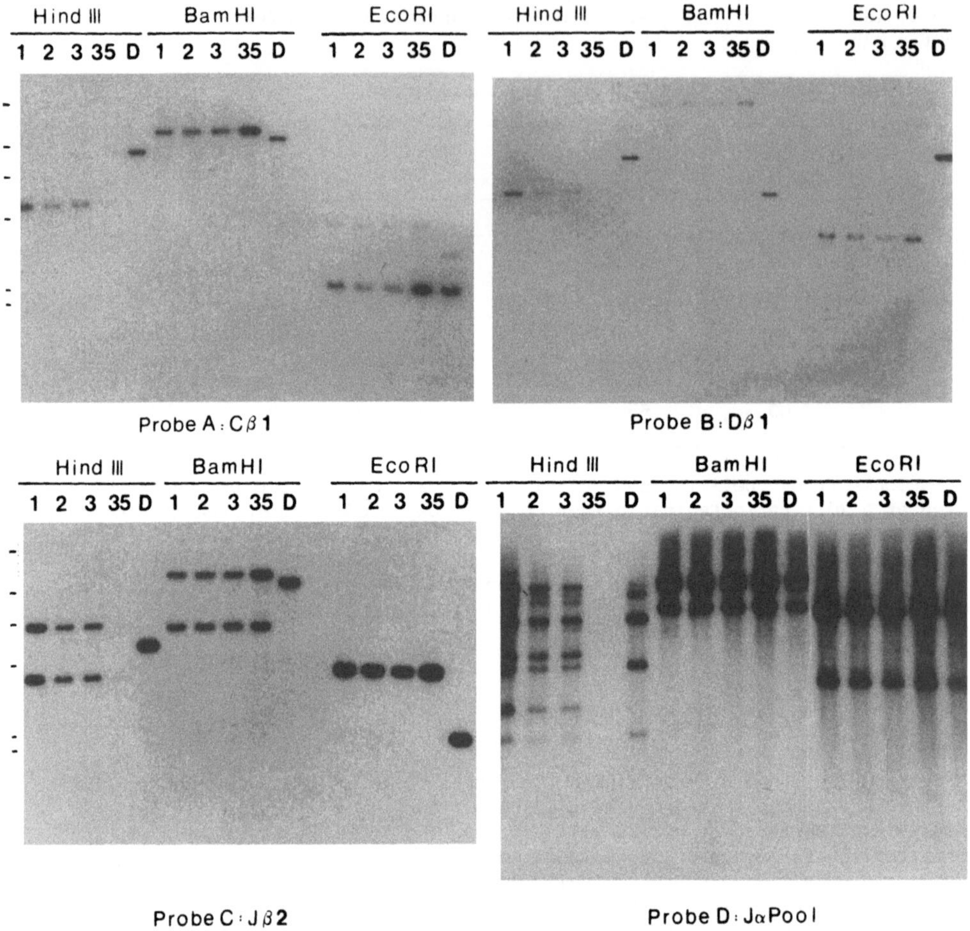

Fig. 2. Southern analysis of T cell receptor α and β gene rearrangements of anti-HLA-Cw3 H-2 unrestricted CTLs. Genomic DNA from 4 non-H-2 restricted anti-HLA-Cw3 CTL clones (lanes 1-4, 6-9, 11-14) and from DBA/2 liver cells (lane 5, 10, 15) were cut with the following endonucleases : Hind III (lanes 1-5), Bam HI (lanes 6-10) or Eco RI (lanes 11-15). Following nitrocellulose blotting, hybridizations were sequentially performed with Cβ1- (A), Dβ1- (B), Jβ2- (C), Vβ14- (not shown) and pooled Jα locus- (D) specific DNA probes. phage DNA cut with Hind III was included on the gel as mw markers (see the scale on the left hand side of the picture) corresponding from top to bottom to DNA fragments of 23.6, 9.4, 6.7, 4.3, 2.2 and 2 kilobases

ACKNOWLEDGEMENTS : This work was supported by grants from Institut National de la Santé et de la Recherche Médicale, Centre National de la Recherche Scientifique, and Association pour la Recherche sur le Cancer.

REFERENCES

Dill O, Kievits F, Koch S, Ivanyi P, Hämmerling GJ (1988) Immunological function of HLA-C antigens in HLA-Cw3 transgenic mice. Proc Natl Acad Sci USA 85:5664-5668

Hämmerling GJ, Hunt T, Dill O, Moreno J (1989) Cytotoxic T lymphocyte recognition of HLA-A2 antigens in normal and HLA-Cw3 transgenic mice. Eur. J. Immunol 19:599-604

Kievits F, Ivanyi P, Krimpenfort P, Berns A, Ploegh HL (1987) HLA-restricted recognition of viral antigens in HLA transgenic mice. Nature 329:447-449

Kronenberg M, Goverman J, Haars R, Malissen M, Kraig E, Phillips L, Delovitch T, Suciu-Foca N, Hood L (1985) Rearrangement and transcription of the β-chain genes of the T-cell antigen receptor in different types of murine lymphocytes. Nature 313:647-653

Le A-XT, Bernhard EJ, Holterman MJ, Strub S, Parham P, Lacy E, Engelhard VH (1989) Cytotoxic T cell responses in HLA-A2.1 transgenic mice. Recognition of HLA alloantigens and utilization of HLA-A2.1 as a restriction element. J Immunol 142:1366-1371

Malissen M, McCoy C, Blanc D, Trucy J, Devaux C, Schmitt-Verhulst A-M, Fitch F, Hood L, Malissen B (1986) Direct evidence for chromosomal inversion during T-cell receptor β-gene rearrangements. Nature 319:28-33

Malissen M, Trucy J, Letourneur F, Rebaï N, Dunn DE, Fitch FW, Hood L, Malissen B (1988) A T cell clone expresses two T cell receptor α genes but uses one αβ heterodimer for allorecognition and self MHC-restricted antigen recognition. Cell 55:49-59

Marrack P, Kappler J (1988) The T cell repertoire for antigen and MHC. Immunol Today 9:308-315

Maryanski JL, Pala P, Corradin G, Jordan BR, Cerottini J-C (1986) H-2-restricted cytolytic T cells specific for HLA can recognize a synthetic HLA peptide. Nature 324:578-579

Pérarnau BM, Gillet AC, Hakem R, Barad M, Lemonnier FA (1988) Human β2-microglobulin specifically enhances cell surface expression of HLA class I molecules in transfected murine cells. J Immunol 141:1383-1389

Schneck J, Lee Maloy W, Coligan JE, Margulies DH (1989) Inhibition of an allospecific T cell hybridoma by soluble class I proteins and peptides : Estimation of the affinity of a T cell receptor for MHC. Cell 56:47-55

Swain SL, Dutton RW, Schwab R, Yamamoto J (1983) Xenogeneic human anti-mouse T cell responses are due to the activity of the same functional T cell subsets responsible for allospecific and major histocompatibility complex-restricted responses. J Exp Med 157:720-729.

H-2-Restricted Recognition of Xeno-MHC Antigens by Primary Mouse Cytotoxic T Cells is the Exception Rather than the Rule

F. KIEVITS, J. WIJFFELS, W. LOKHORST, W. J. BOERENKAMP, and P. IVANYI[1]

[1] Central Laboratory of the Netherlands Red Cross Blood Transfusion Service and Laboratory for Experimental and Clinical Immunology, University of Amsterdam, P.O.Box 9406, 1006 AK Amsterdam, The Netherlands

ABSTRACT

HLA-B27.2 and SLA-PD1 transgenic mice were used to study the question whether the recognition of xenogeneic MHC determinants is restricted by self-MHC and so follows the recognition of other foreign antigens, such as haptens, viral, tumor or minor histocompatibility antigens.

Our results show that this is not the case, but indicate that xenogeneic MHC molecules are recognized in the same way as allelic variants of self-MHC molecules. This was shown by primary bulk CTL cultures as well as in split-well analysis. H-2-restricted recognition of xeno-MHC antigens (Maryanski, 1986a,b; Lemonnier, 1977; Holterman, 1986; Achour, 1986; Burakoff, 1977) can therefore not be generalized as a part of a classical MHC class-I-specific immune response.

INTRODUCTION

Cytotoxic T lymphocytes (CTL) recognize peptide antigens which have been processed by the target cell and then presented in association with the relevant self-class-I molecule (Kindred, 1972; Zinkernagel, 1975; Ziegler, 1982; Townsend, 1985). An intriguing question emerging from this antigen presentation model is whether allo- or xeno-MHC-specific CTL may also recognize a complex between the foreign MHC and some peptides. Such a peptide could be derived from the foreign HLA itself or from another molecule. Alternatively, self-MHC molecules expressed on the membrane may bind peptides derived from processed allo- or xeno-antigens. In that case, the recognition of foreign MHC peptides might be restricted by self-MHC. However, the possibility remains that CTL recognize native allo- or xeno-MHC molecules rather than peptides.

During the last years, the idea has been spread that the recognition of xeno-MHC class-I antigens by mouse CTL is restricted by H-2 (Maryanski, 1986a,b; Lemonnier, 1977; Holterman, 1986; Achour, 1986; Burakoff, 1977). Most of the experiments were done with selected CTL clones and lines raised against HLA-transfected cells in secondary immune responses. Such CTL might represent only a minimal part of the T-cell repertoire. Therefore, we examined whether mouse primary bulk-CTL populations, directed against xeno-MHC molecules (HLA-B27.2 and SLA-PD1), were restricted by the mouse MHC. The experiments were extended by split-well analysis from limiting-dilution experiments. We found no H-2-restricted recognition of HLA class-I antigens by primary CTL. Therefore, we propose the possibility that allo- and xeno-MHC antigens are recognized as peptides, preferentially presented by their native MHC molecules.

I.K. Egorov/C.S. David (Eds.)
Transgenic Mice and Mutants in MHC Research

MATERIALS AND METHODS

The mouse strains were maintained at our Institute. The origin, functional and biochemical analysis of the HLA-B27.2 and SLA-PD1 (human and porcine) xeno-MHC products in transgenic mice have been described elsewhere (Frels, 1985; Bluestone, 1987; Kievits, 1987; Kievits, 1989; Krimpenfort, 1987). $H\text{-}2^b$, $H\text{-}2^q$ and $H\text{-}2^k$ B27dTGM (double transgenic mice expressing the human β_2-microglobulin and HLA-B27.2 genes) were obtained from crosses between B27dTGM mice and B10, B10.Q and B10.BR mice, respectively). The SLA-PD1sTGM (single transgenic mice expressing the PD1 allele of SLA) are of B10 origin (Frels, 1985). $H\text{-}2^k$ SLA-PD1 TGM were obtained from crosses to B10.BR ($H\text{-}2^k$) mice. All mice used in this study were from the fourth backcross generation.

Spleen cells from naive B10 and B10.Q mice were cocultured with the same number of irradiated (2300 rad) spleen cells from $H\text{-}2^b$ or $H\text{-}2^q$ B27dTGM. In the same way, B10 and B10.BR spleen cells were cocultured with spleen cells from $H\text{-}2^b$ or $H\text{-}2^k$ SLA-PD1-transgenic mice. This was done in 50 ml of Iscove's modified Dulbecco's medium, supplemented with 10% (v/v) fetal calf serum (FCS) and 100 IU ml^{-1} penicillin, 100 µg ml^{-1} streptomycin and 5×10^{-5} M 2-mercaptoethanol. Autologous transgenic lymph-node cells were cultured in the same medium and used as target cells. After 5 days, the responding CTL were collected and tested in the ^{51}Cr-release assay on human, transgenic and nontransgenic cells stimulated for 3 days with 2.5 µg ml^{-1} Concanavalin A. Responding CTL were incubated with ^{51}Cr-labeled target cells at the effector:target ratio indicated.

For blocking experiments, target cells were incubated for 30 min on ice with 100 µl of monoclonal antibodies (1:10) and washed three times with medium before incubation with CTL. The percentage of specific ^{51}Cr release was calculated as (release in the presence of CTL spontaneous release) / (release in 10% Triton X-100 spontaneous release) x 100%. The spontaneous release in medium from labeled target cells ranged from 20 to 30% of the maximal release produced by Triton X-100.

For split-well analysis, responder and stimulator cells were cocultured in microtiter plates for 7 days in the presence of 20 U/ml rat IL-2. Responding cultures were defined as those in which the ^{51}Cr-release values exceeded the mean background ^{51}Cr release plus three times the standard deviation.

HLA-typed peripheral lymphocytes were isolated by flotation on Ficoll-Isopaque and stored in liquid nitrogen until used. HLA types of the human cells were: HLA-A2,8; B27.2, B27.2; C2,-; DR2,5 (27,27): HLA-A3,21; B27.2,61; C2,2; DR8,5 (27,-): HLA-2,31; B18,62; C3,-; DR4,8 (-,-): HLA-A2,28; B5,12; C4,- (-,-).

RESULTS AND DISCUSSION

The CTL response against xenogeneic class-I-MHC determinants expressed on cells of two types of transgenic mice (TGM) was studied, the first double transgenic for the human genes HLA-B27.2 and β_2-microglobulin, the second expressing a porcine MHC class-I gene SLA-PD1 (PD1TGM). HLA-B27-specific CTL were elicited by stimulation *in vitro* without prior priming *in vivo* of $H\text{-}2^b$ or $H\text{-}2^q$ spleen cells with irradiated B27dTGM cells with matched H-2 haplotypes. The specificity of the CTL population was tested on HLA-B27.2-positive and -negative human cells (Fig. 1A).

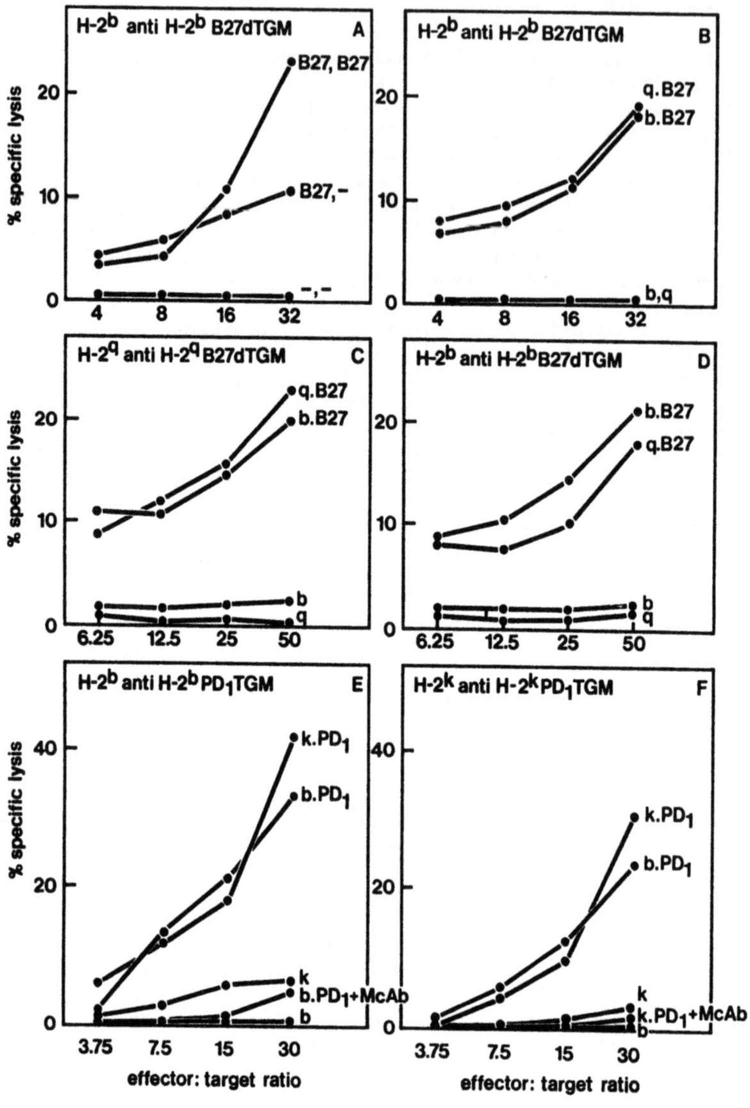

Fig. 1 - Cytotoxic T lymphocytes (CTL) specific for HLA-B27.2 and hu-β_2m double-transgenic mouse cells (B27dTGM) were obtained from H-2^b (A,B,D) and H-2^q (C) mice and tested on HLA-B27-homozygous (B27,B27), heterozygous (B27,-) and -negative (-,-) human cells (A) and on transgenic and nontransgenic mouse cells (B,C,D), matched or mismatched for H-2 at the effector:target ratios indicated. Human and mouse target cells expressing HLA-B27 are lysed. CTL specific for SLA-PD1 were obtained from H-2^b (E) and H-2^k (F) mice and tested on H-2^b- and H-2^k-transgenic and nontransgenic mouse cells. Lysis of all target cells expressing SLA-PD1 was blocked by PD1-specific monoclonal antibody 74-11-10. Figs. 1E and 1F show only the blocking on syngeneic target cells

The $H-2^b$ anti-HLA-B27 CTL cells were able to kill HLA-B27-positive *human* cells and the level of specific lysis was dependent on whether the target cells were homozygous or heterozygous for HLA-B27. These results already show that the recognition of HLA-B27 xenogeneic MHC molecules by mouse CTL occurred in a non-H-2-restricted way. The same CTL population was perfectly able to kill HLA-B27-positive $H-2^b$ cells, whereas HLA-B27-negative cells were not lysed (Fig. 1B). Further, the same $H-2^b$ CTL population was examined on transgenic $H-2^q$ mouse cells (Fig. 1B). Again, cells expressing HLA-B27 were lysed, whereas no lysis was observed on HLA-B27-negative cells. No H-2 restriction in the recognition of HLA-B27 was found.

In following experiments, the approach was extended by HLA-B27-specific CTL elicited in $H-2^q$ mice. Figs. 1C and 1D show that CTL obtained from $H-2^q$ or $H-2^b$ mice were able to kill both the $H-2^b$- and $H-2^q$-transgenic target cells. Thus, on HLA-B27dTGM cells, the mouse class-I antigens are not used as restriction elements. To see whether this finding holds true in general, we extended the experiments with $H-2^b$ and $H-2^k$ SLA-PD1-transgenic mice. Again, cells expressing the xenogeneic MHC molecules were lysed independently of the H-2 haplotypes, whereas normal mouse cells were resistent (Figs. 1E and 1F). Specificity of the CTL population was tested by blocking the lysis on syngeneic transgenic cells by monoclonal antibody 74-11-10 specific for a polymorphic determinant on SLA-PD1 (Sachs, 1976; Pescovitz, 1984). Figs. 1E and 1F show that the anti-SLA-PD1 monoclonal antibody virtually abolish lysis of SLA-PD1-positive $H-2^b$ and $H-2^k$ cells, indicating a specific recognition of SLA-PD1 determinants.

Similar results (i.e. complete blocking of the bulk-CTL population) were obtained by blocking HLA-B27-specific CTL activity by the HLA-specific monoclonal antibody W6-32 and by cold target inhibition experiments (Kievits, 1989b).

It may well be that, using CTL bulk cultures, we have missed a low percentage of CTL precursors which recognize HLA only in the context of H-2. Therefore, we estimated in split-well analysis how many CTL cultures could be generated that were only specific for the syngeneic HLA-B27-transgenic cells. Fig. 2 shows that only few examples of H-2-restricted recognition were found. However, the reactivity pattern of the other cultures indicate that this reactivity is due to technical details, such as reactivity just below or above the cut-off value rather than by an H-2-restricted recognition of the xeno-MHC antigen.

Split-well analysis confirms the observation that the overwhelming majority of primary mouse CTL recognizes xeno-MHC antigens in a non-H-2-restricted way (Kievits, 1989a,b). Taken together, the results show that mice are perfectly able to generate xeno-MHC-specific CTL cells in primary response against class-I MHC-transgenic cells. The recognition of the xenogeneic MHC antigens by mouse CTL appeared not to be restricted by self-H-2 and occurred equally well on cells of different haplotype. Identical observations have been reported for SLA-transfected mouse cells (Bluestone, 1987) and recently also for HLA-transgenic mouse cells (Samaan, 1989; Taurog, 1989). However, a syngeneic H-2-restricted recognition of HLA antigens has frequently been reported by others (Maryanski, 1986a,b; Lemonnier, 1977; Holterman, 1986; Achour, 1986; Burakoff, 1977). These authors have concluded that HLA is generally recognized by mouse T cells as a nominal antigen in the context of a self-H-2 class-I-restricted element. This implies that xenogeneic MHC determinants are not recognized in the same way as allo-antigens. This conclusion has been based on the reactivity pattern of CTL obtained in secondary immune responses after priming *in vivo* with HLA-transfected cells. Xenogeneic responses induced in such a way

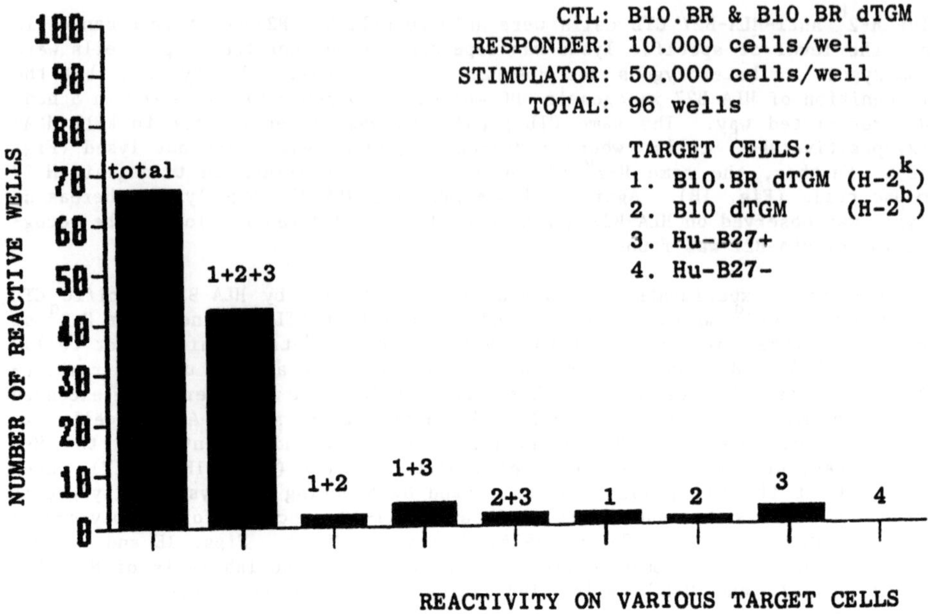

Fig. 2 - Reactivity of individual CTL microcultures on four different target cells defined by split-well analysis. Responder lymph-node cells were cocultured with irradiated stimulator cells in 200 µl culture medium. Forty five µl of each culture were tested on 5000 ^{51}Cr-labeled target cells

are often accompanied by responses against viral, tumor or minor histocompatibility antigens and are therefore not representative for a classical MHC class-I-specific response.

In the experiments described here, the availability of HLA-B27.2- and SLA-PD1-transgenic mice allowed us to test whether xenoreactive CTL, in analogy with alloreactive CTL, preferentially recognize their antigens in a non-H-2-restricted way. Our results show that xeno-MHC antigens are perfectly recognized in primary bulk cultures as alloantigens. These results parallel the primary immune response observed *in vivo* after grafting normal mice with tail skin from HLA-transgenic mice, resulting in a rapid graft rejection (Kievits, 1989a; Dill, 1988). The possibility remains that mouse bulk-CTL cultures recognize xenogeneic MHC as nominal antigens in the context of polymorphic restriction elements shared between various H-2 haplotypes (H-2-public determinants).

This possibility seems unlikely, since a) the induction of xenoreactive mouse CTL occurs without priming; b) the same CTL cultures recognized HLA antigens on human cells; c) recognition of viral antigens by bulk CTL is normally restricted by private H-2 determinants, and d) in split-well analysis almost none of the CTL cultures were reactive only with mouse cells. Furthermore, the finding that anti-SLA-PD1-specific CTL were blocked by a PD1-specific monoclonal antibody that does not crossreact with H-2 antigens also suggests that the recognition of xeno-MHC antigens is neither restricted by private nor by polymorphic MHC determinants. The same holds true for blocking experiments with HLA-specific monoclonal antibodies (Kievits *et al.*, 1989b).

Assuming that, in analogy to H-2-restricted CTL, xeno- and allo-specific CTL recognize also peptides presented by MHC molecules (Bjorkman, 1987a,b), we favor the view that MHC peptides are preferentially presented by their native MHC molecules. This does not rule out that some xenoreactive CTL are responding to foreign MHC, presented by self-MHC molecules. However, the frequency of such CTL will be extremely low in primary responses which makes them the exception rather than the rule.

ACKNOWLEDGEMENTS

We thank Dr. David Sachs for providing us with SLA-PD1 TGM breeding couples and hybridoma cells producing monoclonal antibody 74-11-10. Breeding pairs of B27-transgenic mice were generously provided by Dr. H.L. Ploegh and Dr. A. Berns. We thank Mrs. Jetty Gerritsen for typing the manuscript.

This work was supported by the Foundation for Medical Research (MEDIGON), which is subsidized by the Netherlands Organization for the Advancement of Pure Research (grant no. 509-110) and by the Dutch Multiple Sclerosis Foundation.

REFERENCES

Achour A, Begue B, Gomard E, Paul P, Sayagh B, van Pel A, Levy JP (1986) Eur J Immunol 16: 597-604

Bjorkman PJ, Saper MA, Samraoui B, Bennett WS, Strominger JL, Wiley DC (1987a) Nature 329: 506-512

Bjorkman PJ, Saper MA, Samraoui B, Bennett WS, Strominger JL, Wiley DC (1987b) Nature 329: 512-518

Bluestone JA, Pescovitz MD, Frels WI, Singer DS, Hodes RJ (1987) Eur J Immunol 17: 1035-1041

Burakoff SJ, Ratnofsky SE, Benacerraf B (1977) Proc Natl Acad Sci USA 74: 4572-4576

Dill O, Kievits F, Koch S, Ivanyi P, Hämmerling GJ (1988) Proc Natl Acad Sci USA 85: 5664-5668

Frels WI, Bluestone JA, Hodes RJ, Capecchi MR, Singer DS (1985) Science 228: 577-580

Holterman MJ, Engelhard VH (1986) Proc Natl Acad Sci USA 83: 9699-9703

Kievits F, Ivanyi P, Krimpenfort P, Berns A, Ploegh HL (1987) Nature 329: 447-449

Kievits F, Wijffels J, Lokhorst W, Boerenkamp WJ, Ivanyi P (1989a) Tissue Antigens, in press

Kievits F, Wijffels J, Lokhorst W, Ivanyi P (1989b) Proc Natl Acad Sci USA 86: 617-620

Kindred B, Schreffler DC (1972) J Immunol 109: 940-943

Krimpenfort P, Rudenko G, Hochstenbach F, Guessow D, Berns A, Ploegh HL (1987) EMBO J 6: 1673-1676
Lemonnier F, Burakoff SJ, Germain RN, Benacerraf B (1977) Proc Natl Acad Sci USA 74: 1229-1233
Maryanski JL, Pala P, Corradin G, Jordan B, Cerottini JC (1986a) Nature 324: 578-581
Maryanski JL, Accolla RS, Jordan B (1986b) J Immunol 136: 4340-4347
Pescovitz MD, Lunney JK, Sachs DH (1984) J Immunol 133: 368-375
Sachs DH, Leight G, Gone J, Schartz S, Stuart L, Rosenberg S (1976) Transplantation 22: 559-567
Samaan A, Gillet D, Chopin M, Degos L, Pla M (1989) Immunogenetics 29: 210-212
Taurog JD, Lowen L, Forman J, Hammer RE (1989) J Immunol 141: 4020-4023
Townsend ARM, Gotch FM, Davey J (1985) Cell 42: 457-468
Ziegler K, Unanue ER (1982) J Immunol 127: 1869-1874
Zinkernagel RM, Doherty PC (1975) J Exp Med 141: 1427-1436

Copy Number and the Presence of Human β_2-Microglobulin Control Cell Surface Expression of HLA-B27 Antigen in Transgenic Mice with a 25 kb B27 Gene Fragment

E. H. WEISS[1], G. SCHLIESSER[1], W. KUON[1], M. LANG[1], G. RIETHMÜLLER[1], F. KIEVITS[2], P. IVANYI[2], and G. BREM[3]

[1] Institut für Immunologie, Goethestraße 31, 8000 München, FRG
[2] Central Laboratory of the Netherlands Red Cross Blood Transfusion Service, Plesmanlaan 125, P.O. Box 9190, 1006 AD Amsterdam, The Netherlands
[3] Lehrstuhl für Molekulare Tierzucht, Veterinärstraße 13, 8000 München 22, FRG

INTRODUCTION

Major histocompatibility (MHC) class I antigens serve as receptors for viral or allogeneic peptides, and cytotoxic T lymphocytes only recognize foreign antigens bound to MHC molecules. Due to the outbred population, analysis of the interaction of peptides with individual MHC antigens and the recognition of the resulting complex by T cells, is difficult in the human system.

The association of the HLA-B27 antigen with seronegative spondarthritides renders the HLA-B27 gene unique among the class I gene family. Current evidence strongly argues that the HLA-B27 gene product itself plays a direct role in the pathogenesis of the HLA-B27 associated diseases. In order to study the receptor function of the HLA-B27 antigen for viral or microbial peptides and to identify putative enterobacterial, cross-reactive epitopes, we have introduced the gene encoding the heavy chain of the HLA-B27 molecule (B*2705 variant) into transgenic mice.

MATERIALS AND METHODS

<u>DNA microinjection into mouse embryos</u>, The 25 kb Sal I fragment of the HLA-B27 genomic clone cd2.6 (Weiss 1985) used for microinjection, was excised from low gelling temperature agarose gels (SeaPlaque, FMC), and isolated by electroelution. DNA was dissolved in TE (10 mM Tris/HCl pH7.5; 0.25 mM EDTA) passed through a 0.2 μm filter and diluted with TE to a concentration of 500 molecules/pl. 1-2 pl of DNA solution were microinjected into the male pronucleus of one cell embryos of outbred NMRI mice. Embryos surviving microinjection were reimplanted into oviducts of pseudopregnant female mice. Tail DNA was prepared from 3-wk-old mice by proteinase K digestion and phenol extraction, digested with EcoRI (5-10 μg), and electrophoresed on a 0.7% agarose gel. The Southern blot was hybridized with the genomic 6.7 kb EcoRI fragment (HLA-B27, see Fig.1).

I.K. Egorov/C.S. David (Eds.)
Transgenic Mice and Mutants in MHC Research

Expression of HLA-B27 in transgenic mice, Total RNA from spleen was prepared by extraction with guanidium isothiocyanate and centrifugation through cesium chloride step gradient. 10 μg of total RNA were separated on a formaldehyde gel, transferred to nylon membrane, and hybridized with the HLA-B specific probe (450 bp PvuII/PstI fragment of pMF18, (Pohla 1989)), the H-2 class I probe (2.2 kb 3' BamHI fragment of H-$2D^b$ (Weiss 1984)), and human actin cDNA. Hybridization of DNA and RNA blots was carried out overnight at 65°C according to the protocol of Church and Gilbert (1986).

Indirect immunofluorescence and FACS analysis, Total spleen or lymph node cells were isolated and $1x10^6$ lymphocytes were stained with supernatants of the following monoclonal antibodies at saturating conditions: ME1 (HLA-B7, -B27, -B22, B-14), 42.3.9.8. (rat anti H-2 class I), and BBM.1 (anti human $ß_2$-microglobulin ($ß_2$m) were obtained from the American Type Culture Collection; TM5 (anti HLA class I, (Thurau 1989)). ME1 and TM5 recognize the HLA-B heavy chain independently of the origin of $ß_2$m. FITC-conjugated goat anti-mouseIgG or FITC-conjugated goat anti-rat IgG for 42.3.9.8 were used as second antibodies. Immunofluorescence profiles were obtained by analysing 5,000 cells with a Becton Dickinson FACScan (Mountain View, CA).

Generation and assay of allogeneic and virus-specific cytotoxic T lymphocytes (CTL), For CTL assays mice of the same litter of a single founder were used and tested for syngeneic H-2 haplotype by Southern blot analysis using the 5'flanking K^b fragment (Weiss 1984) as the diagnostic polymorphic probe. Spleen cells of two negative, H-2 matched male siblings (875-5-38/40) were stimulated in vitro with irradiated spleen cells of male B27TGM 875-5-37. After 5 days, the CTL generated in the primary bulk culture were tested in the ^{51}Cr-release assay on different target cells: spleen cells from nontransgenic or B27 transgenic mice stimulated for 3 days with 2.5 μg/ml Concanavalin A (ConA); untransfected P815 cells (H-2^d), HLA-B27 positive P815 transfectant clone, and HLA-B51 positive P815 transfectant clone.
Cytotoxic T lymphocytes specific for influenza virus were generated in H-2 matched 875-5 progeny. A H-2^k/B27TGM was primed in vivo by intraperitoneal injection of 500 haemagglutination units (HAU) of influenza virus A/X79 in 0.5 ml phosphate-buffered saline (PBS). 21 days after influenza priming, the mouse was sacrificed and spleen cells ($5x10^7$) were stimulated with $5x10^7$ virus-infected irradiated autologous spleen cells. After five days, the responding CTL were collected and tested in the ^{51}Cr-release assay on influenza-virus-infected (18h) target cells. Autologous lymph node cells stimulated for 3 days with 2.5 μg/ml ConA were used as target cells. Infection of target cells was performed as described for stimulator cells. Responding CTL were incubated with ^{51}Cr-labeled target cells at the effector/target ratios indicated. The percentage of specific release was calculated as (release in the presence of CTL - spontaneous release)/(release in 10% Triton X-100 - spontaneous release) x 100.

CHARACTERIZATION OF HLA-B27 TRANSGENIC MICE

When producing HLA-B27 transgenic mice, we aimed at obtaining integration of only few copies and a physiological expression of the transgene, in order to avoid artefacts due to integration sites or abberrant expression. We therefore used a genomic 25 kb Sal I fragment encoding 7 kb 5' flanking and 14 kb 3' flanking region for microinjection (see Fig. 1).

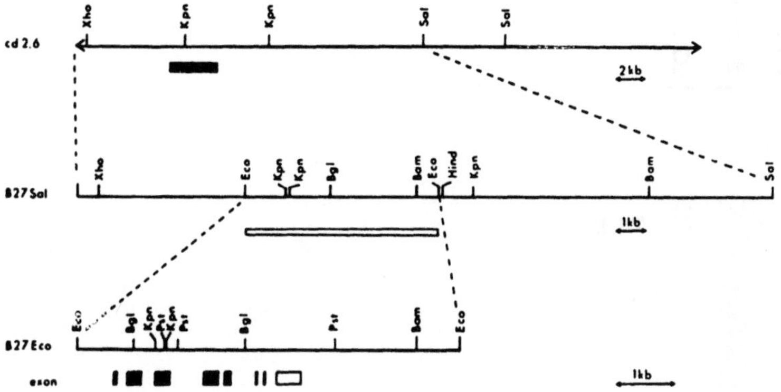

Fig. 1. Map of the HLA-B27 region and origin of the 25 kb Sal I fragment. On the top line is shown the restriction map of cosmid cd2.6, and on the middle line the range of the Sal I fragment. The open box depicts the extend of the 6.7 kb EcoRI fragment used for hybridization, which is shown enlarged on the bottom line.

Southern blot analysis of tail DNA of five independent founder mice and liver DNA of individual offspring indicates that two TGM (865-11, 864-12) carry only one copy, and one TGM each contains two (865-1), four (854-8) and eight (875-5) copies of the HLA-B27 gene (see Fig. 2). The transgene copy number was determined by quantitating the autoradiographic hybridization signal intensities. The homozygous lymphoblastoid B cell line LG2 (HLA-A2, -B27, -Cw1) contains two HLA-B27 gene copies. The larger EcoRI fragment, hybridizing to the HLA-B gene probe, contains the HLA-C gene. No cross-hybridization was obtained with mouse or other human class I genes, even after long exposure. The integrated single Sal I fragment in TGM 864-12 underwent a rearrangement, presumably in the 3' flanking region of the HLA-B27 gene. The founder mice were backcrossed either to human β_2-microglobulin (hβ_2m) transgenic CBA (H-2^k) mice (Krimpenfort 1987), kindly provided by Dr. H. Ploegh, Amsterdam or to C57BL/6 mice.

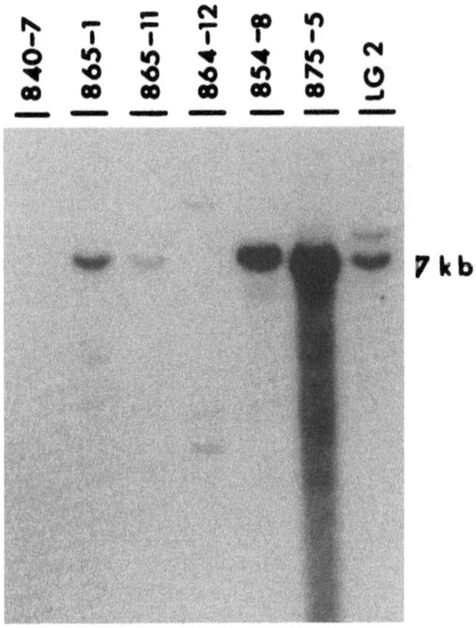

Fig. 2. Detection of HLA-B27 sequences in transgenic founder mice. 7 μg of tail DNA or genomic human DNA, isolated from the HLA-B27 homozygous B LCL LG2, were digested with EcoRI, and analyzed by gel electrophoresis and Southern blotting using the genomic 6.7 kb EcoRI B27 gene fragment as the hybridization probe.

The expression of HLA-B27 RNA correlates with the HLA-B27 gene copy number: Several HLA class I transgenic mice have been reported so far, including HLA-B27 TGM (B*2702, Krimpenfort 1987; B*2705, Taurog 1988), but only the HLA-B7 transgenic mice (Chamberlain 1988) have been analysed in detail for transcription of the HLA-B allele. We examined the tissue specific expression of the transgene in all five HLA-B27 TGM lines by Northern blot analysis of RNA extracted from different tissues of all mice. The expression of the HLA-B27 gene was similar to that of the endogenous murine class I genes. Expression was high in spleen, lung, and liver. As reported previously for HLA-B7 transgenic mice, the amount of HLA-B27 mRNA detected in spleen correlated with the HLA-B27 gene copy number. High copy number TGM (875-5) expressed elevated levels of HLA-B27 RNA comparable to the high transcription of HLA-B27 in the human lymphoblastoid B cell line (B LCL) LG2. The TGM lines containing one or two copies of the transgene, express less HLA-B27 mRNA, an amount similar to that of individual endogenous H-2 class I transcripts. One TGM (864-12) was found to express more HLA-B27 mRNA than expected from the one copy integrated. We do not know whether the rearrangement in the 3' region is responsible for the increased transcription In contrast to the HLA-B7 transgenic mice, HLA-B27 mRNA was not reduced in the liver of any TGM line. The reason for the reduced transcription of the HLA-B7 transgene (as 6.0 kb EcoRI/BamHI or 18 kb Hind III fragment) in the liver is not known. Either the regulation of the liver specific transcription varies between the two HLA-B alleles, which has not been observed in man so far, or the different fragments of HLA-B genomic DNA used for microinjection may account for the disparate transgene expression in the liver.

Thus, additional regulatory DNA elements might be localized 3' to the Hind III site of the HLA-B locus.

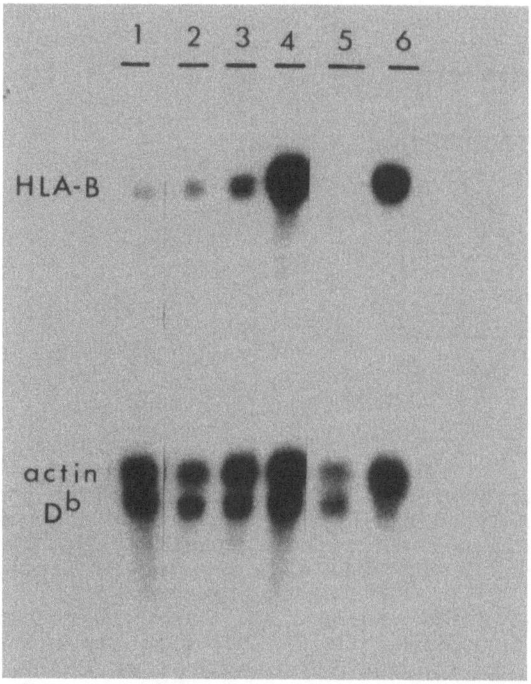

Fig. 3. Transcription of HLA-B27 in the spleen of transgenic mice. 10 μg of total spleen RNA (lane 1-5) and 5 μg of total RNA isolated from the human B LCL LG2 (lane 6) were analysed and hybridized either with the B-locus specific probe (top half) or with a human actin cDNA (1.8 kb band) and with a H-2 class I probe (1.6 kb) on the lower part. Lane 1: offspring of founder 865-11 (one B27 copy); offspring of founder 865-1 (two B27 copies) either heterozygous (lane 2) or homozygous (lane 3) for the transgene; lane 4: offspring of founder 875-5 (eight B27 copies); lane 5: nontransgenic sibling.

<u>Cell surface expression of HLA-B27 in transgenic mice</u>, Much controversy about the requirement of human β_2 microglobulin for cell surface expression of the HLA-B27 heavy chain has existed for some time (Rein 1987), which can now be explained by the usage of anti HLA class I reagents (e.g. W6/32), recognizing the HLA antigen only as complex with human β_2m. We analysed all TGM lines for cell surface expression of the HLA-B27 heavy chain on splenocytes and lymph node cells. We found that all TGM express the HLA-B27 antigen on the cell surface. The amount detected on the lymphocytes correlated with the HLA-B27 gene copy number (see Fig. 4, panel A and C). Even cells from mice containing only one copy of the transgene could be stained with anti HLA class reagents. To our knowledge, no other TGM lines with one or few HLA class I gene copies integrated, have been reported thus far to show cell surface expression of the transgene.

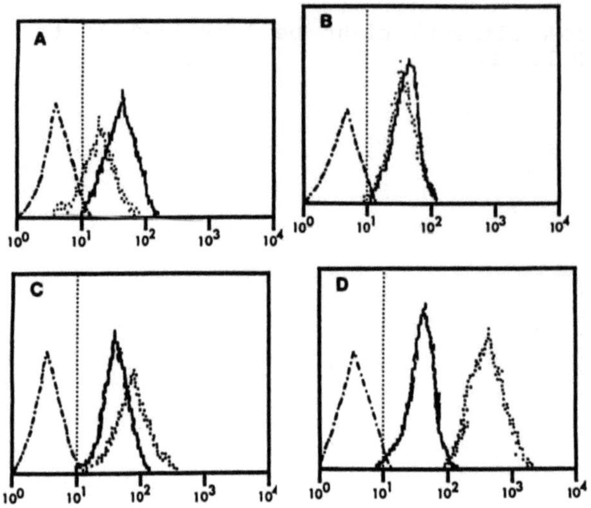

Fig. 4. FACS analysis of cell surface expression of HLA-B27 in transgenic mice. Splenocytes from singly and doubly transgenic and nontransgenic offspring were analysed by indirect immunofluorescence staining with mouse anti HLA class I monoclonal antibody (mAb) TM5 (dotted line) and rat anti H-2 class I mAb 42.3.9.8 (solid line). The fluorescence profile with TM5 of HLA-B27 negative offspring is shown in panel A and C as broken line. Panels A and B are the result for offspring of founder 865-11 encoding one HLA-B27 gene copy, and C/D the result for offspring of founder 854-8 with four HLA-B27 copies, either HLA-B27 singly (A/C) or HLA-B27/hβ_2 m doubly (B/D) transgenic mice. The x axes show the $\log_{10}$ relative fluorescence intensity and the cell number is presented on the ordinates.

<u>Coexpression of human β_2m increases cell surface expression of the HLA-B27 heavy chain</u>, All TGM founder mice were backcrossed to hβ_2mTGM, and singly and doubly transgenic offspring were analysed by FACS. The level of HLA-B27 cell surface fluorescence was 5-10 times greater for lymphocytes from doubly transgenic mice (see Fig. 4, panel B and D). Only the presence of hβ_2m allowed efficient cell surface representation of HLA-B27 in TGM expressing HLA-B27 mRNA at the same relative level as the endogenous H-2 class I genes.

IMMUNOLOGICAL FUNCTIONS OF HLA-B27 IN TRANSGENIC MICE

The availability of HLA-B27TGM allowed us to investigate the immunological function of a single HLA class I restriction element in isolation from the complexity of the human MHC. All functional tests were performed with singly HLA-B27TGM without human β_2m. Cell mediated cytotoxicity assays were performed to determine whether normal mouse lymphocytes could generate in vitro a primary CTL response against HLA-B27-expressing splenocytes (see Fig. 5A).

We observed a low but significant lysis of HLA-B27 positive target cells: either HLA-B27TGM splenocytes, or HLA-B27$^+$P815 and HLA-B27$^+$L929 (data not shown) transfectant clones. The fact that the xenoantigen is recognized on different H-2 haplotypes argues that the mouse T-cell repertoire does not use self-H-2 as restriction element, which has also been observed in other HLA-TGM (Chamberlain 1988; Kievits 1989; Samaan 1989; Hämmerling 1989).

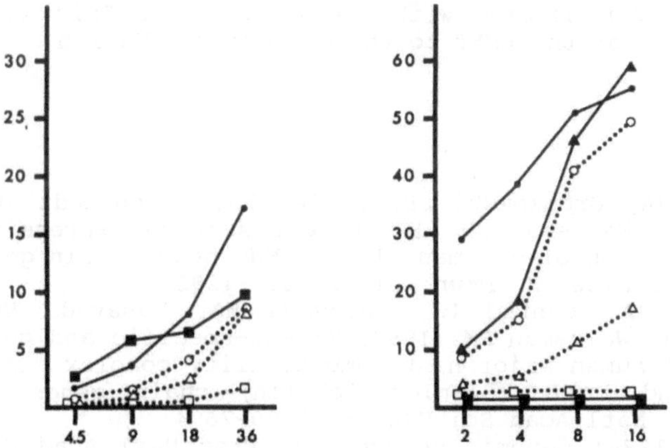

Fig. 5. HLA-B27 specific CTL responses in HLA-B27 transgenic mice. Panel A, primary HLA-B27 allospecific response: Closed symbols mark HLA-B27$^+$ target cells: ● HLA-B27$^+$P815; ■ HLA-B27TGM splenocytes. Open symbols represent HLA-B27$^-$ cells: ○ HLA-B51$^+$P815; △ P815; □ nontransgenic mouse splenocytes. Panel B, influenza virus-specific HLA-B27 restricted response: Closed symbols show the lysis of HLA-B27TGM cells, and the open symbols the corresponding cells of nontransgenic targets, either virus-infected or uninfected: ● H-2^k/B27TGM virus-infected; ▲ H-2^d/B27TGM virus-infected; ■ H-2^k/B27TGM uninfected; ○ H-2^k virus-infected; △ H-2^d virus-infected, H-2^k uninfected □

Moreover, we also investigated whether our HLA-B27TGM carrying the HLA-B*2705 subtype, are capable of using the HLA-B27 variant antigen as a restriction element for virus-specific T cell responses (as reported for other HLA class I TGM: Kievits 1987; Dill 1988; Le 1989; Kievits 1989). If that were the case, we could then address the question whether the human and mouse T cell repertoire recognizes similar peptides in context with the HLA-B27 antigen. As influenza virus recognition has been extensively studied in man and mouse, we used this virus to analyse anti-viral CTL (see Fig. 5B). For both the 875-5 and the 854-8 HLA-B27/hβ_2mTGM lines we obtained anti-viral CTL in vitro from transgenic mice after priming in vivo, which were able to kill virus-infected HLA-B*2705 positive mouse or human target cells (data not shown). No lysis was observed of non-infected cells. Both murine and human class I antigens were used as restriction elements. The absence of hβ_2m in the HLA-B27TGM had no influence on the generation of HLA-B27 restricted virus-specific CTL which could also lyse infected human HLA-B27$^+$ cells (data not shown; Kievits 1989).

The experiments reported here are an important step towards the establishment of HLA-B27 transgenic mice as a possible tool in the study of HLA-B27 disease associations, as the introduction of one genomic 25 kb HLA-B27 Sal I fragment results in a physiological expression of the HLA-B27 antigen.

Acknowledgements, The authors wish to thank Drs. E. Baas and H. Ploegh, Amsterdam, for providing the hβ_2mTGM breeder, and Dr. Judith Johnson and S. Förster for helping with the manuscript. This work was supported by a grant of the BMFT to the Genzentrum, Munich.

REFERENCES

Chamberlain JW, Nolan JA, Gromkowski SH, Kelley KA, Eisenstadt JM, Herrup K, Janeway CA, Weissman SM (1988) Cell surface expression and alloantigenic function of a human class I MHC heavy chain gene (HLA-B7) in transgenic mice. J Immunol 140: 1285-1292

Chamberlain JW, Nolan JA, Conrad PJ, Vasavada HA, Vasavada HH, Ganguly S, Janeway CA, Weissman SM (1988) Tissue-specific and cell surface expression of human major histocompatibility complex class I heavy (HLA-B7) and light (β_2-microglobulin) chain genes in transgenic mice. Proc Natl Acad Sci USA 85: 7690-7694

Church GM, Gilbert W (1984) Genomic sequencing. Proc Natl Acad Sci USA 81: 1991-1995

Dill O, Kievits F, Koch S, Ivanyi P, Hämmerling GJ (1988) Immunological function of HLA-C antigens in HLA-Cw3 transgenic mice. Proc Natl Acad Sci USA 85: 5664-5668

Hämmerling GJ, Hunt T, Dill O, Moreno J (1989) Cytotoxic T lymphocyte recognition of HLA-A2 antigens in normal and HLA-Cw3-transgenic mice. Eur J Immunol 19: 599-604

Kievits F, Ivanyi P, Krimpenfort P, Berns A, Ploegh HL (1987) HLA-restricted recognition of viral antigens in HLA transgenic mice. Nature 329: 447-449

Kievits F, Wijffels J, Lokhorst W, Ivanyi P (1989) Recognition of xeno-(HLA,SLA) major histocompatibility complex antigens by mouse cytotoxic T cells is not H-2 restricted: A study with transgenic mice. Proc Natl Acad Sci USA 86: 617-620

Kievits F, Wijffels J, Lokhorst W, Ivanyi P (1989) HLA expression and function in single and double HLA-B27-transgenic mice. Tissue Antigens 34: 1-14

Krimpenfort P, Rudenko G, Hochstenbach F, Guessow D, Berns A, Ploegh H (1987) Crosses of two independently derived transgenic mice demonstrate functional complementation of the genes encoding heavy (HLA-B27) and light (β_2-microglobulin) chains of HLA class I antigens. EMBO J 6: 1673-1676

Le AXT, Bernhard EJ, Holterman MJ, Strub S, Parham P, Lacy E, Engelhard VH (1989) Cytotoxic T cell responses in HLA-A2.1 transgenic mice. Recognition of HLA alloantigens and utilization of HLA-A2.1 as a restriction element. J Immunol 142: 1366-1371

Pohla H, Kuon W, Tabaczewski P, Dörner C, Weiss EH (1989) Allelic variation in HLA-B and HLA-C sequences and the evolution of the HLA-B alleles. Immunogenetics 29: 297-307

Rein RS, Seemann GHA, Neefjes JJ, Hochstenbach FMH, Stam NJ, Ploegh HL (1987) Association with β_2-microglobulin controls the expression of transfected class I genes. J Immunol 138: 1178-1183

Samaan A, Gillet D, Chopin M, Degos L, Pla M (1989) Mouse cytotoxic T cells can recognize HLA-B27 antigen without H-2 restriction. Immunogenetics 29: 210-212
Taurog JD, Lowen L, Forman J, Hammer RE (1988) HLA-B27 in inbred and non-inbred mice. Cell Surface expression and recognition as an alloantigen in the absence of human β_2-microglobulin. J Immunol 141: 4020-4023
Thurau SR, Wildner G, Kuon W, Weiss EH, Riethmüller G (1989) Expression and immunogenicity of HLA-B27 in high-transfection recipient P815: a new method to induce monoclonal antibodies directed against HLA-B27. Tissue Antigens 33: 511-519
Weiss EH, Golden L, Fahrner K, Mellor AL, Devlin JJ, Bullman H, Tiddens H, Bud H, Flavell RA (1984) Organization and evolution of the class I gene family in the major histocompatibility complex of the C57BL/10 mouse. Nature 310: 650-655
Weiss EH, Kuon W, Dörner C, Lang M, Riethmüller G (1985) Organization, sequence and expression of the HLA-B27 gene: A molecular approach to analyze HLA and disease associations. Immunobiol 170: 367-380

Expression and Function in a Transgenic Mouse of an H-2K^b Variant Gene Obtained by Site-Directed Mutagenesis in a K^{bm1} Mutant Mouse

M. GENDRON[1], M. NAKAGAWA[2], K. HASENKRUG[2], S. NATHENSON[2], and A. W. SEITZ[1]

[1] University of Massachusetts, Amherst, MA 01003, USA
[2] Albert Einstein College of Medicine, Bronx, NY 10461, USA

INTRODUCTION

H-2K^b is a member of the family of highly polymorphic mouse Major Histocompatibility Complex Class I genes. An <u>in vivo</u> mutant of the H-2K^b allele known as bm1, differs from wild type at three amino acid residues: #152 (glu->ala); #155 (arg->tyr); and #156 (leu->tyr) (Schulze et al 1983). These changes are sufficient to elicit mutant to parent skin graft rejections and to alter class I restricted viral presentation (Melief et al 1983). Transgenic mice were used to investigate the <u>in vivo</u> effect on the function of each of these three amino acids by transferring into bm1 mice single amino acid K^b variant genes obtained by site directed mutagenesis. The genes were wild type K^b except for changes at #152 (glu->ala), #155 (arg->tyr) or #156 (leu->tyr). One transgenic mouse line with the K^b #152 (glu->ala) variant gene in B6.C-H-2^{bm1} mice was studied extensively.

In addition to studying the structure-function relationship of the transgene, these transgenic animals are useful for the study of the regulation of transgene expression in the presence of the endogenous gene. By including large flanking sequences such that regulatory sequences are similar to those of the endogenous gene, and by analyzing the presence of the K^{bm1} and the K^b variant gene products on the cell surfaces of lymphocytes, it is possible to determine if there are any mechanisms that may regulate gene expression.

TRANSGENIC MOUSE LINE

Transgenic mice were produced by microinjection of DNA (Gordon et al 1980) into the pronucleus of one cell B6.C-H-2^{bm1} mouse embryos. The K^b variant gene used in this study was obtained by site-directed mutagenesis with a single amino acid change at 152 (glu->ala). The gene construct injected was a 7.0 kb HindIII - EcoRI fragment of the variant H-2K^b gene cloned into the plasmid pTZ19R and linearized with EcoRI.

A transgenic mouse line with this H-2K^b #152 (glu->ala) variant gene was established by mating mice carrying this transgene with B6.C-H-2^{bm1} mice. The integrated gene was in multiple copies and estimated to be 30-60 copies.

EXPRESSION OF THE H-2K^b #152 (glu->ala) VARIANT TRANSGENE

The expression of the transgene on lymphocytes was analyzed by monoclonal antibodies using FACScan analysis. The monoclonal antibody

I.K. Egorov/C.S. David (Eds.)
Transgenic Mice and Mutants in MHC Research

5F1-2-14 recognizes the H-2K^b #152 (glu->ala) variant gene product and not the bm1 gene product (Table 1). Fluorescence was detected in the C57BL/6 positive control with a mean fluorescence intensity (MFI) of 128 and not in the B6.C-H-2^{bm1} negative control. All sixty-four transgenic mice tested were found to express the transgene (MFI= 94). No significant difference can be detected between homozygous (MFI=93) and hemizygous mice (MFI=94) (Table 2).

Further investigations were done to examine the expression of the transgene in relationship to that of the endogenous bm1 gene. Since there is no monoclonal antibody available which is specific for bm1 and does not react with K^b #152 the monoclonal antibody Y-3, which recognizes both, was used (Table 1). Once again, no significant difference can be detected between homozygous (MFI=115) and hemizygous (MFI=123) transgenic lymphocytes. However, there is a significant difference ($p<0.05$) in the mean fluorescence intensity of transgenic lymphocytes, either homozygous or hemizygous, (MFI=119) as compared to bm1 lymphocytes (MFI=106) (Table 2).

While it is known that the expression of transgenes may be affected by parental imprinting (Swain et al 1987), no difference in the expression of this transgene in offspring derived from transgenic males or from a mating of transgenic male and female was detected. It is unlikely that the expression of this transgene in homozygous mice is affected by parental imprinting.

Table 1. Specificity of 5F1-2-14 and Y-3 monoclonal antibodies

	Monoclonal Antibody	
	5F1-2-14	Y-3
bm1	-	+
K^b	+	+
K^b#152 (glu->ala)	+	+

Table 2. Mean fluorescence intensities determined by Y-3 and 5F1-2-14 monoclonal antibodies

	Mean Fluorescence Intensity	
	5F1-2-14	Y-3
h/h	92.9^a	115.4^b
h/-	94.3^a	123.3^b
bm1	2.15^d	106.3^c

Numbers with the same superscript are not significantly different ($p<0.05$).
h/h, homozygous; h/-, hemizygous for transgene.

In summary, the level of transgene expression in lymphocytes is abundant and easily detectable. No difference in the profile of expression in homozygous versus hemizygous transgene expression could be detected. In addition, our data do not indicate that parental imprinting has any effect on transgene expression. When the total level of transgene and endogenous bm1 gene expression was assessed by a monoclonal antibody that recognizes both gene products, transgenic lymphocytes were found to express slightly higher than those from bm1 mice. While the level of H-2K products on lymphocytes is increased in transgenic mice, the lack of increase in homozygous over hemizygous mice indicates that there may be some mechanism that governs the level of this molecule on the cell surface.

FUNCTION OF THE FOREIGN GENE AS A TRANSPLANTATION ANTIGEN

Skin grafts were performed (Billingham and Medawar 1951) to determine whether the integrated K^b #152 (glu->ala) variant transgene is capable of functioning as a transplantation antigen. Tail skin from transgenic mice was grafted on to B6.C-H-2^{bm1} mice. Some recipient mice were primed with lymphocytes from transgenic donors 7-10 days prior to grafting, but no differences were seen between primed and unprimed recipients.

A summary of the skin graft results is presented in Table 3. All skin grafts from these transgenic mice were rejected by bm1 mice. The rejection times of these skin grafts are quite variable and many took longer to reject when compared with B6 (K^b) as the donor. Grafts from homozygous and hemizygous transgenic mice appear to be rejected by the bm1 mice at the same rates.

Table 3. Results of skin grafts

Donor	Recipient	No. of Grafts	Rejection	Time (Mean) Days	FACS
Controls					
B6	bm1	2	+	19-25 (31)	+
B6	B6	2	-	170*	+
bm1	bm1	3	-	170*	-
Transgenics					
h/h	bm1	5	+	13-84 (41)	+
h/-	bm1	6	+	15-84 (40)	+

* grafts not rejected after 170 days; h/h, Homozygous; h/-, Hemizygous for transgene K^b#152(glu->ala).
FACS (+) indicates cell surface expression of the transgene detected by FACScan analysis using 5F1-2-14 monoclonal antibody.

REFERENCES

Billingham RE, Medawar PB (1951) The technique of free skin grafting in mammals. J Exp Biol 28:385-401

Gordon JW, Scangos GA, Plotkin DJ, Barbosa J, Ruddle F (1980) Genetic transformation of mouse embryos by microinjection of purified DNA. Proc Natl Acad Sci 77:7380-7384

Melief CJM, Stukart MJ, deWaal LP, Kast WM, Melvold RW (1983) Specificity and regulation of cytotoxic T-lymphocyte responses analyzed with H-2 mutants. Transplantation Proc, vol XV 4:2086-2089

Schulze D, Pease L, Geier S, Reyes A, Sarmiento L, Wallace R, Nathenson S (1983) Comparison of the cloned H-$2K^{bm1}$ variant gene with the H-$2K^b$ gene shows a cluster of seven nucleotide differences. Proc Nat Acad Sci 80:2007-2011

Swain J, Stewart T, Leder P (1987) Parental legacy determines methylation and expression of an autosomal transgene: A molecular mechanism for parental imprinting. Cell 50:719-727

V. Expression and Function of Class II Genes in Transgenic Mice

Clonal Anergy in Transgenic Mice with Pancreatic Expression of MHC Class II I-E

L. C. BURKLY[1], D. LO[2], O. KANAGAWA[3], R. L. BRINSTER[2], and R. A. FLAVELL[1,4]

[1]Biogen., Inc., 14 Cambridge Center, Cambridge, MA 02142, USA
[2]Laboratory of Reproductive Physiology, School of Veterinary Medicine, University of Pennsylvania, Philadelphia, PA 19104, USA
[3]Lilly Research Laboratory, 3252 Holiday Court, La Jolla, CA 92037, USA
[4]Howard Hughes Medical Institute, Section of Immunobiology, Yale University School of Medicine, 424 FMB, New Haven, CT 06510, USA

INTRODUCTION

Experimental conditions whereby T cell tolerance can be established and maintained have been investigated and defined in many systems. However, the mechanisms of tolerance in these systems and their relevance to tolerance induction during normal T cell development *in vivo* is still not fully understood. The development of monoclonal antibodies (mAbs) to T cell receptor (TcR) Vβ chains, and the observation that certain Vβ chains are associated with reactivity to antigens such as I-E (Kappler, et al. 1987b) and Mls (Kappler, et al. 1988; MacDonald, et al. 1988) has provided *in vivo* evidence for clonal deletion. Thus, Kappler et al. showed that T cells utilizing the Vβ17a TcR gene which is identified by mAb KJ23 are generally reactive to I-E and are present in I-E negative mice but deleted in the thymus of I-E expressing mouse strains (Kappler, et al. 1987ab, Marrack, et al. 1988). The ability to monitor the presence or absence of antigen-reactive T cells offers a valuable alternative to functional measurements and provides a means to determine mechanisms of T cell tolerance. There is now *in vitro* evidence for clonal paralysis of T cells. T cell clones are reportedly inactivated upon exposure to antigen in the presence of human T cell clones (Lamb, et al. 1983), chemically modified spleen cells (Jenkins and Schwartz, 1987), purified MHC class II on planar membranes (Quill and Schwartz, 1987), class II^{+} keratinocytes (Gaspari, et al. 1988), and I-E^{+} islet cells (Markmann, et al. 1988). It has been proposed that antigen presentation to T cells in the absence of appropriate costimulatory signals results in T cell paralysis rather than activation. An *in vivo* situation in which MHC class II molecules are borne on nonlymphoid cell types would allow one to determine the consequences of this presentation for normal T cell function. There is to date, however, no evidence for *in vivo* paralysis of T cells.

We have previously reported INS-I-E transgenic mice (Lo, et al. 1988) which express MHC class II I-E on pancreatic islet β cells and kidney tubule epithelium but not in the thymus or cells of peripheral lymphoid organs. These mice bear no evidence of autoreactive T cell infiltrates, even after *in vivo* priming with I-E^{+} spleen cells (Lo, unpublished). The transgenic T cells were found to be specifically tolerant to the I-E molecule since they failed to mount a primary mixed lymphocyte reaction when cultured with conventional I-E^{+} splenic antigen presenting cells (APC). Here, we address the basis for tolerance to

I.K. Egorov/C.S. David (Eds.)
Transgenic Mice and Mutants in MHC Research

the I-E molecule in Ins-I-E mice by using two independent anti-TcR mAbs which detect I-E reactive T cells.

EXPERIMENTAL PROCEDURES

Mice

SJL/J and C57BL/6 mice were obtained from the Jackson Laboratory (Bar Harbor, ME). Transgenic animals previously described, Ins-I-E mice (Lo, et al. 1988), El-I-E (Lo, et al. 1989), -1.4kb E_{α}^{d} (v^{+}) (Widera, et al. 1987; Burkly, et al. 1989), and 107-1 (Widera, et al. 1987) were bred at the University of Pennsylvania. Transgenic mice were backcrossed several times to the SJL/J strain in order to obtain mice homozygous for the Vβ17a gene (Kappler, et al. 1987b) for analysis of T cells bearing $V\beta17a^{+}$ TcR. These mice were identified by DNA dot hybridization by the absence of Vβ8 TcR genes since Vβ8 genes are lacking in the SJL/J genome. For analysis of $V\beta5^{+}$ T cells, each transgenic line was backcrossed several times to the C57BL/6 strain such that they were Vβ5 $gene^{+}$ and Vβ17a $gene^{-}$. All progeny were also analyzed as previously described for the presence of the transgenes by DNA dot hybridization and only $transgene^{+}$ individuals were used.

Cytofluorographic Analyses

Vβ17a-bearing T cells were detected by immunofluorescent staining of lymph node T cells prepared by treatment with the B cell-specific mAb J11d (Bruce, et al. 1981) followed by rabbit complement (Pel Freez, Brown Deer, Wisconsin). The frequency of positive cells is calculated as the percentage of cells staining with cell culture supernatant containing the KJ23 mAb followed by Fitc-Goat anti-mouse Ig minus the percentage staining with second step alone. Vβ5-bearing T cells were measured by staining total lymph node cells with the MR9-4 mAb (Bill, Kanagawa, Palmer, submitted) followed by Fitc-Goat anti-mouse IgG Fc specific Ab (Jackson Immunoresearch, West Grove, Pa). The frequency of B cells detected with this secondary reagent is <3%. Stained cells were analyzed by flow cytofluorimetry on a Becton Dickinson FACStar and contour plots shown have logarithmic scales.

T Cell Proliferation

Stimulation of T cells with anti-Vβ17a KJ23 mAb was performed with B cell-depleted T cells or $L3T4^{+}$ T cell populations prepared by treatment with J11d and anti-Lyt 2 (53-6.7, rat IgG2a) mAb + complement. T cells were coated with KJ23-containing ascites fluid (1/10) for 30 minutes at 4° C. After washing to remove unbound antibody, the cells were then plated at 2×10^{6} responder cells per well in Costar 24 well plates (3524). 2.5×10^{6} syngeneic, mitomycin c treated splenic stimulators T cell depleted by anti-thy 1.2 (mouse IgM, New England Nuclear, Boston, MA) + complement were added per well. After 4 days, cells were transferred to Falcon 96 well plates (3072) at 2×10^{5} cells/well and triplicate wells pulsed with ^{3}[H]TdR on days 4, 5, or 6 for the final 6 hrs. of culture. Wells were harvested and data reported are shown for day 5. T cell proliferation induced by anti-Vβ5 MR9-4 mAb was performed with total spleen cells as follows. Falcon 96 well plates were coated with Goat anti-mouse IgG at 20μg/ml in Phosphate buffered saline (PBS) for 4 hrs. at room temperature. After washing, TcR-specific mAbs were added (50 µl of undiluted cell culture supernatant or ascites fluid at 1/100) for two hours at room temperature, wells washed again, and 2 x 10^5 spleen cells added per well. Cells were cultured for 5 days with ^{3}[H]TdR added during the final 6 hrs. Cultures were performed in RPMI 1640 medium containing 2 x 10^{-5} 2-mercaptoethanol, L-glutamine, penicillin and streptomycin, and 10% (v/v) fetal calf serum (FCS) (Hyclone Laboratories, Logan, Utah) or 1% (v/v) fresh normal mouse sera where indicated. Stimulation with anti-T3 mAb 145-2C11 (Leo, et al. 1987) was performed either with 1 x 10^5 responder cells and 2 x 10^5 T-depleted mitomycin c treated stimulator cells (Vβ17a experiments) or 2 x 10^5 whole spleen cells (Vβ5 experiments) added to wells of a 96-well plate. Cells were cultured for three days in the presence of 145-2C11 mAb containing cell culture supernatant and pulsed with ^{3}[H]TdR during the final 18 hrs.

RESULTS

To determine if the tolerance in Ins-I-E mice was the result of deletion of I-E reactive T cells, we used the Vβ17a-specific mAb KJ23 and the MR9-4 mAb which recognizes T cells expressing Vβ5.1$^+$ and Vβ5.2$^+$ TcR (Bill, et al. 1988) to measure the frequency of these I-E reactive populations amongst Ins-I-E peripheral T cells. No significant reduction in the frequency of Vβ17a$^+$ T cells was seen when we compared the immunofluorescence staining of lymph node T cells of transgenic mice with the SJL/J positive control strain (Table 1). Furthermore, the density of TcR molecules per cell on Ins-I-E and SJL/J Vβ17a$^+$ T cell populations appears indistinguishable (Fig. 1) since the intensity of Vβ17 staining is comparable. The proportion of KJ23$^+$ T cells which is CD4$^+$ (65-70%) vs. CD8$^+$ (30-35%) is also comparable in the two strains (Fig 1). Transgenic 107-1 mice expressing I-E in a normal fashion on thymus epithelium, mφ, DC and B cells have a low frequency of Vβ17a$^+$ T cells (Table 1) reflecting clonal elimination of these T cells in the thymus (Marrack, et al. 1988). Analysis of Vβ5$^+$ lymph node cells shows that a considerable frequency of Vβ5$^+$ T cells are present in INS-I-E mice as compared to B6 positive controls (Table 1) whereas 107-1

Table 1. T cells bearing I-E specific receptors are not deleted in INS-I-E Transgenic Mice

I-E Expression	Mouse	% Vβ17a$^+$	Mouse	% Vβ5$^+$
none	SJL/J	9.7 +/- 1.0 (5)	C57BL/6	5.7 +/- 0.9 (4)
pancreatic β cells, kidney	INS-I-E	8.3 +/- 0.5 (6)	INS-I-E	4.2 +/- 0.6 (4)
Pancreatic acinar cells	EL-I-E	8.6 +/- 0.9 (5)	EL I-E	6.1 +/- 0.1 (2)
wildtype	107-1	2.2 +/- 0.5 (5)	107-1	0.1 +/- 0.1 (2)

Percentage represents the average +/- SEM of n independent derterminations (n value is given in parenthesis).

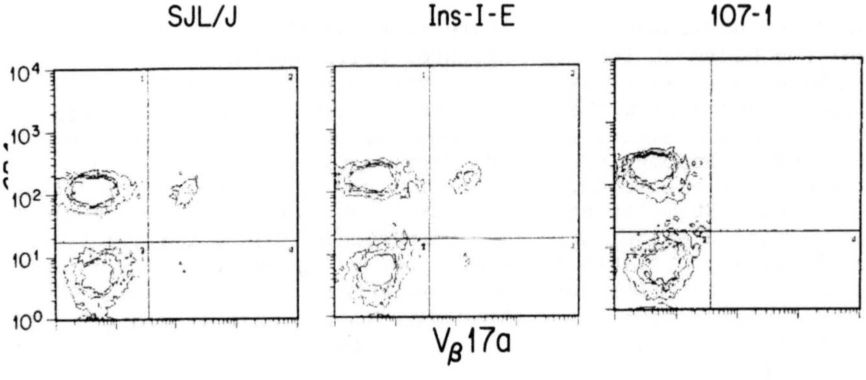

Fig. 1. Log fluorescence intensity of CD4 vs. Vβ17a on lymph node T cells

mice on the C57Bl/6 background exhibit elimination of the $V\beta5^+$ T cell population. Since 107-1 transgenic mice differ from C57BL/6 mice only by expression of the I-E transgene, this result formally demonstrates the deletion of Vβ5-bearing T cells is due to expression of the I-E molecule. Thus, *in vivo* tolerance in INS-I-E mice is not due to clonal deletion of these I-E reactive T cell populations.

Table 2. INS-I-E transgenic T cells are not activated by anti-Vβ17a mAb

Exp.	Responder	^{3}[H]Tdr Incorporation Induced by TcR Specific mAbs			
		Vβ17a-Specific mAb			T3-specific mAb
		-	+		
		cpm x 10^{-3}		cpm (SI)	cpm (SI)
1	SJL/J	3.3	24.9	21.6 (7.5)	
	INS-I-E	0.9	1.1	0.2 (1.2)	
	EL-I-E	2.7	17.5	14.8 (6.5)	
	107-1	0.4	0.3	-0.1 (0.8)	
2	SJL/J	13.0	55.1	42.1 (4.2)	69.2 (139.4)
	INS-I-E	6.5	10.1	3.6 (1.6)	186.4 (144.4)
3	SJL/J	11.5	67.9	56.4 (5.9)	288.2 (388.4)
	INS-I-E	14.7	21.0	6.3 (1.4)	268.5 (355.3)
	EL-I-E	30.0	140.2	110.2 (4.7)	272.2 (273.7)
	107-1	3.3	3.1	-0.2 (0.9)	266.0 (205.6)

Responder cells are B cell-depleted lymph node T cells (Exp. 1) and $L3T4^+$ lymph node T cells (Exp. 2 and 3).

To determine if T cells from the Ins-I-E mice were clonally paralyzed, we attempted to activate these I-E specific T cells. Normally, T cells can be activated by receptor crosslinking with mAbs which recognize TcR in the presence of Fc receptor$^+$ accessory cells, or by insolubilized anti-TcR antibodies in the presence of accessory cells or soluble costimulatory signals (Frelinger, et al. 1984; Kaye, et al. 1983; Manger, et al. 1985; Crispe, et al. 1985). If Ins-I-E T cells were rendered anergic by inappropriate exposure to I-E *in vivo*, then we reasoned that they should not proliferate on reexposure to antigen (Jenkins and Schwartz, 1987; Quill and Schwartz, 1987) or through receptor crosslinking with anti-TcR antibodies. Tables 2 and 3 show representative data generated with the Vβ17a- and Vβ5-specific mAbs and Figure 2 summarizes the results of many experiments. Stimulation of T cells with mAb 145-2C11 (Leo, et al. 1987) specific for the T3 molecule of the T3/antigen receptor complex was used as a positive control to demonstrate the general responsiveness of all T cell populations.

While anti-Vβ17a mAb could induce proliferation in SJL/J control T cells, Ins-I-E transgenic T cells showed little, if any, proliferation under the conditions described at various time points tested (Table 2), as would be expected for paralysed T cells. Only one responsive

Table 3. INS-I-E transgenic T cells are not activated by anti-Vβ5 mAb

Exp.	Responder	^{3}[H]TdR Incorporation Induced by mAb Specific for Vβ17a	Vβ5			T3	
		total cpm x 10^{-3}		cpm	(SI)	cpm	(SI)
1	C57BL/6	0.2	19.5	19.3	(97.5)	144.9	(52.8)
	C57BL/6	0.3	39.2	38.9	(130.7)	265.4	(62.7)
	INS-I-E	3.5	3.9	0.4	(1.1)	115.2	(58.6)
	INS-I-E	3.5	3.2	-0.3	(0.9)	98.1	(110.0)
	107-1	1.0	4.3	3.3	(4.3)	111.5	(30.3)
2	C57BL/6	0.5	11.6	11.1	(23.2)	133.4	(223.3)
	INS-I-E	1.8	4.6	2.8	(2.6)	200.6	(223.9)
	INS-I-E	2.3	18.8	16.5	(8.2)	128.9	(118.2)
	EL-I-E	1.1	5.9	4.8	(4.9)		
	107-1	1.0	1.2	0.2	(1.2)	267.0	(268.0)

In Exp. 1, cultures were supplemented with 1% (v/v) fresh NMS.

individual was found (Fig 2). 107-1 mice that have deleted most Vβ17a$^+$ T cells, should therefore not show T cell proliferation, and, therefore, serve as a negative control.

Stimulation of spleen cells with anti-Vβ5 mAb MR9-4 insolubilized on anti-Ig coated wells (Table 3, Fig. 2) revealed a positive response from C57BL/6 strain spleen cells. In this case, we used insolubilized Vβ17a as a negative control Ab. In contrast to B6 cells, proliferation of Ins-I-E T cells was absent or greatly diminished. Only 1/6 individuals gave a significant response (Table 3, Exp. 2).

We have also analyzed El-I-E transgenic mice which only express I-E extrathymically on pancreatic acinar cells; these mice are tolerant to I-E (Lo, et al. 1989) and have normal percentages of Vβ17a$^+$ T cells

(Table 1). In contrast to Ins-I-E mice, E1-I-E T cells proliferate in response to anti-Vβ17a at levels comparable to SJL/J T cells (Fig. 2). Though E1-I-E mice respond to anti-Vb17a at control levels, anti-Vβ5 induced low levels of proliferation when compared to B6 controls. Thus, Vβ5⁺ and Vβ17a⁺ T cells appear to differ in their ability to be tolerized by the I-E presented in E1-I-E mice.

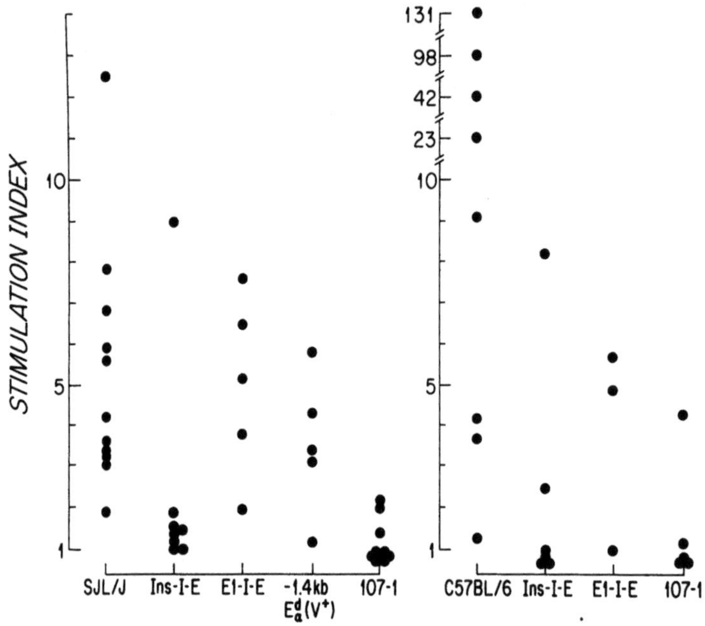

Fig. 2. Summary of TcR crosslinking experiments with anti-Vβ17a (left) and anti-Vβ5 (right). Each dot represents the Stimulation Index (SI) for an individual mouse where SI is the cpm incorporated with KJ23 mAb over incorporation without mAb (left) and incorporation with MR9-4 mAb over incorporation with KJ23 as negative control Ab (right)

DISCUSSION

We have used anti-TcR mAbs which detect I-E reactive T cell populations in order to examine I-E tolerant Ins-I-E mice. Our data indicates that Vβ17a⁺ and Vβ5⁺ T cells are not clonally deleted in Ins-I-E mice. Moreover, these T cells respond poorly, if at all, to receptor crosslinking. We believe, therefore, that exposure to I-E on nonlymphoid cells *in vivo* may have rendered the T cells anergic. Our results parallel those described for the induction of T cell unresponsiveness *in vitro*. In particular, Markmann, et al. (1988) demonstrated that T cell clones exposed to I-E⁺ Ins-I-E transgenic islet cells *in vitro* were clonally paralyzed. Previous indications that clonal paralysis can occur *in vivo* include tolerance established by injection of chemically fixed APC (Jenkins and Schwartz, 1987) and MHC class II L cell transfectants (Madsen, et al. 1988), and induction of *in vivo* tolerance to Mls (Rammensee, et al. 1989). However, our experiments provide, to our knowledge, the first evidence that paralysis of T cells does occur during normal T cell development in a transgenic mouse system.

Low levels of proliferation were rarely observed from Ins-I-E mice on TcR crosslinking with anti-Vβ17a (Fig. 2). This might be expected since individual Vβ17a$^+$ T cells differ in their reactivity to allogeneic forms of the I-E molecule (Marrack, et al. 1988; Kappler, et al. 1987b). It is possible that not all Vβ17a$^+$ T cells recognize the I-E expressed in INS-I-E mice ($E_{\alpha}{}^{d}E_{\beta}{}^{b}$) equally well and thus not all Vβ17a$^+$ T cells will be tolerized. Whether or not the same is true of Vβ5$^+$ T cells has not been determined. Although Vβ5$^+$ T cells are clonally eliminated in I-E expressing strains, Vβ5$^+$ T cell hybrids are not demonstrably reactive to I-E$^+$ spleen stimulator cells (Bill, et al. 1988). This observation raises a second explanation for the proliferation sometimes evident in Ins-I-E mice (Fig. 2). The concept has been put forth that allogeneic I-E molecules are I-E/peptide complexes (Matzinger and Bevan, 1977) with the I-E associated peptides dependent on the I-E$^+$ cell type (Marrack and Kappler, 1988) and thus possibly on the non-MHC background genes (Bill, et al. 1989). I-E/peptide complexes may vary between individual Ins-I-E mice due to the fact that this transgenic line is not inbred. The animals that we analyzed were backcrossed 4 to 6 times onto SJL/J or C57BL/6, respectively. Thus, in some mice, certain Vβ17a$^+$ or Vβ5$^+$ I-E reactive T cells may not be exposed to their appropriate ligand and thereby escape tolerance.

Our experiments with Vβ17a demonstrate clonal anergy of peripheral CD4$^+$ Ins-I-E T cells. Curiously, 30-35% of Vβ17a$^+$ T cells and the vast majority of Vβ5$^+$ T cells are CD4$^-$ CD8$^+$ in both normal and transgenic mice. Unlike normal CD8$^+$ T cells which respond to receptor crosslinking (Crispe, et al. 1985; Kanagawa, unpublished results), our experiments indicate that the CD8$^+$ transgenic T cells are apparently unresponsive as well. Although CD8$^+$ T cells are commonly restricted to class I, their use of particular Vβ genes may confer sufficient reactivity to I-E to be paralyzed. Paralysis could result from contact with I-E$^+$ islet cells which lack APC function (Markmann, et al. 1988) or from exposure to I-E in the absence of T cell help.

We have suggested the possibility that clonal anergy of I-E tolerant Ins-I-E T cells occurs by exposure to I-E$^+$ islet cells *in vivo*. We have previously reported, however, that both peripheral T cells and thymocytes of Ins-I-E mice are tolerant to I-E. This contrasts with studies reported by Morahan, et al. (1989) who showed that peripheral T cells but not thymocytes are tolerant to H-2K^b in transgenic mice expressing K^b only on islet cells. Given the tolerant Ins-I-E thymocytes, we must consider the possibility that I-E was transported from the pancreas or kidney and presented in processed form in the thymus to induce tolerance there. The possibility that an I-E derived peptide may be presented by I-A molecules has recently been reported (Murphy, et al. 1989). Nonetheless, our conclusion that tolerance in INS-I-E mice is not due to clonal deletion is still valid. It is possible that tolerance may occur in the thymus by deletion vs. paralysis depending on the quantity or site of presented antigen.

Ins-I-E mice contain significant numbers of potentially I-E reactive T cells. Yet it is possible that a very small percentage of I-E reactive T cells was deleted which is required for function of the remaining I-E specific population. We consider this possibility unlikely given results obtained with transgenic -1.4kb $E\beta^{d}(v^{+})$ mice which we have shown to have a readily detectable, partial reduction of $V\beta17a^{+}$ T cells (5.9 +/- 0.3 %; Marrack, et al. 1988). These mice express I-E on thymic epithelial cells (Widera, et al. 1987; Burkly, et al. 1989) but not on marrow-derived cells in the thymus or on peripheral cell types. Despite a significant reduction of $V\beta17a^{+}$ cells, T cells of these mice still proliferate in the presence of KJ23 mAb (Fig. 2).

We have also analyzed El-I-E mice for $V\beta17a^{+}$ and $V\beta5^{+}$ T cells and conclude that tolerance to I-E in these mice occurs by a non-deletional mechanism (Table 1). However, El-I-E $V\beta17a^{+}$ T cells do proliferate on receptor crosslinking (Table 2). The difference between INS-I-E and El-I-E $V\beta17a^{+}$ T cells (both lines backcrossed 4-6 times to SJL/J) indicates I-E expression in these two transgenic strains is not equivalent. Since both lines express I-E at high levels (Lo, et al. 1988; Lo, et al. 1989), our data argue in favor of a qualitative difference in I-E. We suggest that $V\beta17a^{+}$ T cells do not recognize the particular I-E/peptide complexes presented in El-I-E mice. In contrast, the $V\beta5^{+}$ population appears to be functionally impaired in El-I-E mice (Table 3 and Fig. 2).

In summary, we have analyzed two transgenic lines with nonlymphoid expression of MHC class II I-E. We demonstrate that tolerance to I-E occurs in these mice by a mechanism other than clonal deletion and provide functional evidence for clonal paralysis of normal I-E specific T cells *in vivo*.

ACKNOWLEDGEMENTS

We thank K. Hughes for technical assistance and M. Pomeroy for operating the flow cytofluorimeter. Work by R. Flavell was supported by the Howard Hughes Medical Institute, NIH, and Biogen; by L. Burkly was supported by Biogen and the American Cancer Society; by D. Lo and R. Brinster was supported by NIH.

REFERENCES

Bill, J, Appel, VB, Palmer, E (1988) An analysis of T-cell receptor variable region gene expression in major histocompatibility complex disparate mice. Proc Natl Sci USA 85: 9184

Bill, J, Kanagawa, O, Woodland, DL, Palmer, E (1989) The MHC molecule I-E is necessary but not sufficient for the clonal deletion of Vβ11-bearing T cells. J Exp Med 169: 1405

Bruce, J, Symington, FW, McKearn, TJ, Sprent, JJ (1981) A monoclonal antibody discriminating between subsets of T and B cells. J Immun 127: 2496

Burkly, LC, Lo, D, Brinster, RL, Flavell, RA (1989) I-E transgenic mice: A model system to dissect the regulation and function of MHC class II genes *in vivo*. Immunologic Research (in press)

Crispe, IN, Bevan, MJ, Staerz, U (1985) Selective activation of Lyt 2^{+} precursor T cells by ligation of the antigen receptor. Nature 317: 627

Frelinger, J, Sing, A, Infante, A, Fathman, CG (1984) Clonotypic antibodies which stimulate T cell clone proliferation. Immunol Rev 81: 21

Gaspari, AA, Jenkins, MK, Katz, SI (1988) Class II MHC-bearing keratinocytes induce antigen-specific unresponsiveness in hapten-specific TH1 clones. J Immunol 141: 2216

Jenkins, MK, Schwartz, RH (1987) Antigen presentation by chemically modified splenocytes induced antigen-specific T cell unresponsiveness *in vitro* and *in vivo*. J Exp Med 165: 302

Kappler, JW, Roehm, N, Marrack, P (1987) T cell tolerance by clonal elimination in the thymus. Cell 49: 273

Kappler, JW, Staerz, U, White, J, Marrack, P (1988) Self-tolerance eliminates T cells specific for Mls-modified products of the major histocompatibility complex. Nature 332: 35

Kappler, JW, Wade, T, White, J, Kushnir, E, Blackman, M, Bill, J, Roehm, N, Marrack, P (1987) A T cell receptor Vβ segment that imparts reactivity to a class II major histocompatibility complex product. Cell 49: 263-271

Kaye, JA, Porcelli, S, Tite, J, Jones, B, Janeway, CAJr (1983) Both a monoclonal antibody and antisera specific for determinants unique to individual cloned helper T cell lines can substitute for antigen and antigen-presenting cells in the activation of T cells. J Exp Med 158: 836

Lamb, JR, Skidmore, BJ, Green, N, Chiller, JM, Feldmann, MJ (1983) Induction of tolerance in influenza virus-immune T lymphocyte clones with synthetic peptides of influenza hemagglutinin. J Exp Med 157: 1434

Leo, O, Foo, M, Sachs, DH, Samelson, LE, Bluestone, JA (1987) Identification of a monoclonal antibody specific for a murine T3 polypeptide. Proc Natl Acad Sci USA 84: 1374

Lo, D, Burkly, LC, Flavell, RA, Palmiter, RD, Brinster (1989) Tolerance in transgenic mice expressing class II MHC on pancreatic acinar cells. J Exp Med (in press)

Lo, D, Burkly, LC, Widera, G, Cowing, C, Flavell, RA, Palmiter, RD, Brinster, RL (1988) Diabetes and tolerance in transgenic mice expressing class II MHC molecules in pancreatic beta cells. Cell 53: 159

MacDonald, HR, Schneider, R, Lees, RK, Howe, RC, Acha-Orbea, H, Festenstein, H, Zinkernagel, RM, Hentgartner, H (1988) T cell receptor use predicts reactivity and tolerance to MLs^a-encoded antigens. Nature 332: 40

Madsen, JC, Superina, RA, Wood, KJ, Morris, PJ (1988) Immunological unresponsiveness induced by recipient cells transfected with donor MHC genes. Nature 332: 161

Manger, B, Weiss, A, Weyand, C, Goronzy, J, Stobo, JD (1985) T cell activation: Differences in the signals required for IL-2 production by nonactivated and activated T cells. J Immunol 135: 3669

Markmann, J, Lo, D, Naji, A, Palmiter, RD, Brinster, RL, Heber-Katz, E (1988) Antigen presenting function of class II MHC expressing pancreatic beta cells. Nature 336: 476

Marrack, P, Kappler, J (1988) T cells can distinguish between allogeneic major histocompatibility complex products on different cell types. Nature 332: 840

Marrack, P, Lo, D, Brinster, R, Palmiter, R, Burkly, L, Flavell, RA, Kappler, JW (1988) The effect of thymus environment on T cell development and tolerance. Cell 53: 627

Matzinger, P, Bevan MJ (1977) Hypothesis: Why do so many lymphocytes respond to major histocompatibility antigens? Cell Immun 29: 1

Morahan, G, Allison, J, Miller, JFAP (1989) Tolerance of class I histocompatibility antigens expresses extrathymically. Nature 339: 622

Murphy, DB, Lo, D, Rath, S, Brinster, RL, Flavell, RA, Slanetz, A, Janeway, CAJr (1989) A novel MHC class II epitope expressed in thymic medulla but not cortex. Nature 338: 765

Quill, H, Schwartz, RH (1987) Stimulation of normal inducer T cell clones with antigen presented by purified Ia molecules in planar lipid membranes: specific induction of a long-lived state of proliferative nonresponsiveness. J Immunol 138: 3704

Rammensee, H-G, Kroschewski, R, Frangoulis, B (1989) Clonal anergy induced in mature $V\beta6^+$ T lymphocytes on immunizing $Mls\text{-}1^b$ mice with $Mls\text{-}1^a$ expressing cells. Nature 339: 541

Widera, G, Burkly, LC, Pinkert, CA, Boettger, E, Cowing, C, Palmiter, RD, Brinster, RL, Flavell, RA (1987) Transgenic mice selectively lacking MHC class II (I-E)antigen expression on B cells: An *in vivo* approach to investigate Ia gene function. Cell 51: 175

Role of Ia Antigens in Clonal Deletion of T Cells: Studies with Recombinant and Transgenic Mice

C. S. DAVID and G. ANDERSON[1]

[1] Department of Immunology, Mayo Clinic, Rochester, MN 55905, USA

The Mls locus was discovered nearly 20 years ago by Hilliard Festenstein (1973). In the beginning it was merely a curiosity as to what other genes besides the major histocompatibility complex were involved in the mixed lymphocyte reactions. In the 1970's, more laboratories started looking at this locus because incompatibility at the Mls locus activated much higher frequency of T cells than the MHC (Janeway and Katz, 1985). A major breakthrough came in the early 1980's in the elucidation of the Mls locus when two laboratories showed that specific T cell receptors can be clonally deleted in the context of certain MHC class II molecules and Mls locus (Kappler, Staerz, White and Marrack 1988; MacDonald, Schneider, Lees, Howe, Acha-Orbea, Festenstein, Zinkernagel and Hengartner 1988). T cells bearing $V_\beta 6$ and $V_\beta 8.1$ T cell receptors were reactive to Mls and in Mls-1^a positive animals these T cells were clonally deleted. A hierarchy of response to Mls was found in H-2 haplotype both in vitro and in vivo.

We initiated studies in our laboratory to determine the role of the various MHC class II genes in the clonal deletion of $V_\beta 6$ positive T cells. Congenic strains of mice in the C57BL/10 background expressing the various MHC class II genes were mated to DBA/1 and F_1 mice were analyzed for the percent of V_β positive T cells among the Thy-1 positive T cells. DBA/1 (H-2^q haplotype) is of the Mls-1^a genotype but since the H-2^q haplotype does not present Mls, the $V_\beta 6$ positive T cells are not deleted. In the first set of experiments strains with various alleles at the I-A region but expressing an intact I-E molecule were studied. As shown in Table 1, all of the strains expressing an intact I-E molecule completely eliminated $V_\beta 6$ positive T cells.

Table 1. Role of I-A and I-E in $V_\beta 6$ Deletion

Strain		I-A	I-E	$V_\beta 6$ in F_1, %
DBA/1 (Mls-1^a) x	B10.K	k	+	0.3 ± 0.1
	B10.D2	d	+	0.3 ± 0.1
	B10.P	p	+	0.1 ± 0.1
	B10.RIII	r	+	0.3 ± 0.1
	B10.PL	u	+	0.2 ± 0.1
	B10.SM	v	+	0.2 ± 0.1
	B10.A(5R)	b	+	0.2 ± 0.2
	B10.S(9R)	s	+	0.3 ± 0.1

I.K. Egorov/C.S. David (Eds.)
Transgenic Mice and Mutants in MHC Research

Next we looked at strains which are I-E negative expressing the different I-A genes. A^k and A^b showed the highest clonal deletion of $V_\beta6^+$ T cells whereas A^f and A^s were intermediate and A^q as previously known (MacDonald, Lees, Schneider, Zinkernagel and Hengartner 1988) showed no clonal deletion (Table 2).

Table 2. Role of I-A in $V_\beta6$ Deletion

Strain	I-A	I-E	$V_\beta6$ in F_1, %	$V_\beta6$ in parent, %
DBA/1 x B10.RKB	k	-	1.6 ± 0.2	7.4
B10	b	-	1.3 ± 0.3	8.5
B10.M	f	-	2.4 ± 0.1	6.3
B10.S	s	-	3.1 ± 0.1	8.8
B10.Q	q	-	4.7 ± 0.1	4.8

Next we looked at recombinants involving the f haplotype within the I region (Table 3). The recombinant strains having an E_α^q gene which does not generate an intact I-E molecule gave intermediate deletion like the parent strains, while two recombinants, B10.RFB2 with the E_α^k gene and B10.RFB3 with E_α^p gene deleted $V_\beta6$ cells more completely (Lafuse, Savarirayan, McCormick and David 1987).

Table 3. Role of E_α (A_β^f) Molecule in $V_\beta6$ Deletion

Strain	A_β	A_α	E_β	E_α	$V_\beta6$ in F_1, %	V_β in parent, %
DBA/1 x B10.M	f	f	f	f	2.4 ± 0.1	6.3
B10.RFB2	f	f	f	k	0.6 ± 0.0	7.1
B10.RFDB	f	f	f	q	2.4 ± 0.1	5.5
B10.RFB3	f	f	f	p	0.7 ± 0.1	7.5

B10.RFB2 and B10.RFB3 do not express intact I-E molecule since the E_β^f gene product is not expressed on the cell surface (Tacchini-Cottier and Jones 1988). Previous studies by cytotoxicity and immunoprecipitation have shown that the E_α polypeptide is not expressed in these strains (Lafuse, Savarirayan, McCormick and David 1987). When cells from these strains were analyzed by cell sorter, to our surprise we found expression of the E_α chain using the E_α specific antibody 14-4-4 (Ozato, Mayer and Sachs 1980). We believe that E_α^k and E_α^p in these recombinants associate with A_β^f to form a mixed isotype hybrid dimer, which is not only expressed on the cell surface but can function in vivo in the clonal deletion of $V_\beta6$ positive T cells in the context of $Mls\text{-}1^a$ locus.

Figure 1.

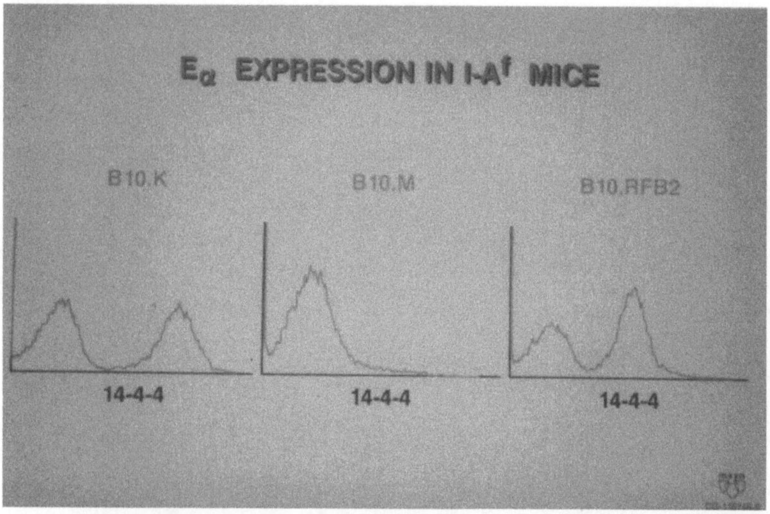

Next we looked at some recombinants involving the nonpermissive $H\text{-}2^q$ haplotype (Table 4). We found a recombinant mouse, B10.RQB3, with the E_α^k gene which clonally deleted $V_\beta6$ positive T cells, while B10.Q and B10.RQB2 with nonfunctional E_α^q gene showed no deletion.

Table 4. Role of E_α (A_β^q) in $V_\beta6$ Deletion

Strain	A_β	A_α	E_β	E_α	$V_\beta6$ in F_1, %	$V_\beta6$ in parent, %
DBA/1 x B10.Q	q	q	q	q	4.6 ± 0.2	4.8
B10.RQB2	q	q	q	q	4.9 ± 0.1	5.1
B10.RQB3	q	q	q	k	1.7 ± 0.1	6.8

E_β^q is also a nonfunctional gene (Tacchini-Cottier and Jones 1988). We again found expression of the E_α chain in these mice by cell sorter analysis using 14-4-4 (Figure 2). These mice were negative with monoclonal antibody 17-3-3 (Ozato, Mayer and Sachs 1980) which is specific for E_β chain indicating that E_β is not expressed in these mice. E_α^k associates with A_β^q to form a mixed isotype hybrid dimer.

Figure 2.

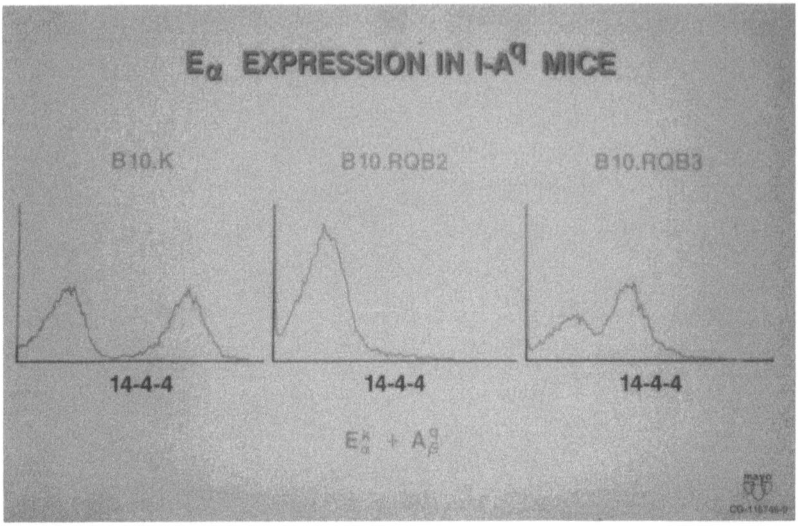

In order to confirm that this phenomena is due to the E_α chain and not some closely linked gene product we generated transgenic mice in the H-2^q haplotype expressing the E_α gene. The E^K_α transgene from B6.E^k_α transgenic mice (courtesy of Chris Benoist) was introduced into the B10.Q strain. These mice were then crossed to DBA/1 to produce q/q homozygous mice positive and negative for the E_α transgene and were tested for the expression of the $V_\beta 6$ positive T cells. The E_α positive mice showed clonal deletion of the $V_\beta 6$ positive cells while their E_α negative full sibs had the same number of $V_\beta 6$ positive T cells as the parent B10.Q strain (Fig. 3). This proved the role of E_α gene in the clonal deletion of $V_\beta 6^+$ T cells.

Figure 3.

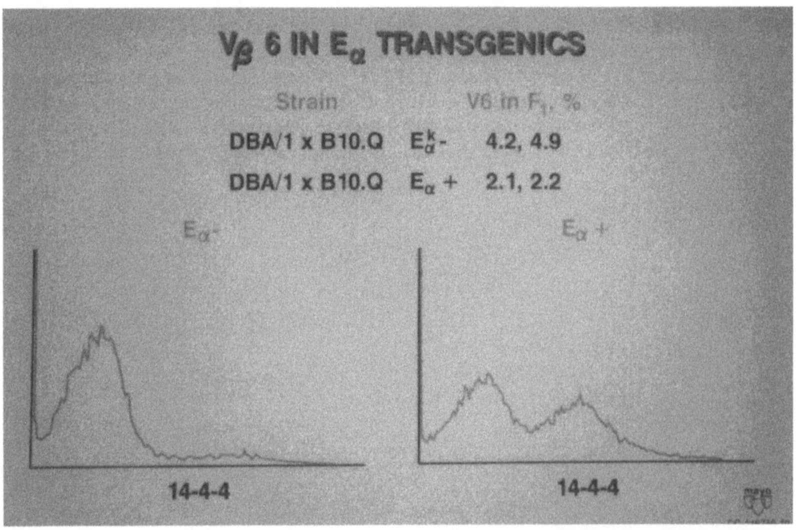

Next, we wanted to see whether H-2^q mice expressing an intact I-A^k molecule coded by transgenic A_α^k and A_β^k genes can delete $V_\beta 6$ positive T cells. I-A^k positive H-$2^{q/q}$ mice deleted 50% of the $V_\beta 6$ positive T cells as compared to normal I-A^k negative H-$2^{q/q}$ mice (Table 5).

Table 5. $V_\beta 6$ in I-A^k Transgenics

Strain	A_α^k $A_\alpha^k + A_\beta^{k*}$	$V_\beta 6$, %
(SWR x SJL)F_2	H-$2^{q/q}$, I-A^k -	6.3 ± 0.3
(SWR x SJL)F_2	H-$2^{q/q}$, I-A^k +	2.6, 2.7
SWR	H-2^q, Mls-1^a	7.1 ± 0.1
SJL	H-2^s, MlS-1^b	10.7 ± 0.1
(SWR x SJL)F_1	H-2^s, Mls-1^b	5.2 ± 0.3

On the other hand, H-2^q mice with the transgenic A_α^k gene do not show any deletion of $V_\beta 6$ positive T cells. The A_α^k gene is not expressed on these mice since A_α^k does not associate with the A_β^q. Thus, free alpha chains cannot present Mls to delete T cells (Table 6). B10 mice which express A_α^k transgene as a hybrid A_α^k A_β^b molecules can clonally delete $V_\beta 11$ positive T cells.

Table 6. $V_\beta 6$ TCR Expression in A_α^k Transgenics

Mice	A_α^k Gene SB	A_α^k Gene FACS	% $V_\beta 6$ of Thy-1+ T Cells
DBA/1 - H-2^q	-	-	3.8
[DBA/1 x B10-A_α^k+] x DBA/1			
24 - q/q	-	-	4.6
10 - q/q	-	-	4.4
11 - q/q	+	-	4.3
15 - q/q	+	-	4.3

MHC class II antigens also showed a clear hierarchy in positive selection. Strains expressing the intact I-E and permissive I-A genes showed the highest percent of $V_\beta 6$+ T cells in the context of Mls-1^b whereas I-E negative strains and nonpermissive I-A positive strains showed lower number of $V_\beta 6$ positive T cells. E_α in the context of nonpermissive q showed higher number of $V_\beta 6$ positive cells than the parent strains (Table 7).

Table 7. $V_\beta 6^+$ T cells in Mls-1^b Mice

Mice	H-2	% $V_\beta 6^+$ T Cells
B10.K	k	11.0 ± 0.1
B10.D2	d	11.0
B10.P	p	10.5
B10.RIII	r	10.1
B10.SM	v	9.6
B10.PL	u	10.0
B10.S	s	7.9 ± 0.4
B10	b	7.8 ± 0.1
B10 - A_α^k	b, b/k	9.0 ± 0.4
B10.M	f	6.4 ± 0.1
B10.Q	q	5.0 ± 0.2
B10.Q - E_α^-	q	5.5 ± 0.3
B10.Q - E_α^+	q, q/E_α	6.5 ± 0.3

Similarly, A_α^k transgene increased the number of $V_\beta 11$ positive T cells in I-E negative mice. B10 mice showed 5.5% $V_\beta 11$ positive T cells in their CD4+ population while B10 mice expressing A_α^k showed 50% higher $V_\beta 11$ positive T cells (Table 8).

Table 8. $V_\beta 11$ TCR Expression in A_α^k Transgenics

Mice	A_α^k	CD4/CD8	% $V_\beta 11$ of CD4+
B10	-	2.8	5.2
B10 - A_α^k (b/b)			
391	+	4.5	9.5
392	+	5.0	8.1
458	+	-	9.0
467	+	-	9.8
525	-	3.0	5.5
526	+	-	8.6
527	+	-	8.7
528	-	-	4.8

In conclusion, we have shown that mixed isotype hybrid dimers of E_α and A_β can be expressed in vivo in mice. This hybrid molecule is easily detectable in haplotypes which lack a functional E_β chain. In strains where E_β is expressed, E_α preferentially associates with E_β (Sant and Germain 1989). In the absence of the E_β chain, some E_α chains associate with A_β chains. In strains which lack the E_α expression, no $E_\beta A_\alpha$ hybrids have been found. It is possible that the number of A_β chains synthesized are higher than the A_α chains such that there are excess A_β chains to associate with E_α, but there is not excess A_α to associate with E_β. We have also shown that in strains expressing both the A and E genes, there is complete deletion of $V_\beta 6$ positive T cells. Among the different A alleles, there is a hierarchy of k being the most efficient, b, f and s intermediate and q nonpermissive. In the intermediate deleters such as f haplotype, addition of the E_α chain causes total deletion. In the nonpermissive A^q strains, addition of E_α results in 50% clonal deletion. This shows clearly that I-E molecule is not the dominant molecule in clonal deletion of $V_\beta 6$ or any other T cells. The A and the E molecules work synergistically in the clonal deletion. Further, we have shown that E_α plus A_β can cause clonal deletion of $V_\beta 6$ positive T cells, whereas A_α^q plus A_β^q cannot clonally delete T cells, indicating that in the $H\text{-}2^q$ haplotype the A_α chain is deficient in presenting the Mls antigen and by analogy we conclude that presentation of Mls antigen is the property of E_α and A_α chains. We propose that the V_β T cell receptor recognizes the Mls antigen in the context of the alpha chain of the class II molecules. The binding between Mls and class II probably occurs at a site outside the antigen binding pocket (Figure 4).

Figure 4.

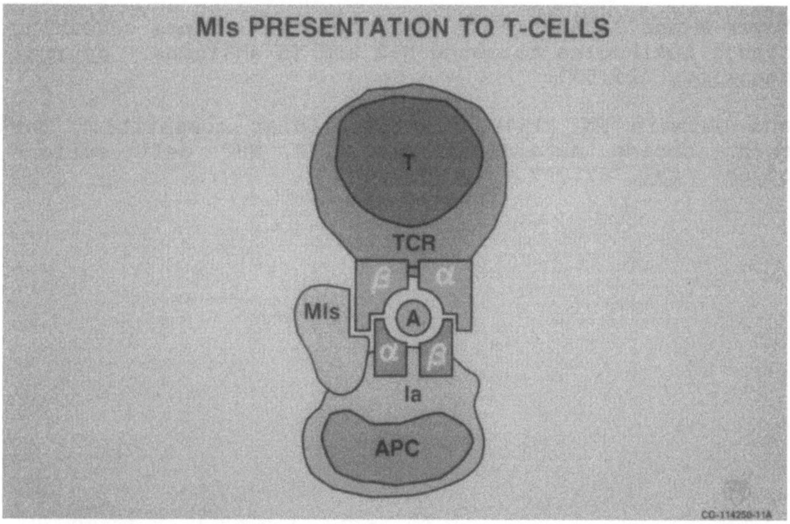

REFERENCES

Festenstein H (1973) Immunogenetic and biological aspects of in vitro lymphocyte allotransformation (MLR) in the mouse. Transplant. Rev. 15:62.

Janeway CA and Katz ME (1985) Immunobiology of the T cell response to Mls-locus-disparate stimulator cells. Unidirectionality, new strain combinations, and the role of Ia antigens. Journal of Immunology 134:2057.

Kappler JW, Staerz U, White J and Marrack PC (1988) Self-tolerance eliminates T cells specific for Mls-modified products of the major histocompatibility complex. Nature 332:35.

MacDonald HR, Schneider R, Lees RK, Howe RC, Acha-Orbea H, Festenstein H, Zinkernagel RM and Hengartner H (1988) T-cell receptor V_{β} use predicts reactivity and tolerance to Mls^{a}-encoded antigens. Nature 332:40.

MacDonald HR, Lees RK, Schneider R, Zinkernagel RM and Hengartner H (1988) Positive selection of $CD4^{+}$ thymocytes controlled by MHC class II gene products. Nature 336:471.

Lafuse WP, Savarirayan S, McCormick JF and David CS (1987) Identification of I-E alpha genes in H-2 recombinant mouse strains by F_1 complementation. Transplantation 43:297.

Tacchini-Cottier FM and Jones PP (1988) Defective E_{β} expression in three mouse H-2 haplotypes results from aberrant RNA splicing. Journal of Immunology 141:3647.

Ozato K, Mayer N and Sachs (1980) Hybridoma cell lines secreting monoclonal antibodies to mouse H-2 and Ia antigens. Journal of Immunology 124:533.

Sant AJ and Germain RN (1989) Intracellular competition for component chains determines class II MHC cell surface phenotype. Cell 57:797.

Transgenic Mice with MHC Class II Genes: The Use in the Study of Allelic α/β Chain Pairing and the Production of Engineered Mice with Mutant I-A Genes

B.-Y. WEI, J. MARTIN, S. SAVARIRAYAN, R. LITTLE, and C. S. DAVID[1]

[1] Department of Immunology, Mayo Clinic, Rochester, MN 55905, USA

Class II major histocompatibility molecules, or Ia antigens, are restriction elements for the regulation of immune response. Ia antigens also play a crucial role in the induction of immune tolerance, as well as in the susceptibility to certain diseases. The murine Ia molecules consist of A and E heterodimers, each composed of an α and a β polypeptide, both of which exhibit high degree of polymorphism. Sequence analysis of allelic Aα and Aβ chain revealed that most polymorphic residues are located in the N-terminal of α1 and β1 domains (Klein, 1986). These polymorphic residues may not only decide the binding of peptide, but also determine the pairing between α and β chains (Braunstein and Germain, 1987; Buerstedde, Pease, Nilson, Bell, Chase, Buerstedde and McKean, 1988a). Analyses of L cell and B lymphoma cell transfected with different α and β alleles have indicated that Aα and Aβ chain did not associate freely to form heterodimer on the cell surface. Haplotype-matched Aα:Aβ pairs gave high level of expression in transfected cells. Some haplotype-mismatched combinations (e.g. $A_\alpha^k A_\beta^b$) can be expressed on the cell surface, while other combinations were not detected (Germain and Malissen, 1986). Studies using hybrid gene constructs suggest that amino acid sequence in the N-terminal half of the β1 domain is responsible for the pairing defects of these α/β chain combinations (Sant, Braunstein and Germain, 1987; Buerstedde, Pease, Nilson, Bell, Chase, Buerstedde and McKean, 1988a).

Ia molecules are mainly expressed in B cells, macrophages and thymic epithelial cells. The expression of Ia on B cells is constititive, while the expression on macrophages and other types of cells can be modulated by a variety of agents, such as gamma-IFN, IL-4, Lipopolysaccharide, TNF, and prostaglandin (Sullivan, Claman, Nakamishi, Tsang, Wang, and Perlin, 1987). In the past few years, new data on DNA sequence of class II genes and transfection of cloned genes of both wild-type and mutated forms have allowed us further understanding of the structure/function relationship of the class II molecules. However, the ultimate evaluation of the significance of the class II molecules on immune response and establishment of a direct relationship between Ia structure and certain diseases requires the introduction of class II genes in the germline cells. Mouse strains with defined mutation or defect of class II genes, such as bm12 have contributed immensely to our understanding of the immune system. The development of transgenic technology has allowed us to produce engineered animals which will further reveal intricate cell interaction in the immune system.

I.K. Egorov/C.S. David (Eds.)
Transgenic Mice and Mutants in MHC Research

In this paper, we will describe our initial efforts to produce a transgenic mice with a wild-type A_α^k and a mutated A_β^k gene with specific d-allele substitution at position 63, 65-67 (Aβ MB gene).

MATERIALS AND METHODS

DNA Microinjection into Mouse Embryos

A_α^k genomic DNA obtained from cos H-2^k 8.1 Hind III fragment (>9 kb) and mutant A_β^k DNA (A_α^{k*} MB) in puc18 linearized with Hind III were used for microinjection (Hogan, Constantini and Lacy, 1986). Briefly, the DNA bands were excised from agarose gels, purified by glass powder and dialyzed against 10 mM Tris-HCl; 0.25 mM EDTA. DNA fragment was diluted to 1-2 μg/ml, passed through a 0.45 μm filter and injected into the male pronucleus of fertilized embryos from (SJL x SWR)F_2 mice. Embryos which had survived microinjection were implanted into oviducts of pseudopregnant foster mothers.

Identification of Transgenic Mice

Tail skin DNA was prepared from 2-3 week old mice (Hogan, Constantini and Lacy, 1986). Briefly, tails were minced in small pieces in a tube containing 500 μl NETS buffer (100 mM NaCl, 50 mM EDTA, 50 mM tris-Cl, pH 7.6, 1% SDS). The preparation was digested with proteinase K and RNase, extracted with phenol:chloroform and extensively dialyzed against TE (10 mM Tris, pH 7.6, 1 mM EDTA) for at least 24 hours at 4°C. Fifteen μg of DNA is digested with Pst I or Eco RI, electrophoresed in 0.7% agarose and transferred to nylon membrane. Hybridization of DNA blots were carried out in 65°C in 6x SSC/5x Denhart/0.5% SDS for 24 hours with A_α^k cDNA and A_β^k cDNA probes labelled by oligolabelling kit (Pharmacia).

Flow Cytometry

Cell surface expression of A_α^k and A_β^{k*} MB genes was detected with biotinylated mAbs 39J (anti-A_α^k) and 25-9-17 (anti-A_β^d) (ATCC), respectively. Spleen and lymph node cells were incubated with LPS at final concentration of 80 μg/ml (for 72 hours in medium), RPMI 1640 medium containing 5% horse serum, 5 x 10^{-5} M ß-mercaptoethanol, 100 μg/ml penicillin, 10% streptomycin, 2 mM glutamine. Lymphocytes from peripheral blood and cultured spleen cells were separated over a Ficoll-hypaque gradient, washed twice with 1% bovine serum albumin in PBS (pH 7.2) containing 0.1% sodium azide. The lymphocytes were incubated with mouse anti-mouse biotinylated mAbs 39J or 25-9-17 for 30 minutes at 4°C, washed twice and further incubated with fluorescein-streptoavidin (Becton Dickinson Monoclonal Center, Mountain View, CA) for 30 minutes at room temperature. After washing, the cells were run on FACS IV flow cytometer (Becton Dickinson, Mountain View, CA).

Skin Grafting

Skin grafts were performed using an adaption of the method by Billingham and Medawar (1951). Briefly, tail skin (approx. 1 cm^2) was taken from either a founder transgene positive mouse or a transgene negative mouse and transplanted on the backs of sex matched (SJL x SWR)F_1 mice. Bandages were removed after seven days and grafts were monitored daily until rejection.

Mixed Lymphocyte Reaction (MLR) After Skin Grafting

Mice were sacrificed on day 16 after the skin grafts were transplanted. At this time point the skin grafts from the founder transgenic mice, #20 (female) and #29 (male) were rejected by the age and sex matched (SJL x SWR)F_1 recipient. Control mice grafted with negative full sib tail skin showed no signs of rejection. The inguinal, axillary, brachial and mesenteric lymph nodes were removed from the recipient mice and single cell suspensions were prepared in RPMI 1640 medium plus .05 mM ß-mercaptoethanol, 10 U/ml penicillin, 1% streptomycin, 2 mM glutamine and 10% fetal bovine serum (Hyclone). The contaminating red blood cells were removed by ammonium chloride and the lymph node cells (LNC) were then resuspended to a concentration of 5 x 10^6 cells/ml. The splenic lymphoid cells (SPC) removed from donor mice were resuspended to 10 x 10^6 cells/ml and irradiated with 3300 rads and were used as stimulator cells.

For measurement of mixed lymphocyte reacton (MLR), 100 μl of the irradiated SPC were used per well and 100 μl of the LNC were used per well in a 96-well NUNC microtiter plate. Each condition was assayed in triplicate. Proliferation was quantitated by adding 1 μCi ^{3}H-methyl thymidine 48 hours after the start of the MLR. On day 3 of the MLR, the plates were harvested onto glass microfibre filters using a Skatron Cell Harvester. The filters were placed into tubes and 2 ml of Packard Ultima Gold scintillation fluid was added to the tube. The tubes were counted on a Beckman model LS7500 scintillation counter and the data are expressed as counts per minute (cpm).

RESULTS

1. Production of A^k_α, A^k_β MB, and A^k_α + A^k_β MB Transgenic Mice

The A^k_α genomic DNA used for microinjection is a Hind III fragment of cos H-2 8.1. The A^k_β MB gene is a mutant Aβ genomic DNA with d-allele substitution at position 63, and 65-67 (polymorphic region "B" of β1 domain). Region B of Aβ polypeptide has been confirmed to be a dominant serologic epitope for Aβ chain by previous studies (Buerstedde, Pease, Nilson, Buerstedde, Murphy and McKean, 1988b). We selected this mutant gene for production of engineered mice because of two reasons: (1) The substitution of d-allele at polymorphic region B will destroy all the binding of A^k_β-reactive mAbs so far tested, while at the same time the mutant A^k_β MB chain gained the reactivity with most A^d_β mAbs. Therefore, it would provide us with an easy method to follow the expression of the Aβ MB polypeptide on the cell surface. (2) Since the N-terminal part of the β1 domain, designated as the polymorphic region "A", is confirmed to be responsible for the α/β chain pairing (Braunstein and Germain, 1987; Buerstedde, Pease, Nilson, Bell, Chase, Buerstedde and McKean, 1988a), the d-substitution at region "B" did not impair the α/β chain pairing. Therefore, it was expected that A^k_β MB chain could form heterodimer on the cell surface with A^k_α transgene product. Both A^k_α and A^k_β MB genes have been successfully used in transfection assays. For production of transgenic mice, we microinjected A^k_α, A^k_β MB genes separately or in combination into the fertilized oocytes. The A^k_α genomic DNA was injected into the male pronucleus of (C57BL/6 x SJL)F2, the A^k_β MB gene into those of (B10.M x CBA/n)F2, and A^k_α + A^k_β MB into (SWR x SJL)F2 mice. The choice of recipient mice is based on the availability of the embryos and the consideration of possible chain pairing occurring in vivo.

Integration of the transgenes into the genome of recipient mice was confirmed by hybridization of Pst I - or ECoR I digested tail DNA with labeled A^k_α cDNA or A^k_β cDNA as probes. The copy number of transgenes was determined by comparison of the intensities of the bands representing the transgene to bands representing the endogeneous genes by densitometry.

2. A^k_α Transgenic Mice

Fifteen transgenic mice were obtained from 105 live offspring, as confirmed by Southern blots of tail DNA. The copy number of transgene ranged from 1-4 copies per cell. When crossed back to B10 mice, most of them except one line transmitted the gene in a Mendelian fashion, suggesting the injected genes had integrated into one chromosome. The offspring inherited the same copies of genes as the founder mice.

Cell surface expression of A^k_α transgene was analyzed in peripheral blood and spleen cells from several different lines of Aα-transgenic mice, using Aα-specific mAbs. The A^k_α expression was negative on spleen cells and peripheral blood cells in all the cases. However, if the spleen cells and lymph node cells were stimulated with LPS in vitro for 72 hours prior to FACS analyses, A^k_α transgene expression in H-2^b mice could be detected by Aα-reactive mAb 39J, indicating that A^k_α can pair with A^b_β chain, if there is abundant A^k_α chain or endogeneous A^b_β chain. The expression and percentage of Aα-positive cells after LPS stimulation varied from line to line and there seeemed no correlation between the level of expression and gene copy number.

To further study the allelic restriction of Aα/Aβ assembly, Aα transgenic mice derived from founder mice #26 were backcrossed with mice of b-haplotype with different background, such as B10.A, A.BY, and C3H.SW. Splenic lymphocytes from these mice were stimulated with LPS and then tested by cytofluorometry using mAb 39J. The results showed that LPS-stimulated spleen cells become A_α^k-positive regardless of background (Table 1). By crossing the A_α^k-transgenice mice to SJL (H-2^s), DBA/1 (H-2^q), B10.PL (H-2^u), B10.M (H-2^f), and B10.SM (H-2^v) to homozygosity, A_α^k expression was examined by immunfluorescence on LPS-stimulated spleen cells. A_α^k expression was only found on the cell surface of B10.PL mice (Table 1). It should be noted that the specificity of 39J used in this study was confirmed on a panel of independent haplotypes (k,b, s, q, v and u) using unstimulated as well as LPS-stimulated spleen cells. To exclude the possiblility that A_α^k was associated with free Eβ chain in H-2^b mice, LPS-stimulated spleen cells from A_α^k transgenic C57BL/6 were analyzed with 17.3.3, an Eβ-specific mAb. No staining with 17.3.3 was observed.

Table 1. Allelic Restriction on $A_\alpha A_\beta$ Pairing

Strain	Haplotype	DNA Integration	Cell Surface Expression*
B10	H-2^b	+	+
A.BY	H-2^b	+	+
C3H.SW	H-2^b	+	+
B10.PL	H-2^u	+	+
SJL	H-2^s	+	-
DBA/1	H-2^q	+	-
B10.M	H-2^f	+	-
B10.SM	H-2^v	+	-

*FACS analysis after LPS stimulation.

3. A_α^k + A_β^k MB Transgenic Mice

By previous in vitro transfection studies into M12.C3 B lymphoma cells, it was demonstrated that the mutations existing at the polymorphic region "B" did not impair the pairing between A_α^k and A_β^k MB. These d-allele substitution destroyed the binding with all of the A_β^k-reactive mAbs so far tested, but at the same time gained the binding with some of the A_β^d-reactive mAbs including 25-9-17. For production of the double transgenic mice, the A_α^k and A_β^k MB genes were mixed and then injected into the pronuclei of fertilized oocytes of (SJL x SWR)F2. We hoped that the expressed transgene products would preferentially pair with each other, instead of pairing with the endogeneous counterparts.

By Southern blot analysis of tail DNA, six transgenic founder mice were generated and in each case it was demonstrated that both genes were integrated. The copy number of the transgenes as measured by densitometry in different founder mice was found to be 1-4 for A_α^k gene and 2-5 for A_β^k MB gene, respectively. The results of Southern blots of second and third generation of the founder mice suggest that A_α^k and A_β^k MB genes were integrated at the same chromosome, since segragation patterns consistently showed that the two genes always segregated together (data not shown).

By cytofluorometry with Aα-reactive mAbs such as 39J and Aβ-reactive mAbs such as 25-9-17, we failed to detect the expected heterodimer Aα/Aβ MB on the surface of spleen cells or peripheral lymphoid cells of several founder mice. To investigate whether the A_α^k or A_β^k MB genes were transcribed in these transgenic mice, Northern blot analyses were performed. Unfortunatley, we failed to distinguish the endogenous and exogeneous Aα mRNA using allele-specific oligonucleotide. It was then that we tried to stimulate spleen cells from transgenic mice with LPS in order to enhance the expression of transgenes.Stimulation with LPS resulted in a large increase in the proportion of Ia-positive cells in CBA (H-2^k) and DBA/2 (H-2^d) mice, detected by 39J (anti-A_α^k) and 25-9-17 (anti-A_β^d), respectively. Spleen cells from transgenic mouse #35 (s/f) which was obtained by crossing one of the founder mice (s haplotype) with B10.M(f) were stimulated with LPS for 72 hours and then tested for the expression of Aα or Aβ MB genes. The spleen cells of #35 became surface-positive for both Aα and Aβ MB chains, detected by mAb 39J and 25-9-17, respectively. The spleen cells from negative littermates of transgenic mice remained negative for staining by 39J or 25-9-17 after LPS stimulation, although they had the same s/f background as #35 transgenic mouse. It should be mentioned that B10.M and SJL mice remain negative for staining by 39J or 25-9-17.

The offspring of these transgenic mice were screened with both Southern blotting of tail DNA and cytofluorometry of spleen cells after incubation with LPS. The transgenes-positive offspring identified by Southern blot consistently expressed both A_α^k and A_β^k MB on spleen cell surface after stimulation with LPS whereas the negative littermates did not show signs of expression of transgenes after LPS stimulation. Limited experiments have been done on gamma-INF induction of transgene expression on peritoneal cells; no significant enhancement of expression of transgene product was observed.

Tail skin from A_α^k + A_β^{k*} MB founder mice #20 (female) and #29 (male) was engrafted to sex matched (SJL x SWR)F_1 recipient mice. These grafts were both rejected with a survival time of approximately 15 days. The skin from a negative littermate was not rejected. Other control grafts from (SJL x SWR)F_1 mice were also retained by the recipients. Thus, the tail skin from the transgenic mice express $A_\alpha^k A_\beta^{k*}$ and function as an allograft in parental (SJL x SWR)F_1 recipients.

To complement the results of skin transplantation, mixed lymphocyte reactions (MLR) were performed in the above mice. As shown in Table 2, spleen cells of the A_α^k + A_β^{k*} MB founder transgenic mice were shown to induce a significant proliferation of lymph node

Table 2. Day 3 MLR Between A_{α}^{k} + MB Transgenic and Parental Strain (x ± SEM)

Founder #20 ♀			Founder #29 ♂		
	SPC+ only	83 ± 10		SPC+ only	236 ± 95
	SPC- only	158 ± 56		SPC- only	238 ± 193
#1+	LNC only	900 ± 162	#1+	LNC only	1,114 ± 113
	LNC + SPC-	5,141 ± 859		LNC + SPC-	14,590 ± 1,350
	LNC + SPC+	14,489 ± 1,276		LNC + SPC+	47,000 ± 1,019
#2+	LNC only	1,654 ± 863	#2+	LNC only	540 ± 78
	LNC + SPC-	7,413 ± 1,100		LNC + SPC-	5,253 ± 183
	LNC + SPC+	30,136 ± 1,541		LNC + SPC+	9,247 ± 352
#3+	LNC only	387 ± 54	#3+	LNC only	726 ± 48
	LNC + SPC-	3,198 ± 864		LNC + SPC-	6,741 ± 1,095
	LNC + SPC+	13,134 ± 2,584		LNC + SPC+	22,060 ± 3,028
#1-	LNC only	4,245 ± 2,205	#1-	LNC only	835 ± 179
	LNC + SPC-	5,538 ± 6		LNC + SPC-	7,839 ± 934
	LNC + SPC+	5,357 ± 578		LNC + SPC+	7,877 ± 482
#2-	LNC only	1,723 ± 219	#2-	LNC only	939 ± 166
	LNC + SPC-	5,854 ± 434		LNC + SPC-	9,218 ± 1,091
	LNC + SPC+	3,780 ± 409		LNC + SPC+	9,224 ± 357
#3-	LNC only	2,193 ± 383			
	LNC + SPC-	7,306 ± 532			
	LNC + SPC+	4,256 ± 226			

Lymph node cells from the skin-transplant recipients (SJL x SWR)F_1 were incubated with the donor spleen cells for mixed lymphocyte reaction. For details, see Materials and Methods. #1+, #2+ and #3+ are the results of MLR between recipient lymph nodes and spleen cells of transgenic mice #20 and #29. #1-, #2- and #3- are the results of MLR between the recipient lymph nodes and spleen cells of normal (SJL x SWR)F_1 mice.

cells from the engrafted (SJL x SWR)F_1 animals (#1+, #2+, and #3+), indicating that the transgene antigens can be successfully recognized as alloantigens by the primed T lymphocytes from (SJL x SWR)F_1 mice; whereas the lymph node cells failed to proliferate to the stimulation of spleen cells from control mice (#1-, #2-, and #3-). The MLR results suggest that the transgene products are expressed on spleen cells when primed T lymphocytes are used as an indicator. However, primary MLR, i.e. lymph node cells from ungrafted animals were unable to detect transgenic products on the stimulator cells (data not shown).

DISCUSSION

The transgenic mouse model offers us an unique opportunity to evaluate the allelic α/β heterdimer assembly in vivo. Our results indicate that in A_α^k-transgenic C57BL/6 mice, there is minimal A_α^k expression on the spleen cells and peripheral blood cells under normal condition. This seems to contrast with in vitro transfection results on L cells. However, 'in vivo' the trans-pairing ($A_\alpha^k A_\beta^b$) faces the competition of cis-pairing ($A_\alpha^b A_\beta^b$). Only in the case of abundant production of A_α^k and/or A_β^b, after LPS stimulation, the trans-pairing between A_α^k and A_β^b will occur.

Transcomplementation studies were performed also with several other H-2^b strains and the results indicate the association of A_α^k chain with an A_β^b chain, regardless of the strain background. It was also shown that A_α^k can pair with A_β^u, but not with A_β^s, A_β^f , A_β^q or A_β^v chains, even after LPS stimulation.

Although A_α^k and A_β^k MB genomic DNA were demonstrated to be integrated into the chromosome of the host, we failed to detect either A_α^k or A_β^k MB chains on the lymphoid cell surface prior to LPS stimulation. The A_α^k and A_β^k MB genomic DNA contain all the known regulatory regions at the 5'end (3kb and 3.5kb, respectively). This seems to contradict the data that in A_α^k + A_β^k double transgenic mice the lymphoid cells were strongly positive for A^k molecules, although the single A_α^k or A_β^k transgenics were very weakly positive (Le Meur, Waltzinger, Gerlinger, Benoist and Mathis, 1989). The lack of obvious existence of A_α^k/A_β^k MB heterodimer on the lymphoid cell surface in our double transgenic mice may have the following reasons: 1. The transcription rate of A_α^k or A_β^k MB genes was low prior to stimulation with LPS because of unknown reasons, such as position effect, low copy number, etc. Preliminary immunoprecipitation studies have shown that prior to LPS stimulation, little A_α^k/A_β^k MB heterodimers was detected even intracellularly. 2. In our A_β^k MB genomic DNA, the puc 18 sequences was not excluded for microinjection. This is because when the gene was prepared for in vitro transfection assay, the Hpa I restriction site at the 3'end of A_β^k genomic DNA was destroyed when cloned into puc 18 during a blunt end ligation process. In many cases it was reported that the plasmid sequence was "toxic" in suppressing the expression of transgene (Palmiter and Brinster, 1986). 3. The results of assembly of α/β heteroduplex in transfection assay may not parallel those obtained from transgenic mice.

LPS is a known B cell stimulant, and can induce a number of trans-acting factors (Sen and Baltimore, 1986; Blanar and Medawar, 1989; Dorn, Benoist and Mathis, 1989) which may override the above deficit to induce transgene expression. The results of skin transplantation suggested that transgene product (A_α^k and/or A_β^k MB) were indeed expressed on the cell surface and could induce allograft rejection in parental mice. In addition, we have observed deletion of $V_\beta 6^+$ T cells in these transgenic mice in the context of $Mls-1^a$, suggesting expression of transgene in the thymus.

The transgenic mice with defined mutations at either α or β chain will be a valuable tool in studies of Ia structure/function as well as on certain disease models.

REFERENCES

Billingham RE, Medawar PB (1951) The technique of free skin grafting in mammals. J Exp Med 28:235-239.

Blanar MA, Buekly LC, Flavell RA (1989) NF-KB binds within a region required for B-cell specific expression of the major histocompatibility complex class II gene E_α^d. Mol Cell Biol 9:844-846.

Braunstein NS, Germain RN (1987) Allele-specific control of Ia structure-function relationship. Proc Natl Acad Sci USA 84:2921-2925.

Buerstedde J-M, Pease LR, Nilson AE, Bell MP, Chase C, Buerstedde G, McKean DJ (1988a) Regulation of murine MHC class II molecule expression: Identification of A_β residues resonsible for allele-specific cell surface expression. J Exp Med 168:823-837.

Buerstedde J-M, Pease LR, Nilson MP, Buerstedde G, Murphy D, McKean DJ (1988b) Identification of an immunodominant region on the I-A_β chain using site-directed mutagenesis and DNA-mediated gene transfer. J Exp Med 167:473-487.

Dorn A, Benoist C, Mathis D (1984) New B lymphocyte-specific enhancer-binding protein. Mol Cell Biol 9:312-320.

Germain RN, Malissen B (1986) Analysis of the expression and function of class II major histocompatibility complex-encoded molecules by DNA-mediated gene transfer. Ann Rev Immunol 4:281-315.

Hogan B, Constantini F, Lacy E (1986) Manipulating the mouse embryo: A laboratory Manual. Cold Spring Harbor Laboratory.

Klein J (1986) Natural history of the major histocompatibility complex. John Wiley and Sons, New York.

LeMeur M, Waltzinger C, Gerlinger P, Benoist C, Mathis D (1989) Restricted assembly of MHC class II molecules in transgenic mice. J Immunol 142:323-327.

Palmiter RD, Brinster RL (1986) Germline transformation of mice. Ann Rev Genet 20:465-499.

Sant AJ, Braunstein HS, Germain RN (1987) Predominant role of amino-terminal sequences in dictating efficiency of class II major histocompatibility complex αβ dimer expression. Proc Natl Acad Sci USA 84:8065-8069.

Sen R, Baltimore D (1986) Multiple nuclear factors interact with immunoglobulin enhancer sequences. Cell 46:705-716.

Sullivan KE, Claman AF, Nakamishi M, Tsang SY, Wang Y, Peterlin BM (1987) A model for the transciptional regulation of MHC class II genes. Immunol Today 8:289-293.

T Cell Recognition of Major Histocompatibility Complex Antigens in HLA Class II Transgenic Mice

S. K. LAWRANCE, L. KARLSSON, J. PRICE, V. QUARANTA, Y. RON, J. SPRENT, and P. A. PETERSON[1]

[1]Department of Immunology, IMM8, Research Institute of Scripps Clinic, 10666 North Torrey Pines Road, La Jolla, CA 92037, USA

INTRODUCTION

The development of the self-tolerant/foreign antigen reactive immune system requires the selection of regulatory T helper cells such that those with specificity for self determinants are deleted or inactivated. In contrast, cells with specificity for foreign antigens mature and circulate in the periphery until activated (figure 1). A peculiar feature of these latter cells, however, is their requirement for simultaneous recognition of gene products encoded in the class II region of the major histocompatibility complex (MHC). The specificity of this recognition is determined by the particular forms of MHC antigens which are expressed in the thymus. Thus, thymic MHC molecules participate both in the negative selection of autoreactive T cells, as demonstrated by analyses of T cells bearing V beta 17a gene products (Kappler et al., 1987), and in the positive selection of MHC restricted antigen reactive T cells, as demonstrated in bone marrow chimera and thymus graft experiments (von Bohemer et al., 1975; Fink and Bevan, 1978) and, more recently, in experiments with T cell receptor (TCR) transgenic mice (Kisielow et al., 1988). Together, these processes generate a repertoire of mature T cells with receptors with exquisite specificity for MHC and antigen. These observations raise the questions: i) what portions of MHC molecules are recognized during these processes and ii) what are the influences of structural variations in MHC molecules on the make-up and functional properties of the mature T cell receptor repertoire. To address these questions we have generated in vivo mutations in MHC molecules.

Molecular studies of the class II region of both the murine and human MHC have demonstrated that each encodes a series of related alpha and beta genes. The products of these genes form heterodimers at the cell surface which serve both to present peptides derived from foreign antigens and to restrict interactions to those T cells expressing appropriate TcR and accessory molecules. In the more than eighty million years since their divergence from a common ancestor, the orthologous human and murine molecules have acquired replacements at between twenty and forty four per cent of their membrane distal domain amino acid residues. With the advent of the ability to transfer genes into the mouse germline by microinjection, these observations raised the possibility of determining the impact of these replacements on T cell recognition of MHC in positive and negative selection and, thus, on the functional characteristics of the mature T cell repertoire.

I.K. Egorov/C.S. David (Eds.)
Transgenic Mice and Mutants in MHC Research

HUMAN MHC TRANSGENIC MICE

Human MHC transgenic mice were generated by microinjection of CsCl gradient purified inserts cleaved from cosmid clones using the appropriate restriction enzymes. The clones employed were, for DP4 alpha/beta, cos SB-2 (Lawrance et al., 1985), for DQ3.2 alpha/beta, cosII-5407 (Jonsson et al., 1987), for DR alpha, cosmid T9C (Spies et al., 1985), for DR3 beta I, cosmid 4-1 (Rollini et al., 1985), and, for DRw53 beta III, cosmid B16A (Spies et al., 1985). Thymus and spleen cell RNAs from DNA positive animals were tested for transgene expression by Northern blotting. RNA positive animals were tested by two-dimensional gel electrophoresis and FACS analysis for expression of transgene protein products. Figure 2 shows the patterns obtained upon staining spleen cell suspensions from C57BL/10 DQ3.2 alpha/beta transgenic mice with monoclonal antibodies directed against a DQ3 specific epitope (IV-D-12) and an IA^b epitope (25-9-17). In both cases, a bimodal distribution of staining is observed, consistent with expression of DQ and IA molecules on

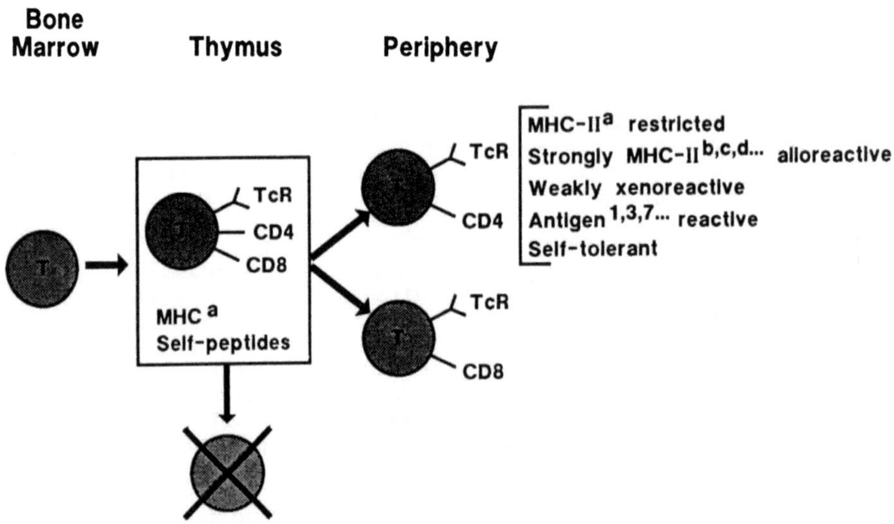

Figure 1. MHC molecules and self-peptides expressed in the thymus play a central role in the selection of the T cell repertoire. Pre-T cells from the bone marrow migrate to the thymus, express T cell receptors and accessory molecules, and interact with MHC and self-peptides. A small percentage survive this process and, in the case of class II restricted cells, mature and migrate to the periphery as $CD4^+$ $CD8^-$ TcR alpha-$beta^+$ T cells, predominantly of the T helper phenotype. At the population level, these cells are restricted in antigen specific responses to interactions with APC's bearing the MHC alleles expressed in the thymus. Other MHC alleles cause apparently antigen independent alloreactive or xenoreactive responses. Finally, the repertoire is tolerant to self determinants and responsive to a subset of the universe of foreign antigens. MHC also plays a central role in the selection of this subset of antigens.

splenic macrophages, B lymphocytes, and dendritic cells and their absence on splenic T cells. Different levels of expression were observed in lines derived from various transgenic founders. These appear to correlate with levels of RNA expression (resulting from effects of copy number and/or position of integration) and thus to reflect quantitative rather than qualitative distinctions in transgene expression.

Numerous attempts were made to detect immunological function mediated by the DQ molecules. These included antibody blocking of immune responses to complex protein antigens, antigen presentation to transgenic lymphocytes by human DQ3.2 cells and mixed lymphocyte culture (MLR) between DQ transgenic and normal C57BL/10 animals, and between human peripheral blood lymphocytes and DQ transgenic cells. In all cases, either no or exceedingly minimal effects were observed. These results, taken in light of analogous observations in human class I MHC transgenic mice (Ai-Xuan et al., 1989; Hammerling et al., 1989) suggest that the expression of human MHC gene products alone are not sufficient for the generation of human MHC restricted transgenic murine T cells. Considerations of the molecules involved in antigen presenting cell (APC): T cell interactions (figure 3) suggest the possibility that divergence of MHC determinants required for interaction with one or more accessory molecule (e.g. antigen, TcR, invariant chain, CD4) may be the cause of the DQ molecules failure to functionally interact with the murine immune system.

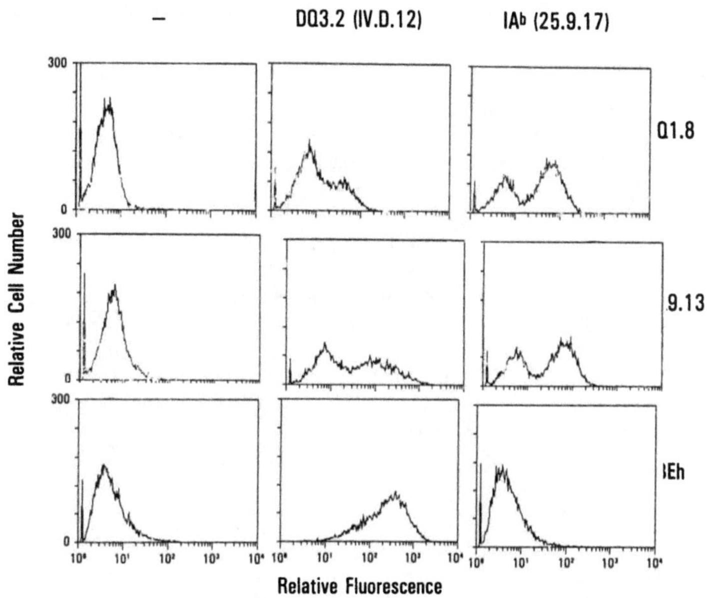

Figure 2. DQ3.2 heterodimers are expressed on DQ alpha beta transgenic spleen cells. Spleen cell suspensions from two DQ transgenic lines were stained with IV-D-12 (mouse IgG anti-DQ3) or 25-9-17 (mouse IgG anti IA^b) and FITC conjugated goat anti-mouse Fc of IgG. BEh is a DQ3.2 positive human lymphoblastoid cell line.

Experimental designs to test this possibility include a) the stepwise addition of the appropriate human accessory molecules and b) the stepwise replacement of human domains of MHC class II molecules with the corresponding murine domains. Both approaches, in addition to providing an avenue back to the original hypothesis regarding the influence of human molecules on the development of the murine T cell repertoire, provide potentially very informative means of experimentally manipulating the immune response, e.g. by modifying domains controlling the species incompatibility or by targeting expression of required accessory molecules to particular cell types involved in T cell differentiation and function.

HYBRID MURINE/HUMAN MHC CLASS II HETERODIMERS

Amongst the extensive series of H-2 haplotypes which have been characterized there exist haplotypes in which mutations have occurred which inactivate either the class II IE alpha gene ($H\text{-}2^{b,s}$), the IE beta gene ($H\text{-}2^{ap5}$), or both genes ($H\text{-}2^{f,q}$), (Mathis et al., 1983; Tacchini-Cottier et al., 1988). In the strains which express only one of the IE genes, the resulting alpha or beta chain polypeptides are expressed only intracellularly. These observations raised the possibility of generating hybrid molecules in which the product of the defective alpha or beta chain gene is supplied by the orthologous human DR gene. By this procedure we anticipated that it might be possible to generate hybrid heterodimers consisting of a

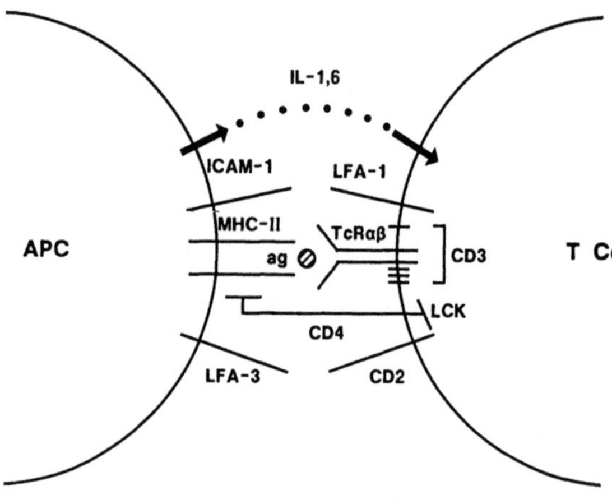

Figure 3. The interaction of T cells and APC involves a complex group of molecular interactions and signal transductions. Activation of T cells by APC's requires or is enhanced by the interaction of intercellular adhesion molecules; LFA-1 and ICAM-1 and LFA-3 and CD2; the secretion and uptake of interleukins IL1 and IL6; the interaction of CD4 (associated with the LCK protein kinase) with MHC class II; as well as the interaction of the CD3 TcR alpha beta complex with the antigen MHC complex.

human alpha chain combined with a murine beta chain and vice versa. If species conserved determinants, as discussed above, are confined to either the alpha or the beta chain, only one of the resulting hybrid molecules should be immunologically functional.

DR alpha RNA positive progeny of (C57BL/6 x SJL)F2 (IE alpha non expressing [IE^-]) founder animals were tested for expression of hybrid DR alpha:E beta molecules by 2d gel electrophoresis of immunoprecipitates of metabolically labelled spleen cells and FACS analyses of spleen cells stained with a panel of monoclonal antibodies directed against epitopes located on IE molecules. As shown in figure 4, DR alpha transgenic mice exhibit a unique pattern of staining distinct both from (B6 x SJL)F1 IE^- and B10.A IE^+ animals. Specifically, DR alpha transgenic IE^- spleen cells are stained by the 17.3.3 and ISCR-3 antibodies. In contrast, they are not stained by the 14.4.4 and Y17 antibodies. These data indicate that hybrid DR alpha:E beta molecules are expressed on the cell-surface of DR alpha transgenic spleen cells. Secondly, consistent with the substitution of 25% of amino acids in DR alpha relative to E alpha, they demonstrate that DR alpha:E beta molecules are serologically distinct from E alpha:E beta molecules.

To assess the immunological function of the hybrid DR alpha:E beta molecule, we took advantage of the immune response (IR) gene defects which are associated with mutations in IE genes. As shown in figure 5, IE^- mice are nonresponders to either moth cytochrome c (MCC) or random polymers of glutamic acid, lysine, and phenylalanine (GLØ). In contrast, both IE^+ and DR alpha transgenic mice give strong proliferative responses to these antigens. These results indicate that DR alpha:E beta molecules are functional in antigen

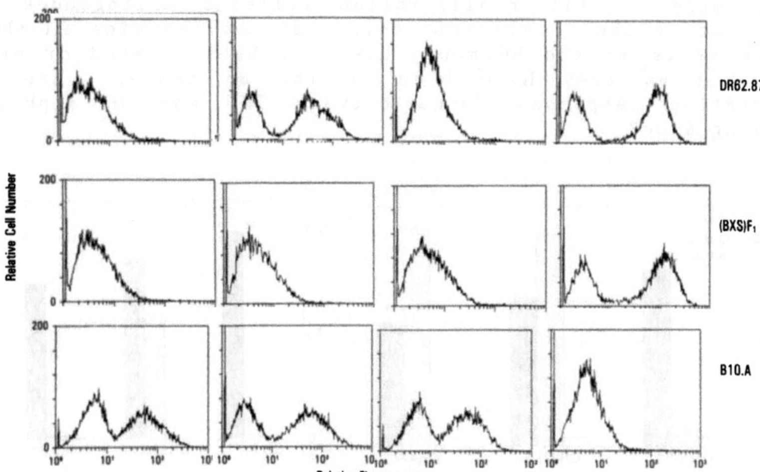

Figure 4. Hybrid DR alpha:E beta molecules are expressed on IE $alpha^-$ DR alpha transgenic spleen cells. Spleen cells from DR transgenic, (B6 x SJL)F1 (IE^-), and B10.A ($I\text{-}E^+$) animals were stained with IE specific monoclonal antibodies and FITC conjugated goat anti-mouse Fc of IgG.

presentation and that T cells are able to functionally interact with these hybrid molecules. Clearly, no species conserved determinants required for T cell recognition are located on the IE and DR alpha chains.

Having demonstrated the expression and function of the hybrid human:mouse class II MHC heterodimer, we may return to our original hypothesis dealing with the effects of structural mutations in MHC class II molecules on the development and function of T lymphocytes. The availability of monoclonal antibodies directed against variable region determinants on TcR's provides a direct means of testing this. As discussed above, it has recently been demonstrated that expression of IE molecules results in the clonal elimination in the thymus of T cells bearing receptors containing the V beta 17a gene product (Kappler et al., 1988) or the V beta 11 gene product. In contrast, those cells bearing the V beta 8 gene product are not deleted. We stained thymocytes and peripheral lymph node cells from DR alpha transgenic animals with monoclonal antibodies directed against each of these V gene products. The results of these experiments demonstrated identical deletions of V beta 17a and V beta 11 but not V beta 8 positive T cells from the periphery as are found in IE positive animals. Therefore, the structural distinctions between IE and DR alpha:E beta moelcules do not influence the expression of these elements of the T cell repertoire.

To more generally assess the composition of the T cell repertoires of IE as compared to DR alpha:E beta expressing mice we assessed the ability of each molecule to stimulate in reciprocal mixed lymphocyte cultures. It has previously been demonstrated that limited amino acid substitutions in the membrane distal region of the IA beta chain (IA^{bm12}) cause very strong proliferative responses in reciprocal primary MLR cultures with cells from otherwise syngeneic C57BL/6 mice. This proliferation reflects differences in the composition of the T cell repertoires of the two mice in that bm12 specific cells in the B6 mouse have not been deleted or otherwise inactivated as they have been in the B6 mouse. Are similar proliferative responses found between IE^+ and DR alpha:E beta expressing mice?

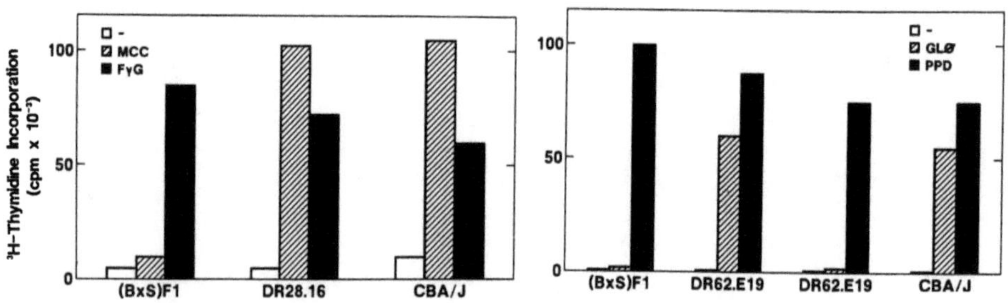

Figure 5. DR alpha:E beta molecules mediate antigen presentation of moth cytochrome c and glu-lys-phe antigens. Lymph node cells from immunized mice were restimulated in vitro with antigen. ^{3}H-Thymidine incorporation was measured on day 3. FGG = fowl gamma globulin, PPD = purified protein derivative.

To answer this question, mice which differ only in the expression of IE alpha and DR alpha were generated from the following crosses: (C57BL/6 x B10.A(18R))F1 (I-E negative), (C57BL/6 x B10.A(5R))F1 (I-E positive) (DR alpha $H2^{bb}$ x B10.A(18R))F1 (DR alpha:E beta positive). As shown in the left panel of figure 6, both I-E positive and DR alpha:E beta positive spleen cells act as strong alloantigens for I-E negative peripheral lymph node responders. In contrast, as shown in the center panel, while strongly stimulated by H2 incompatible SJL stimulators, IE positive responders are tolerant with respect to DR alpha:E beta positive stimulators. Conversely, as shown in the right panel of figure 6, DR alpha:E beta positive responders are equally tolerant to IE expressing stimulators. We conclude from these analyses that, with respect to alloreactivity and self tolerance induction, there are no significant differences in the peripheral T cell repertoires of DR alpha transgenic and IE positive animals.

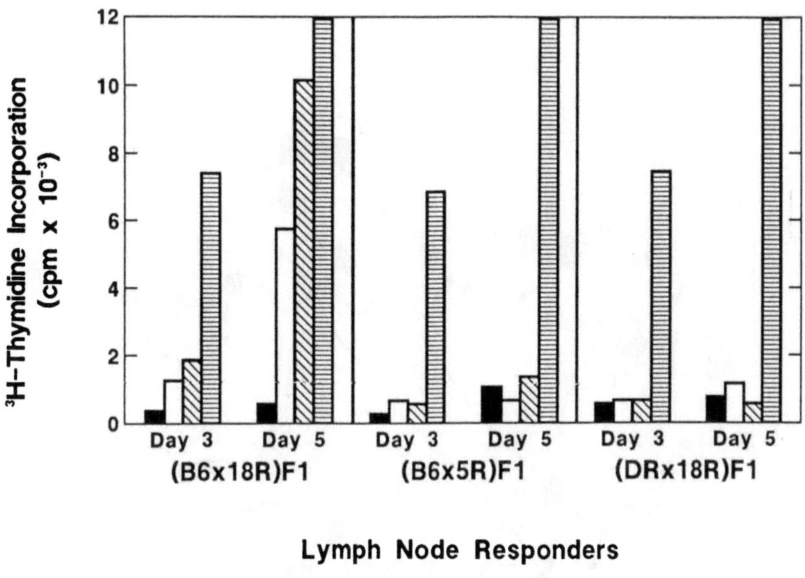

Figure 6. DR alpha transgenic and IE^{+} animals are cross-tolerant in reciprocal mixed lymphocyte culture. Total lymph node cell responders and irradiated spleen cell stimulators from each animal were co-cultured in all combinations. Stimulators and H2 haplotypes (K, A, [E alpha: E beta], D) were: closed, IE^{-} (B6 x 18R)F1 (b, b, [b:-], b/d); open, IE^{+} (B6 x 5R)F1 (b, b, [b:k], b/d); diagonal, DR alpha:E $beta^{+}$ (DR alpha transgenic x 18R)F1 (b, b, [b:-], b/d + DR); and horizontal, H2 incompatible SJL (s, s, [s:-], s). Values for day 5 SJL stimulators exceed the scale of the figure (average value = 59,000).

DISCUSSION

To summarize, we have investigated the influence of structural mutations of MHC class II molecules as generated by the expression of human MHC molecules in transgenic mice on the development and function of the T cell repertoire. Analyses of intact human alpha/beta MHC class II heterodimers suggest that the molecules are functionally inert. To dissect the nature of this incompatibility between human MHC and the murine immune system, we have generated hybrid murine/human heterodimers. Analyses of mice expressing a DR alpha:E beta heterodimer indicate that the molecule is functional in that it can mediate immune responsiveness to foreign antigens. These results suggest that any species specific T cell recognition elements (e.g. the CD4 recognition site) relevant to the inactivity of intact transgenic human heterodimers are likely to be located on the beta chain.

I-E molecules and DR alpha:E beta molecules differ at 25% of the 230 amino acids comprising the alpha chain. Many of these replacements are located in the membrane distal domain which is thought to contribute to the alpha-helical and beta-sheet structures which form a pocket in which antigenic peptides are bound and presented to T

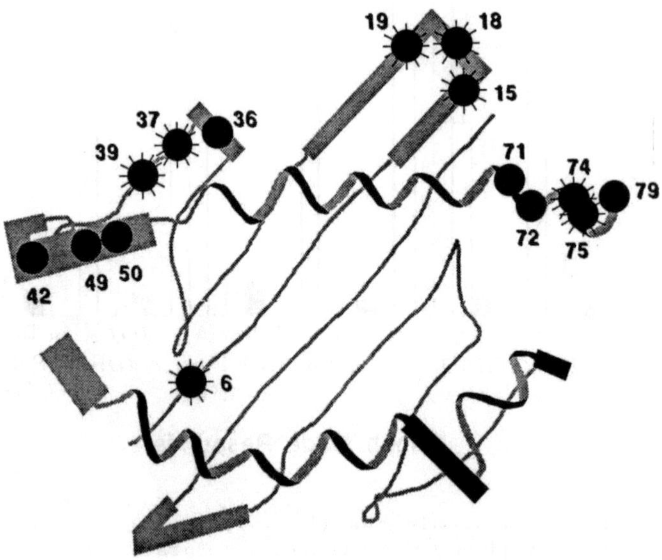

Figure 7. Hypothetical structure of the DR alpha:E beta molecule. Amino acid replacements in DR alpha relative to E alpha are indicated according to the model of the antigen binding site of class II MHC molecules proposed by Brown et al. (1988). Numbering is for IE and DR alpha. Replacements are 6:T->V, 15:L->N, 18:K->Q, 19:R->S, 36:I->M, 37:E->A, 39:S->K, 42:I->V, 49:A->G, 50:K->R, 71:D->E, 72:V->I, 74:K->T, 75:E->V, 79:N->Y, 82:D->I, and 83:A->T. Nonconservative substitutions are highlighted.

cells (Brown et al., 1988), (figure 7). The serological data presented above reinforce this notion that IE and DR alpha:E beta molecules are structurally quite distinct. In spite of these distinctions, however, we have found no functional differences in the two molecules with respect to antigen presentation, tolerance induction by clonal elimination, and alloantigenicity.

These observations have implications regarding the nature of T cell recognition of MHC molecules. It has been debated at some length whether the restriction of T cells to interaction with a particular MHC molecule involves a direct recognition of the MHC molecule itself or, rather, recognition only of the antigen as selected and/or presented by the MHC molecule (e.g. see Werdelin, 1987). The data we have presented do not resolve this issue. They do, however, impose severe constraints on the regions of MHC molecules which T cells may recognize. Specifically these regions are those which are conserved between IE molecules and DR alpha:E beta molecules. Inspection of figure 7 suggests the alpha chain alpha-helical region between residues 51 and 70 as a likely candidate recognition structure.

These conclusions can also be extended to shed light on the mechanism of alloreactivity. Does this interaction represent a direct recognition of polymorphisms in MHC molecules? If so, recognition, again, must be limited to those regions conserved between DR alpha and E alpha. The results argue strongly against alloreactivity as T cell recognition either of intact regions or processed peptides derived from regions of MHC molecules other than the previously mentioned alpha helical stretch. In contrast, the observation of concordance in antigens presented (i.e. MCC, GLØ, antigens involved in V beta 17a and V beta 11 elimination) by IE and DR alpha:E beta molecules, together with the observation of cross-tolerance between the two mice, are consistent with alloreactivity as the recognition of different subsets of self-peptides as selected by MHC molecules. It remains to be seen whether the two phenomena (antigen presentation and tolerance) can be separated. The present data are, however, consistent with the notion that T cells either do not recognize MHC molecules directly or recognize only very limited regions through the T cell receptor.

Finally, some mention should be made regarding the unique characteristics of the DR alpha and E alpha genes. Both genes exhibit extremely limited intraspecies polymorphism suggesting strong pressure for structural and functional conservation. The observation that DR alpha can faithfully replace E alpha indicates that this function is not only conserved within species but has also been conserved over 80 million years since the divergence from the common ancestor of man and mouse. Paradoxically, however, multiple independent mutations inactivating IE expression are tolerated both by inbred and wild mice. Furthermore, expression of IE molecules causes the somatic elimination of a significant portion of the V beta TcR repertoire. It remains unclear how these observations are to be reconciled. It seems likely, however, that additional experimentation with mutant and transgenic animals will shed further light on these questions.

REFERENCES

Ai-Xuan TL, Bernhard EJ, Holterman MJ, Strub S, Parham P, Lacy E, Engelhard VH (1989) Cytotoxic T cell responses in HLA-A2.1 transgenic mice. J. Immunol. 142:1366-1371.

Brown JH, Jardetsky T, Saper MA, Samraoui B, Bjorkman PJ, Wiley DC (1988) A hypothetical model of the foreign antigen binding site of class II histocompatibility molecules. Nature 332:845-850.

Fink PJ, Bevan M (1978) H-2 antigens of the thymus determine lymphocyte specificity. J. Exp. Med. 148:766-775.

Hammerling, GT, Hunt T, Dill O, Moreno J (1989) Cytotoxic T lymphocyte recognition of HLA-A2 antigens in normal and HLA-Cw3-transgenic mice. Eur. J. Immunol. 599-604.

Jonsson AK, Hyldig-Nielson JJ, Servenius B, Larhammar D, Andersson G, Jorgensen F. Peterson PA, Rask L (1987) Class II genes of the human MHC. J. Biol. Chem. 262:8767-8777.

Kappler JW, Roehm N, Marrack P (1987) T cell tolerance by clonal elimination in the thymus. Cell 49:273-280.

Kilielow P, Teh HS, Bluthmann H, von Boehmer H (1988) Positive selection of antigen-specific T cells in thymus by restricting MHC molecules. Nature 335:730-733.

Lawrance SK, Das HK, Pan J, Weissman SM (1985) The genomic organization and nucleotide sequence of the HLA-SB (DP) alpha gene. Nucl. Acids Res. 13:7515-7528.

Mathis DJ, Benoist C, Williams VE, Kanter M, McDevitt HO (1983) Several mechanisms can account for defective E alpha gene expression in different mouse haplotypes. Proc. Natl. Acad. Sci. 80:273-277.

Rollini P, Mach B, Gorski J (1985) Linkage map of three HLA-DR beta-chain genes: evidence for a recent duplication event. Proc. Natl. Acad. Sci. 82:7197-7201.

Spies T, Sorrentino R, Boss JM, Okada K, Strominger JL (1985) Structural organization of the DR subregion of the human major histocompatibility complex. Proc. Natl. Acad. Sci. 82:5165-5169.

Tacchini-Cottier FM, Jones PP (1988) Defective E beta expression in three mouse H-2 haplotypes results from aberrant RNA splicing. J. Immunol. 141:3647-3653.

von Boehmer H, Hudson L, Sprent J (1975) Collaboration of histoimcompatible T and B lymphocytes using cells from tetraparental bone marrow chimeras. J. Exp. Med. 142:989-997.

Werdelin O (1987) T cells recognize antigen alone and not MHC molecules. Immunology Today 8:80-84.

VI. Transgenic and Mutant Models of Disease

HLA-B27 and Arthritis

H. S. Luthra, C. L. Nickerson, S. Savarirayan, and C. S. David[1]

[1]Mayo Medical School, Mayo Clinic, Rochester, MN 55905, USA

In 1973 two studies reported almost simultaneously the highly significant association between HLA-B27 (B27) and ankylosing spondylitis (AS) (Schlosstein 1973; Brewerton 1973a). In the former study 88% of the patients (all Caucasion) were B27 positive compared to only 8% of the controls. Across the Atlantic the results were similar. Brewerton et al (1973a) found that 96% of AS patients were B27 positive compared to 4% of the controls. Since then there have been several reports reconfirming these findings (James 1975; Calin 1975, 1977; van der Linden, SM 1975; Sonoraki 1975; Dewar 1983; Good 1976; Gofton 1975). This strong association between a gene product and disease assigns a critical role to the gene in the pathogenesis of this disease. Studies performed in other racial groups continue to show the high association between B27 and AS. Japanese, who have an incidence of B27 of 1% in the general population, have about 70% positivity in AS patients (Sonoraki 1975; Dewar 1983). Among Afro-American Blacks 60% of AS patients are B27 positive compared to 4% of the general population. In a study involving Haida Indians the association of B27 in AS patients was 100% compared to 50% in the population (Good 1976). One ethnic group in whom the relationship of AS and B27 does not hold is the Pima Indians. Gofton (1975) observed a 20% incidence of radiographic sacroiliitis and an 18% incidence of B27. Interestingly, 50% of males and 9% of female patients were B27 positive. Others have reported association in males only (Calin 1977) and there are reports of discordance between B27 and AS (van der Linden, JMJP 1975). Such observations are rare and will hopefully be explained when the mechanism(s) involved in the pathogenesis of this disease is discovered. For the present time B27 and AS continues to be the strongest association between a Class 1 gene product and disease.

Because of the development of spinal involvement similar to AS in other spondyloarthropathies, several studies have investigated the association between these diseases and B27. A high association of B27 has been observed with Reiter's syndrome (Brewerton 1973b), spondylitis associated with psoriasis (Eastmond 1977), inflammatory bowel disease (Mallas 1976; Huaux 1977), juvenile chronic polyarthritis (Schaller 1977), as well as with post-infectious reactive arthritis following infection with Shigella, Salmonella, Yersinia (Aho 1975, 1985), and other organisms.

Although the association between a genetic marker - HLA-B27, the exposure to an infectious agent (known or presumed) and the occurrence of disease has been known for some time, only recently has there been some evidence to help explain this.

I.K. Egorov/C.S. David (Eds.)
Transgenic Mice and Mutants in MHC Research

The prevalence of AS is only 1-1.5% (Calin 1975), while 8% of the Caucasion population have the B27 gene. This would suggest that either the B27 gene is different in the diseased population or other factors are involved. Using serological reactivity with monoclonal antibodies, isoelectric focusing, and cytotoxic T lymphocytes, at least eight subtypes of HLA-B27 have been identified and six of these have been sequenced. Interestingly, there is no subtype which appears that was found to preferentially associate with disease. This suggests that some nonpolymorphic regions of the molecule must be involved (Khan 1988) in antigen presentation.

Molecular mimicry has been proposed as the mechanism involved in the development of these diseases (Bodmer 1975). Ebringer et al (1978) observed that fecal shedding of Klebsiella was more frequent in AS patients than in a control population, and that this increased in patients with active disease. This however was not substantiated in subsequent studies (Eastmond 1980, Warren 1980). van Bohemen et al (1984) reported cross reactivity of two HLA-B27 reactive monoclonal antibodies -M1 and M2- with bacteria. M1 antibody was found to cross react with a 16 kDa antigen on Klebsiella pneumoniae (types K21 and K43) and Yersinia enterocolitica (type 9). The second antibody -M2- cross reacted with a 20 kDa antigen on Shigella flexneri. Similar studies by Raybourne et al (1988), however showed that the M1 and M2 antibodies reacted wtih 36 kDa and 23 kDa envelope proteins of both Shigella flexneri and Shigella sonnei.

Seager et al (1979) raised rabbit antisera against certain strains of Klebsiella, which was found to cross react with lymphocytes of B27 positive patients with AS but not with B27 positive or negative lymphocytes from healthy individuals. The same group (Edmonds 1981) further showed that B27 positive lymphocytes from patients with other related diseases, e.g., Reiter's syndrome, also reacted with this antisera. Incubation of B27 positive cells from normal donors with Klebsiella culture supernatant imparted susceptibility of these cells to this antisera (Geczy 1980). Family studies showed that this factor on B27 positive cells of AS patients was not inherited (Edmonds 1981). It is nondialysable, heat labile, and about 35-50,000 dalton in size (Geczy 1980). In one of the isolates these investigators observed the presence of a plasmid which coded for the ability of the bacteria to produce the modifiable factor (Cameron 1983). This ability was transferable to another bacteria (E. coli) by transfer of the plasmid. DNA sequence of this plasmid has not been reported so far. Geczy et al (1986) have subsequently reported development of cytotoxic T lymphocytes against B27 positive targets from ankylosing spondylitis patients or normal B27 positive cells which have been modified by Klebsiella supernatants. What role this may have in the development or perpetuation of disease remains to be elucidated.

Recently Schwimmbeck et al (1987) have reported the finding of a six amino acid sequence homology between the hydrophilic domain of HLA-B27 (amino acids 72-77) and Klebsiella pneumoniae nitrogenase (amino acids 188-193). In addition, they observed the presence of autoantibodies in ankylosing spondylitis and Reiter's syndrome patients to synthetic peptides of B27 (AA 69-84, QTDRED) and Klebsiella pneumoniae nitrogenase (AA 184-196, QTDRED). This observation strengthens the concept of molecular mimicry - thus suggesting that the immune response initiated by K. pneumoniae infection is perpetuated by the self B27 molecule, thus leading to development of disease long after the original infection has cleared. Since disease development is not limited to a particular subtype of B27 and the 6AA homology has been observed only with B27.1 the question as to whether there is similar homology of different B27 subtypes with bacterial proteins of other bacteria can be asked. The answer to this can only be available when additional bacterial proteins are sequenced.

Schwimmbeck et al (1987) have shown that subtypes B27.2 and B27.3 have a 5AA homology with Klebsiella pneumoniae nitrogenase (QTDRE). This homology was also observed for B27 and B40 (B7-CREG antigens). Although 4 or 5AA homology can play a role in molecular mimicry (Schwimmbeck 1987), the finding that HLA-B8, an antigen not associated with these diseases, also has the same 5AA homology questions the significance of this observation (Weiss 1988). Stieglitz et al (1988) studied strains of arthritogenic Shigella flexneri and non-arthritogenic Shigella sonnei and identified a plasmid containing a 5 amino acid sequence (AQTDR) which is homologous with AA 71-75 of the alpha-1 domain of HLA-B27.1, B27.2, B27.3 and B27.4.

The other organism which is of interest is Yersinia enterocolitica (Cover 1989). This bacteria, although known to cause a diarrheal illness, can also cause an acute or chronic arthritis. The latter preferentially is observed in B27 positive individuals (Aho 1974). These patients tend to have persistent IgG and IgM antibodies to Yersinia enterocolitica (Granfors 1983) and lymphocytes from Reiter's syndrome patients show increased stimulation to Yersinia in vitro (Brenner 1984). Using live bacteria experimental models of reactive arthritis have been reported in both rats (Hill 1987, Toivanen 1988) and mice (Yong 1988).

The approach we have taken is to use HLA-B27 transgenic mice to develop an animal model of reactive arthropathy, using Yersinia enterocolitica as the inciting infection. If disease is induced, we would have an animal model of reactive arthritis which would enable us to study the mechanisms responsible for disease.

The HLA-B27 transgenic mice were produced by injecting cloned HLA-B27 gene (gift from H. Coppin, Paris) (Coppin 1986) into fertilized single cell embryos of B6/SJL F1 mice. Two founder mice were obtained (Savarirayan 1988) which expressed the B-27 antigen in conjunction with mouse ß2 microglobulin. Northern blot analysis showed (Nickerson et al 1989a) that B27 was expressed in the thymus, salivary glands, lymph nodes, skeletal muscle, lung, liver, spleen, and brain. In addition, B-27 antigen functioned as a transplantation antigen as evidenced by skin grafts among B-27 positive and negative siblings. The two founder mice are being backcrossed to B10 background and offspring of both founder mice were used in these studies.

Yersinia enterocolitica 0:8 WA, (donated by DTR Yu) which was isolated from a patient with reactive arthropathy and known to cause an inflammatory arthritis in certain strains of mice (Yong 1988), and Yersinia enterocolitica 0:3 another strain known to be associated with reactive arthropathy, were used in our studies (Nickerson et al 1989b). The organisms were stored in double strength skim milk at -70°C. The organisms were cultured on blood agar plates overnight at 37°C and then an isolated colon was inoculated into trypticase soy broth for 24 hours at 24°C. This was then recultured to late-log phase, washed and resuspended in phosphate-buffered saline to appropriate dilutions before inoculation into the mice. Initial experiments were aimed at determining the lethal dose of Yersinia in these mice. Intragastric, intraperitoneal, and intravenous routes were used. Animals (B-27 positive and B-27 negative siblings) were given doses of 10^2 to 10^8 by the various routes. Animals were observed daily for occurrence of any evidence of sickness and/or arthritis. Animals receiving bacteria intragastrically or intraperitoneally remained well. Animals given 10^2 and 10^3 intravenously remained well with no obvious disease. Those animals given doses of 10^6 to 10^8 all died within two weeks. The animals receiving 10^4 showed differences in mortality which were of interest (Table 1). The B-27 positive animals receiving 10^4 organisms showed a mortality of 73% compared to 8% in the B-27 negative group ($p=0.005$). There was no difference in the mortality in the animals given 10^5 bacteria.

Table 1

B-27	YERSINIA DOSE	PARALYSIS (%)	MORTALITY (%)
+	10^4	14/30 (47)	22/30 (73)
-	10^4	2/25 (8)	2/25 (8)
+	10^5	8/30 (27)	25/30 (83)
-	10^5	1/24 (4)	20/24 (83)

Within two to three weeks after inoculation several of these animals started to limp. This progressed to complete paralysis of the hind limbs (Fig. 1). This was not limited to a particular haplotype (Table 2), but was increased in B-27 positive animals.

Table 2

B-27	H-2	PARALYSIS (%)	DAY OF ONSET	MORTALITY (%)	DAY
+	b/b	7/15 (47)	9.3 + 1.2	14/17 (82)	21.8 + 7.05
-	b/b	0/11 (0)	-	6/9 (67)	9.2 + 1.24
+	s/s	5/22 (23)	10.0 + 1.3	17/22 (77)	11.0 + 1.88
-	s/s	1/12 (8)	12.0	3/12 (25)	8.3 + 1.16
+	b/s	7/16 (44)	15.3 + 2.2	10/16 (62)	18.1 + 2.16
-	b/s	1/18 (6)	8.0	7/18 (39)	10.1 + 1.58
+	s/f	3/04 (75)	5.3 + 0.3	4/4 (100)	15.8 + 1.11
-	s/f	1/7 (14)	14.0	4/8 (50)	13.5 + 3.60
+	f/f	0/01 (0)	-	1/1 (100)	5.0
-	f/f	0/03 (0)	-	2/2 (100)	7.0 + 3.03
Total					
B-27 +		22/60 (37)		47/60 (78)	
B-27 -		3/49 (6)		22/49 (45)	

Figure 1

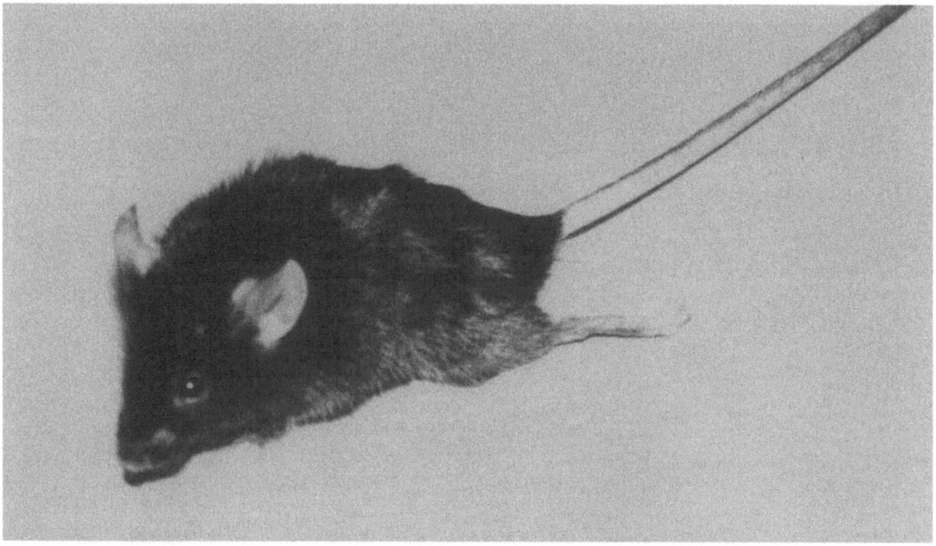

HLA-B27 positive animal showing paralysis of the hind limbs.

On dissection although abscesses were observed in multiple organs including the liver, spleen, lung, and other tissues, all the animals which showed paralysis clinically had abscesses located in the paraspinal and spinal region. These tissues were examined histologically and showed infiltration of the spine with destruction of the bony elements (Fig. 2a and 2b). The infiltration was by polymorphonuclear leukocytes and lymphocytes. There were areas with abscess formation and edema. The infiltrate involved the intraspinal region but there was no infiltrate in the nervous tissues. The spinal cord had the appearance of infarction as seen in vascular thrombosis. There was no clinical or histologic evidence of peripheral arthritis. No radiological changes of joint damage or new bone formation was observed.

Figure 2a

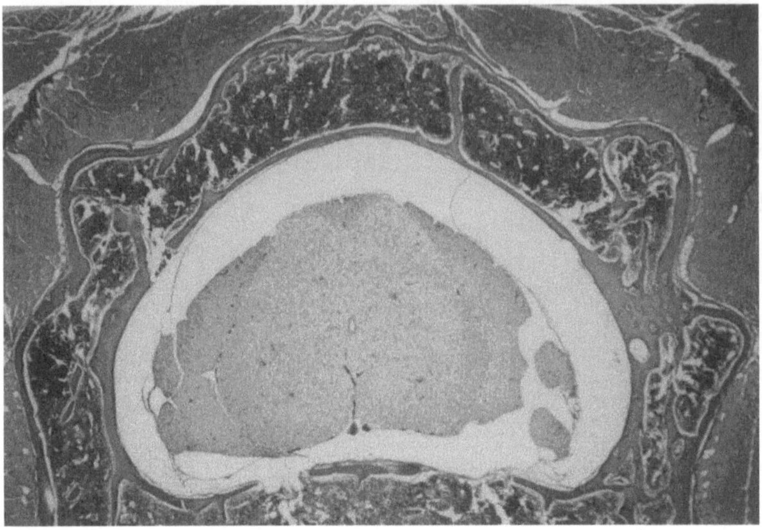

Cross section through the spine showing normal architecture

Figure 2b

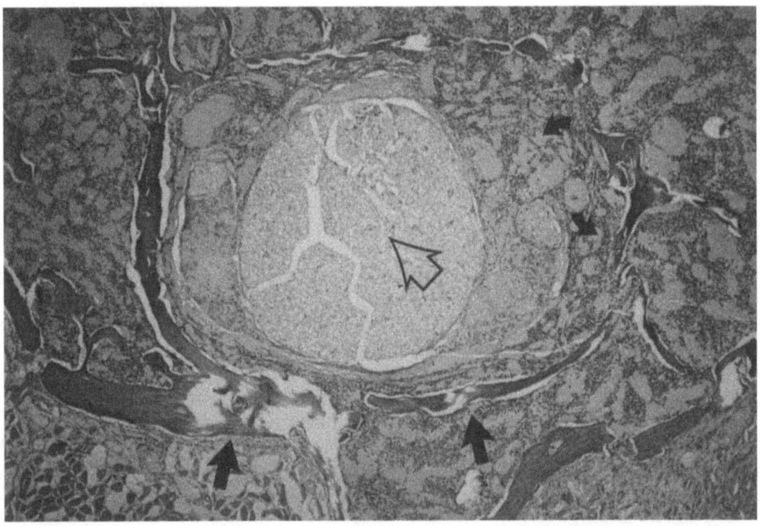

Cross section through the region of the paravertebral abscess involving the spine.
Solid arrow shows the destruction of the bones of the spine small solid arrow shows the infiltrating cells, inflammatory edema, and surrounding the neural tissue.
Open arrow shows the infarcted spinal cord. Note lack of any inflammatory cells.

In summary, we have observed a difference in the manner in which Yersinia enterocolitica 0:8 are handled by HLA-B27 positive and negative transgenic mice. Although we have not developed an animal model of spondyloarthropathy, the above finding is very encouraging and hopefully will allow us to develop such a model. The localization of the abscesses in and around the spinal tissues could likely be explained on the presence of a rich perivertebral venous plexus, but this does not explain the increased susceptibility of the HLA-B27 transgenic mice. Could there be increased adhesion of the bacteria to the endothelial cells during first passage after injection into the tail vein? This could occur if there is increased expression of the transgene in the vascular endothelium since it has been shown that (Klareskorg et al 1978, Maeda et al 1984) HLA Class I molecules can bind to bacterial proteins. How does one resolve the observation that several of these diseases are associated with a HLA Class I molecule, which has limited homology with at least two bacteria - namely Klebsiella and Shigella. One can postulate that the binding of a bacterial protein to the B-27 molecule initiates an immune response which cross-reacts with a native molecule/epitope leading to an autoimmune response and thus damage. Another possibility is that the bacterial antigen/s can lead to selective deletion of certain T cell clones which are important for suppression of the pathogenic response.

REFERENCES

Aho K, Ahvonen P, Lassus A, Sievers K, Tiilikainen A (1974) HL-A27 in reactive arthritis: a study of Yersinia arthritis and Reiter's syndrome. Arth Rheum 17:521-526

Aho K, Ahvonen P, Alkio P, Lassus A, Sairanen E, Sievers K, Tiilikainen A (1975) HL-A27 in reactive arthritis following infection. Ann Rheum Dis 34:29 (Supplement)

Aho K, Leirisalo-Repo M, Repo H. (1985) Reactive arthritis. Clin Rheum Dis 11:25-40

Bodmer WF (1975) The HL-A system and its association with immune response and disease. Ann Rheum Dis 34:13-16

Brenner MB, Kobayashi S, Wiesenhutter CW, Huberman AK, Bales P, Yu DTY (1984) In vitro T lymphocyte proliferative response to Yersinia enterocolitica in Reiter's syndrome: lack of response in other HLA-B27 positive conditions. Arth Rheum 27:250-257

Brewerton DA, Caffrey M, Hart FD, James DCO, Nicholls A, Sturrock RD (1973a) Ankylosing spondylitis and HL-A27. Lancet i:904-907

Brewerton DA, Caffrey M, Nicholls A, Walters D, Oats JK, James DCO (1973b) Reiter's disease and HL-A27. Lancet ii:996-998

Calin A, Bennett PH, Jupiter J, Terasaki PI (1977) HLA-B27 and sacroiliitis in Pima Indians - association in males only. J Rheum 4:44-48

Calin A, Fries JF (1975) Striking prevalence of ankylosing spondylitis in <<healthy>> W27 positive males and females. New Eng J Med 293:835-839

Cameron FH, Russell PJ, Sullivan J, Geczy AF (1983) Is a Klebsiella plasmid involved in the aetiology of ankylosing spondylitis in HLA-B27 positive individuals? Mol Imm 20:563-566

Coppin H, McDevitt HO (1986) Absence of polymorphism between HLA-B27 genomic exon sequences isolated from normal donors and ankylosing spondylitis patients. J Immunol 137:2168-2172
Cover TL, Aber RC (1989) Yersinia enterocolitica. New Eng J Med 321:16-24
Dewar PJ (1983) HLA antigens. Clin Rheum Dis 9:93-116
Eastmond CJ, Willshaw HE, Burgess SEP, Shinebaum R, Cooke EM, Wright V (1980) Frequency of fecal Klebsiella aerogenes in patients with ankylosing spondylitis and controls with respect to individual features of the disease. Ann Rheum Dis 39:118-123
Eastmond CJ, Woodrow JC (1977) The HLA system and the arthropathies associated with psoriasis. Ann Rheum Dis 36:112-120
Ebringer RW, Cawdell DR, Cowling P, Ebringer A (1977) Sequential studies in ankylosing spondylitis: association of Klebsiella pneumoniae with active disease. Ann Rheum Dis 37:146-151
Edmonds J, Macauley D, Tyndall A, Liew M, Alexander K, Geczy A, Bashir H (1981) Lymphocytotoxicity of anti-Klebsiella antisera in ankylosing spondylitis and related arthropathies. Arth Rheum 24:1-7
Geczy AF, Alexander K, Bashir HV, Edmonds J (1980a) A factor(s) in Klebsiella culture filtrates specifically modifies an HLA-B27 associated cell surface component. Nature 283:782-784
Geczy AF, Alexander K, Bashir HV, Edmonds JP (1980b) Characterization of a factor(s) present in Klebsiella culture filtrates that specifically modifies an HLA-B27 associated cell surface component. J Exp Med 152:331s-340s
Geczy AF, McGuigan LE, Sullivan JS, Edmonds JP (1986) Cytotoxic T lymphocytes against disease-associated determinant(s) in ankylosing spondylitis. J Exp Med 164:932-937
Gladman DD, Urowitz MB, Anhorn KAB, Chalmers A, Mervont H (1986) Discordance between HLA-B27 and ankylosing spondylitis: a family investigation. J Rheum 13:129-136
Gofton JP, Chalmers A, Price GE, Reeve CE (1975) HL-A27 and ankylosing spondylitis in BC Indians. J Rheum 2:314-318
Good AE, Kawanishi H, Schultz JS (1976) HLA-B27 in Blacks with ankylosing spondylitis or Reiter's disease. New Eng J Med 294:166-167
Granfors K, Viljanen MK, Tiilikainen A, Toivanen A (1980) Persistence of IgM, IgG, and IgA class Yersinia antibody in Yersinia arthritis. J Infect Dis 141:424-429
Hill JL, Yu DTY (1987) Development of an experimental animal model for reactive arthritis induced by Yersinia enterocolitica infection. Infect Immunity 55:721-726
Huaux JP, Fiasse R, Le Bruyere M, Nagaut de Deuxchaisnes C (1977) HLA-B27 in regional enteritis with and without ankylosing spondylitis or sacroiliitis. J Rheum 3:60-63 (Supplement 4)
James DCO (1975) Ankylosing spondylitis and HL-A27. Ann Rheum Dis 34:1-3 (Supplement)
Khan MA (1988) Ankylosing spondylitis and heterogeneity of HLA-B27. Sem Arth Rheum 18:134-141
Klareskorg L, Banck G, Forsgren A, Peterson PA (1978) Binding of HLA antigen-containing liposomesto bacteria. Proc Natl Acad Sci USA 75:6197-
Maeda K, Kono D, Kobayashi S, Brenner MB, Yu DTY (1984) A study of the specificity of the direct binding between bacteria and HLA antigens. Clin Exp Immunol 57:694-702
Mallas EG, Mackintosh P, Asquith P, Cooke WT (1976) Histocompatibility antigens in inflammatory bowel disease. Their clinical significance and their association with arthopathy with special reference to HLA-B27 (W27). Gut 17:906-910

Nickerson CL, Luthra HS, David CS (1989a) The role of enterobacteria and HLA-B27 in spondyloarthropathies. Ann Rheum Dis (in press)

Nickerson CL, Luthra HS, Savarirayan S, David CS (1989b) HLA-B27 transgenic mice are more susceptible to Yersinia enterocolitica infection than HLA-B27 negative littermates. (Submitted).

Raybourne RB, Bunning VK, Williams KM (1988) Reaction of anti HLA-B27 monoclonal antibodies with envelope proteins of Shigella species. J Immunol 140:3489-3495

Savarirayan S, Prakash S, Banerjee S, Haqqi T, Little R, Hanson J, McCormick J, Nickerson C, David CS (1988) Expression of HLA-B27 in association with mouse B2 in transgenic mice. FASEB Jour 2:A889

Schaller JG (1977) Juvenile rheumatoid arthritis series. Arth Rheum 20:165-170 (Supplement 2)

Scholosstein L, Terasaki PI, Bluestone R, Pearson CM (1973) High association of an HL-A antigen, W27, with ankylosing spondylitis. New Eng J Med 288:704-706

Schwimmbeck PL, Yu DTY, Oldstone MBA (1987) Autoantibodies to HLA-B27 in the sera of HLA-B27 patients with ankylosing spondylitis and Reiter's syndrome. Molecular mimicry with Klebsiella pneumoniae as potential mechanism of autoimmune disease. J Exp Med 166:173-181

Seager K, Bashir HV, Geczy AF, Edmonds J, de Vere-Tyndall (1979) Evidence for a specific B27-associated cell surface marker on lymphocytes of patients with ankylosing spondylitis. Nature 277:68-70

Sonoraki H, Seki H, Chang S, Okuyama M, Juji T (1975) Human lymphocyte antigen HL-A27 in Japanese patients with ankylosing spondylitis. Tissue Antigens 5:131-136

Stieglitz H, Fosmire S, Lipsky PE (1988) Bacterial epitopes involved in the induction of reactive arthritis. Am J Med 85:56-58 (supplement)

Toivanen BP, Merilahti-palo R, Gripenberg C, Soderstrom KO, Jaakkola UM (1988) Experimental Yersinia-associated arthritis in spontaneously hypersensitive rat. Br J Rheum 27:52-54 (supplement II)

van Bohemen OG, Grumet FC, Zanen HC (1984) Identification of HLA-B27 M1 and M2 crossreactive antigens in Klebsiella, Shigella, and Yersinia. Immunology 52:607-610

van der Linden JMJP, Cats A, Keuning JJ, van Rood JJ, Wuisman JHC (1975) HL-A27 and ankylosing spondylitis: a family study. Ann Rheum Dis 34:53 (supplement)

van der Linden SM, Valkenberg HA, deJongh BM, Cats A (1975) The risk of developing ankylosing spondylitis in HLA-B27 positive individuals. Arth Rheum 27:241-249

Warren RE, Brewerton DA (1980) Faecal carriage of Klebsiella in patients with ankylosing spondylitis and rheumatoid arthritis. Ann Rheum Dis 39:37-44

Weiss E, Bloemer K, Doerner C, Kuon W, Lang M, Pohla H, Schattenkirchner M, Riethmuller G (1988) Molecular biology of the HLA-B27 locus. Br J Rheum 27:12-18 (supplement II)

Yong Z, Hill JL, Hirofuji T, Mander M, Yu DTY (1988) An experimental mouse model of Yersinia-induced reactive arthritis. Microbial Pathogenesis 4:305-310

HLA-B27 Transgenic Mice as Potential Models of Human Disease

J. D Taurog[1], R. E. Hammer[1], S. D. Maika[1], K. L. Sams[1], F. A. K. El-zaatari[1], S. A. Stimpson[2], and J. H. Schwab[3]

[1] University of Texas, Southwestern Medical Center, Dallas, TX 75235, USA
[2] Glaxo Laboratories, Research Triangle Park, NC 27709, USA
[3] University of North Carolina, Chapel Hill, NC 27514, USA

INTRODUCTION

HLA-B27 is a serologically identified group of human class I MHC molecules that is very closely associated with the human diseases ankylosing spondylitis, reactive arthritis, and acute anterior uveitis. A number of infectious agents are known to trigger reactive arthritis in HLA-B27-positive individuals, but the mechanism by which this occurs is unknown. Elucidation of this mechanism may provide insight into a number of fundamental processes in the pathogenesis of joint inflammation in particular and inflammatory diseases in general. We have therefore sought to investigate the role of HLA B27 in the pathogenesis of arthritis in HLA-B27 transgenic animals. Our initial endeavors in this regard are reported here.

PRODUCTION OF HLA-B27 TRANSGENIC MICE

An HLA-B27 gene was introduced into non-inbred (C57BL/6 x SJL/J)F_2 and inbred C57BL/6 (B6) mice. The lines established from the (B6 x SJL/J)F_2 founders have been maintained by crossing with (B6 x SJL/J)F_1 at each generation; the background of these lines is designated BXS. The B27 gene product was expressed on spleen cell surfaces and stimulated both primary and secondary anti-B27 cytolytic T lymphocyte (CTL) responses in non-transgenic mice (Taurog, et al. 1988b). Because in humans the B27 heavy chain is associated with human β_2-microglobulin (hβ_2-m), which is only 70% homologous with the mouse β_2-m, it is possible that reproduction of human reactive arthritis in mice might require the presence of two transgenes, HLA-B27 and hβ_2-m. We therefore introduced a genomic clone encoding hβ_2-m (Guessow, et al. 1987; the gift of Dr. H. L. Ploegh, Amsterdam) into both BXS mice and inbred B6 mice, and crossed progeny of these mice with those of the B27 transgenics. In addition, we produced a line of BXS mice, designated 56-3, in which both the B27 and hβ_2-m genes were coinjected into the same fertilized egg and became cointegrated into the same chromosomal location such that they are transmitted together as a single Mendelian entity. The cell surface expression of the B27 molecule in all of the B27+, hβ_2-m+ mice is approximately 10 fold higher than in the mice with B27 only, as assessed by FACS analysis of spleen cells.

I.K. Egorov/C.S. David (Eds.)
Transgenic Mice and Mutants in MHC Research

HLA-B27 AND EXPERIMENTAL DISEASE

Conventional Models of Arthritis in Mice

One potential role of B27 in the pathogenesis of arthritis might be to amplify an independently initiated inflammatory response (Rosenbaum 1981). To investigate this possibility, we compared the responses of B27 transgenic and non-transgenic mice to two stimuli known to produce arthritis in mice, immunization with type II collagen and intraarticular injection with streptococcal cell wall peptidoglcyan.

Type II collagen-induced arthritis: Type II collagen arthritis is normally inducible only in mice expressing the class II MHC antigens I-A^q or I-A^r (Wooley, et al. 1981). For this experiment, B27+, hβ_2-m+ line 56-3 mice were crossed with DBA/1 (H-2^q) mice, and the offspring were immunized with chick type II collagen (the gift of Dr. M. Cremer, Memphis, TN) emulsified in CFA. No difference was observed between the B27+, hβ_2-m+ transgenic mice and the non-transgenic mice in either the incidence or severity of the ensuing arthritis (Figs. 1 and 2). Although additional experiments would be needed to determine if the B27 transgene influences the incidence or course of the disease in mice given a dose of type II collagen that is normally subarthritogenic in H-2^q mice or in mice lacking any of the susceptible H-2 haplotypes, these results suggest that the presence of B27 has no effect on conventional collagen-induced arthritis in mice.

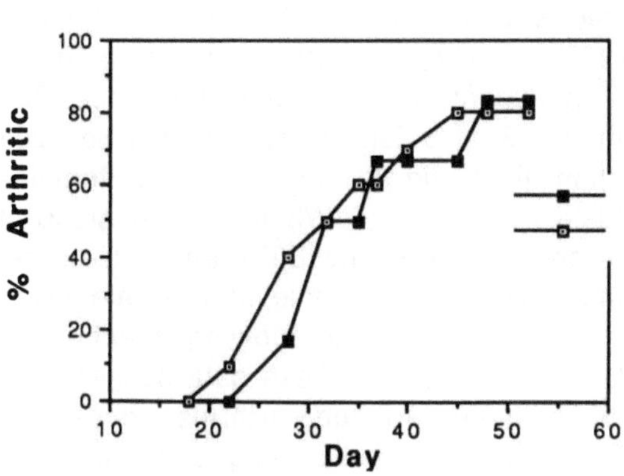

Fig. 1. Incidence of collagen-induced arthritis in transgenic and non-transgenic mice. Six B27+, hβ_2-m+ and 10 non-transgenic (BXS line 56-3 x DBA/1)F_1 mice were injected on day 0 with 200 μg of chick type II collagen in complete Freund's adjuvant.

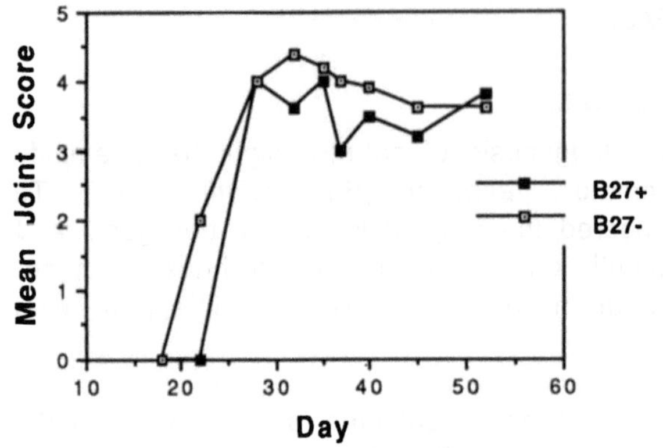

Fig. 2. Severity of collagen-induced arthritis in the transgenic and non-transgenic mice described in the legend of Fig. 1. Arthritis was scored by visual inspection on a scale of 0-3 for each extremity, with a maximum score of 12 per mouse per day.

Streptococcal cell wall-induced arthritis: Streptococcal cell wall (SCW) arthritis has been well characterized in rats (Cromartie, et al. 1977). In this model, injection i.p. of sonicated group A streptococcal cell walls or purified SCW peptidoglycan-polysaccharide (PG-PS) in saline into rats of susceptible strains results in an initial acute polyarthritis, subsidence of which is followed by a chronic, T cell-dependent erosive polyarthritis. A variant of this model involves intraarticular injection of SCW PG-PS to induce a transient monoarthritis, which then is reactivated several weeks later by i.v. injection of the same material (Esser, et al. 1985). Mice are susceptible only to the second type of arthritis. In a pilot experiment to examine whether the B27 transgene influences this process in mice, 3 of 3 female B27 transgenic mice showed significantly greater reactivation than did 6 non-transgenic females or 2 B27 and 4 nontransgenic males (data not shown). When a larger group of mice was examined, no difference was seen between the B27+ and B27- mice, but there was again a striking difference between the male and female mice, with 14 of 20 female mice and 1 of 12 male mice showing reactivation following i.v. injection of 300 μg SCW PG-PS (Fig. 3). These data suggest that SCW-induced arthritis in mice, like collagen-induced arthritis, does not appear to be influenced by the B27 transgene. The striking sex difference observed has not been previously reported and merits further investigation.

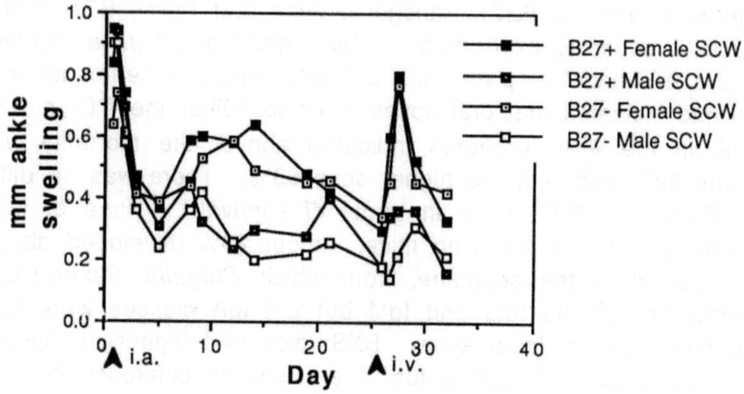

Fig. 3. Streptococcal cell wall peptidoglycan-induced arthritis in transgenic and non-transgenic mice. Twenty female (11 B27+ and 9 B27-) and 12 male (7 B27+ and 5 B27-) BXS mice were injected i.a. in the left ankle on day 0 with 5 µg of SCW PG-PS, and reinjected i.v. on day 26 with 300 µg of SCW PG-PS, and the swelling of the injected ankle measured with digital calipers. The right ankle was injected with saline and showed no significant swelling in any of the animals (data not shown).

Effect in B27 Transgenic Mice of Bacteria Arthritogenic in Humans

A number of bacteria have been shown to trigger reactive arthritis in humans, with B27+ individuals showing a relative risk of approximately 40 fold compared with B27- individuals (Lahesmaa-Rantala and Toivanen 1988). We have undertaken to test whether any of these bacteria has an effect in B27 transgenic mice that differs from their effect in non-transgenic mice.

Studies with *Shigella flexneri*: The microorganism chosen for initial studies was a *Shigella flexneri* isolate, CDI, that was known to have induced reactive arthritis in a B27+ individual, and to contain a plasmid encoding a peptide with a 5 amino acid homology to B27 amino acids 71-75 (Stieglitz, et al. 1989). Although *Shigella* is not a normal pathogen for mice, enteric infection can be established by eradicating the normal gut flora with antibiotics to which the *Shigella* strain is resistant (Cooper 1959). Enteric infection was chosen for initial examination in order to most closely mimic the human situation.

A total of 96 BXS and B6 B27+ transgenic mice and their B27- littermates were studied in 4 separate experiments. Two doses of *Shigella flexneri* CDI were given, 2 x 10^9 and 10^8 p.o. Although preliminary experiments in (B6 x SJL)F_1 mice had indicated that oral doses of up to 10^8 of the CDI strain were tolerated well, in the 4 experiments indicated above, the mortality with the lower dose was 32% and with the higher dose 69%. There was no difference between the B27+ and B27- mice in terms of survival. None of the mice developed arthritis. Of the surviving mice, about 20% developed abscesses, predominantly dorsal to the scapulae, from which *Shigella flexneri* CDI was cultured. Antibodies of the IgG and IgM but not IgA classes were found in serum within one week of inoculation. BXS mice had higher antibody levels than B6 mice, irrespective of B27 status. There was no difference between the B27+ and B27- mice in terms of abscess formation or antibody titers. None of the mice developed diarrhea, although the organism could be cultured from the stool for several weeks after the original inoculation, and histologic evidence of mucosal damage was evident in the cecum. In summary, to date we have not identified any B27-related effect of enteric infection with *Shigella flexneri* in mice.

Studies with *Yersinia enterocolitica*: Another organism known to trigger B27-associated reactive arthritis in humans is *Yersinia enterocolitica*. Recently, Nickerson et al. (1989) reported that B27 transgenic mice injected i.v. with *Y. enterocolitica* strain WA, serotype O:8 (YeWA O:8), showed significantly higher mortality than identically treated non-transgenic littermate controls. In addition, the B27 mice developed hindlimb paralysis, apparently due to the formation of abscesses along the spine, impinging on the spinal cord. In an effort to duplicate this finding, the same organism (the gift of Dr. D. T. Yu, UCLA) was injected i.v. into B27+, hβ_2-m+ and non-transgenic littermate control mice at the dose reported by Nickerson et al. to produce the greatest difference between B27 and non-transgenic mice. In the first experiment, shown in Fig. 4, 10 transgenic and 10 non-transgenic mice were injected i.v. with 10^4 YeWA O:8 organisms. No significant difference in mortality was seen. Several mice developed intraabdominal abscesses, but no spinal abscesses or hindlimb paralysis was seen. A second experiment, in which 5 x 10^2 organisms were given i.v. to 18 B27+, hβ_2-m+ or non-transgenic BXS mice, showed similar results (data not shown). The difference between these results and those of Nickerson et al. may be due to differences in the genetic backgrounds of the mice used, the presence of the hβ_2-m gene, or to environmental factors. Differences in the organism itself are possible, although care was taken to use the same conditions for growing the bacteria as those described by Nickerson et al.

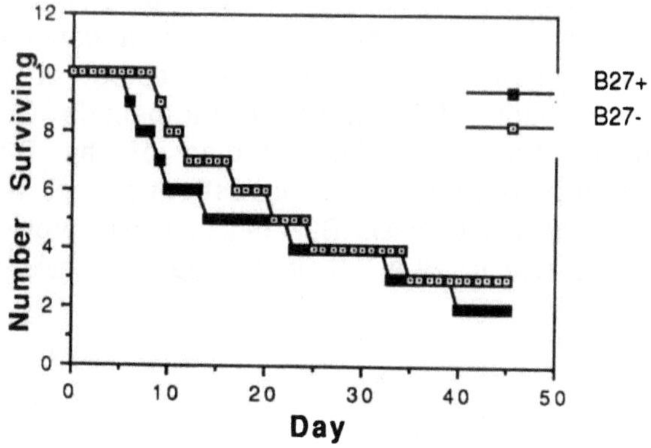

Fig. 4. Survival of transgenic and non-transgenic mice following i.v. injection of 10^4 viable *Yersinia entercolitica* WA O:8 organisms. Ten B27+, hβ_2-m+ and 10 non-transgenic male (line 56-3 x BALB/c) mice were injected on day 0 and observed daily for morbidity and mortality.

DISCUSSION

The data presented here are from initial experiments in which an effect of HLA-B27 on four different experimental disease models was sought in transgenic mice expressing the B27 gene. Although no definite conclusions can be drawn, there is no evidence from the studies to date that B27 exerts an effect on the conventional models of arthritis in mice, whereas evidence from another laboratory suggests that an effect of B27 might be found on the host response to *Y. enterocolitica* (Nickerson et al. 1989). Although we have so far not been able to confirm these findings, the factor(s) responsible for the difference in results in the two laboratories may provide an important clue to non-B27 factors involved in the pathogenesis of B27-associated disease.

In addition to studies of potential models of post-infectious reactive arthritis, the B27 transgenic mice provide an opportunity to examine the function of B27 as a restricting element for CTL. Because experimental priming in vivo is impractical in humans, studies of B27-restricted human CTL are limited primarily to a few pathogens such as influenza. The availability of recombinant vaccinia viruses expressing a large number of well characterized proteins offers the opportunity to explore the restriction by B27 of CTL to a large number of potential peptide antigens. In combination with a panel of mutant B27 molecules created by site-directed mutagenesis (Taurog, et al. 1988a, El-Zaatari, et al. 1989), a murine B27-restricted anti-peptide CTL system promises to be useful in structure-function studies of the B27 molecule as a restricting element. Together, the investigation in the transgenic mice both of disease models and of basic B27 function may provide the insights necessary to identify the role of HLA-B27 in the pathogenesis of rheumatic disease.

Supported by NIH grant AR09989.

REFERENCES

Cooper G (1959) Experimental shigellosis in mice. Austral J Exp Biol 37: 193-200.

Cromartie WJ, Craddock JG, Schwab JH, Anderle SK, Yang CT (1977) Arthritis in rats after systemic injection of streptococcal cells or cell walls. J Exp Med 146:1585-1602.

El-Zaatari FAK, Sams KL, and Taurog JD (1989) In vitro mutagenesis of HLA-B27. Amino acid substitutions at position 67 disrupt anti-B27 monoclonal antibody binding in direct relation to the size of the substituted side chain. Submitted for publication.

Esser RE, Stimpson SA, Cromartie WJ, and Schwab JH (1985) Reactivation of streptococcal cell wall-induced arthritis by homologous and heterologous cell wall polymers. Arthritis Rheum 28: 1402-1411.

Guessow D, Rein R, Ginjaar I, Hochstenbach F, Seemann G, Kottman A, Ploegh HL (1987) The human β2-microglobulin gene. Primary structure and definition of the transcriptional unit. J Immunol 139: 3132-3138.

Lahesmaa-Rantala R and Toivanen A (1988) Clinical spectrum of reactive arthritis. In: Tovianen A, Toivanen P (eds) Reactive arthritis. CRC Press, Boca Raton, FL, p 1-15.

Nickerson C, Luthra H, David C (1989) Increased mortality of HLA-B27 mice during infection with Yersinia enterocolitica. Arthritis Rheum 32:S37 (abstract).

Rosenbaum JT (1981) Why HLA-B27: An analysis based on two animal models. Ann Int Med 94:261-263.

Stieglitz, H, Fosmire S, Lipsky PE (1989) Identification of a 2 mDa plasmid from Shigella flexneri associated with reactive arthritis. Arthritis Rheum, in press.

Taurog JD, El-Zaatari FAK (1988a) In vitro mutagenesis of HLA-B27. Substitution of an unpaired cysteine residue in the α_1 domain causes loss of antibody-defined epitopes. J Clin Invest 82:987-992.

Taurog JD, Lowen L, Forman J, Hammer RE (1988b) HLA-B27 in inbred and non-inbred transgenic mice. Cell surface expression and recognition as an alloantigen in the absence of human β2-microglobulin. J Immunol 141: 4020-4023.

Wooley PH, Luthra HW, Stuart JM, David CS (1981) Type II collagen-induced arthritis in mice. I. Major histocompatibility complex (I region) linkage and antibody correlates. J Exp Med 154: 688-700.

Mouse Mutants Affecting Growth and Metastasis of a Syngeneic Tumor

O. S. EGOROV[1], E. SARGENT[1], G. A. CARLSON[2], and I. K. EGOROV[1]

[1] The Jackson Laboratory, Bar Harbor, ME 04609, USA
[2] The McLaughlin Research Institute, Great Falls, MT 59401, USA

INTRODUCTION

Tumor-specific antigens (TSA) are antigens peculiar to the tumor cells. In theory, TSA can be effectively recognized and reacted against by the host or can be used as targets for an effective immunotherapy. Although such antigens were identified long ago in several animal models of chemically or virally induced tumors (Prehn and Main 1957; Klein 1966), the presence of TSA in spontaneous animal tumors and in human tumors remains controversial (Hewitt et al 1976; Hellstrom and Hellstrom, 1989). One of the reasons for apparently weak or non-existing reactivity of patients against their tumors could be that tumors grow only in genetically unresponsive individuals while those who are able to mount an effective immune response against the tumor simply do not become ill. The information on genes controlling the mechanisms of host defense against the growing syngeneic tumor and its metastases can be useful for designing the optimal strategy of immunotherapy in cancer patients. However such genes have not been studied in animal models because an inbred animal outcrossed to any genetically nonidentical individual produces progeny that are no longer syngeneic with the parental tumor. One way to overcome this difficulty is to produce and study mutants that resist a parental tumor transplant while remaining histocompatible (and isogenic) with the parental strain mice. For the last few years we have been looking for spontaneous mutations of the C57BL/6J (hereafter B6) and C57BL/10SnJ (hereafter B10) strain mice resisting syngeneic tumor transplants using specifically designed screening procedures (Egorov and Egorov, 1988). Several mutants have been found. Here we present some results obtained with these mutants.

RESULTS

We have tested immunological reactions of mutants SH-10, SH-24, S-26, S-27 (two lines were used, one on B6 genetic background, designated B6-S-27, the other one on B10 genetic background, designated B10-S-27), S-28, and S-31. Of these, the SH-10 and the SH-24 mutants have been shown to be histocompatibility gene mutations (they rejected parental skin grafts) not useful for the purpose

I.K. Egorov/C.S. David (Eds.)
Transgenic Mice and Mutants in MHC Research

outlined above. Mutants S-26, S-27, S-28, and S-31 have been shown to reject moderate cell doses of a parental strain lymphoma E2G2, a clone of EL-4, 100 to 200 times higher than the dose that kills parental mice (one tumor cell per mouse kills about 30% of mice of both B6 and B10 strains; mutants resist injections of 100 to 250 cells per mouse). However, no skin graft rejection has been observed between each mutant and its parent (Table 1; data for mutants S-26 and S-28 is not shown). The S-26 and S-28 mutants appear to be not linked with the H-2 complex while the S-27 and S-31 seem to be linked with it. They may be useful for the study of cancer resistance.

Table 1. Parental skin grafts are not rejected by S-27 and S-31 mice that survived E2G2 tumor transplantation

Mouse strain	Number of mice: Total	Dead	Survived	Survival range, days#
	Tumor injection results, S-27 mice			
B10	5	5	0	24-28
B10-S-27*	20	14	6	24-38, >100
	Skin grafting results, S-27 mice injected with tumor			
B10 grafts, total		9	6	
accepted		9	6	
rejected		0	0	
bm23 grafts, total		13	6	
accepted		1**	0	
rejected		12	6	9-24
	Tumor injection results, S-31 mice			
B6	5	5	0	22-28
S-31*	18	2	16	26-28, >100
	Skin grafting results, S-31 mice injected with tumor			
B6 grafts, total		1	15	
accepted		1	15	
rejected		0	0	
bm3 grafts, total		1	15	
accepted		0	0	
rejected		1	15	11-22

Survival time of mice injected i.p. with the tumor (222 cells/mouse) or, alternatively, skin graft survival time in days.

* All S-27 mice received parental C57BL/10SnJ (B10) and H-2 mutant bm23 skin grafts on the day of injection of E2G2 cells. Eight S -31 mice received parental C57BL/6J (B6) and H-2 mutant bm3 skin grafts three days before tumor cell injection and another ten S -31 mice were skin grafted on the day of injection. Some of the grafts were technical failures and are not included in the Table.

** This mouse died of tumor before it rejected the bm23 graft.

We have tested NK (natural killer) activity of spleen cells of the S-27 and S-31 mutants and found no significant difference with the strain of origin (Table 2). Natural resistance to E2G2 leukemia transplantation in S-31 and B6 mice has been tested using [125-I]dUrd prelabelled cells and found to be similar in both of these strains (Table 3). In mixed lymphocyte cultures of responding spleen cells of mutants S-27, S-31 and M560 (an H-2 mutant used as a positive control) with parental stimulating cells, weak, but positive, proliferative responses have been observed indicating the existence of antigenic differences between normal lymphoid cells of these mice (Table 4). No killing of parental 51-Cr labelled target spleen cells by mutant cytotoxic T lymphocytes have been detected, while they strongly killed E2G2 tumor target cells (data not shown).

Table 2. NK activity in spleen cells of S-27 and S-31 mutants and parental mice

Mouse Strain	Percent specific 51-Cr release from target*					
	Target cells: YAC-1			Target cells: L1210		
	Effector to target ratio			Effector to target ratio		
	50:1	17:1	5:1	50:1	17:1	5:1
B6**	9.7±0.8	10.3±1.1	3.8±0.7	1.1±0.9	1.9±0.8	1.3±2.1
B10	10.6±1.2	11.2±1.2	4.6±1.6	1.1±2.4	2.1±0.6	3.3±2.7
S-27	6.0±0.9	5.6±0.6	1.7±0.7	1.5±1.1	0.8±2.0	1.8±3.7
S-31	12.3±1.3	12.5±0.9	6.9±1.2	0.9±1.5	1.6±0.6	0.7±2.4

* Spleen cells were used as effector cells. Target cells were labelled with 51-Cr for 3 hours. Six-hour chromium release test.

** Mean of three to five mice in each strain.

In the course of this study a highly metastatic clone of the E2G2 tumor designated E2M has been derived. Using this clone we have found pronounced differences between the mutants in their ability to resist formation of metastases of the tumor. A modification of the Prehn and Main (1957) assay for tumor immunity has been used to study resistance to metastases. A transplanted tumor growing intradermally for 6 to 8 days and then surgically removed can metastasize into regional lymph nodes and kill the animal. About 50% of the B6 mice are killed by the metastases on day 18 to 25 after the tumor excision (Table 5). A Kb gene mutation, bm3, fully protects mice from metastases in this system and served as a control. The S-27 mutation protects only partially: 18% of these mice are killed by metastases of the tumor which appear on day 45 to 50 and indicate that the resistance to metastases increased in this mutant in comparison with the parental mice. The S-31 mutation, however, increased the sensitivity of the mice to E2M tumor metastases because a higher proportion of them are being killed (about 89%) and they are being killed in a shorter time than B6 mice (12 to 18 days). Since both of these mutants show comparable degree of resistance to intraperitoneal E2G2 cell transplants, it is presently unclear why susceptibility to metastasis has increased in the S-31 mutant.

Table 3. Natural resistance to E2G2 leukemia transplantation in S-31 and C57BL/6J (B6) mice as measured by the use of [125-I]dUrd prelabelled cells

Mouse strain	No of mice	Route of injection	Percent of activity retained, mean On day 1	Percent of activity retained, mean On day 2
	Experiment 1, $1x10^6$ cells per mouse			
B6*	3	i.p.	10.30±1.11	7.63±0.91
S-31**	5	i.p.	51.83±3.25	10.98±2.43
B6**	5	i.p.	50.22±6.86	13.72±2.27
S-31	4	i.v.	6.66±1.17	4.41±0.72
B6	5	i.v.	11.73±0.85	7.14±0.83
	Experiment 2, $2x10^6$ cells per mouse			
S-31**	5	i.p.	47.22±11.0	35.74±6.72
B6**	5	i.p.	52.28±4.40	42.15±5.20
S-31	5	i.p.	21.68±1.77	16.47±2.00
B6	5	i.p.	22.95±5.75	17.10±1.55
S-31	5	i.v.	10.90±6.08	8.77±0.60
B6	5	i.v.	11.73±2.38	8.83±1.43

* These mice were injected with cells that were killed by UV irradiation.

** These mice were injected with labelled E2G2 cells twice, first time in Experiment 1 and second time in Experiment 2 that was performed on day seven after the first injection.

DISCUSSION

Mice that survive a syngeneic tumor inoculation may carry recently arisen mutations for resistance to the tumor transplant. In our series, overall frequency of inheritable variants (mutants) found among the B6 and B10 mice showing resistance to the EL-4 tumor was 1 in every 550 mice tested. For reasons not to be discussed here, this figure is not an estimation of the mutation rate of the genes involved, rather it is a characteristic of the detection system used in our laboratory. The mutants found were resistant to the inoculation of a moderate cell dose of the EL-4 tumor while being susceptible to a higher cell dose of the same tumor. Among them, histocompatibility (H) mutants capable of rejecting parental skin grafts were most frequent. One H-2 mutant (originally designated SH-24, later renamed bm26) and several minor H mutants not linked with the H-2 complex (SH-10 being one of them) have been identified. Some mutants resisting the parental tumor inoculation were incapable of rejecting parental skin grafts. They have been designated "S-mutants". Several S-mutants showed no linkage with the H-2 complex, but have not yet been studied any further. We focused our attention on two mutants, S-27 and S-31, that appear to be linked with H-2.

Table 4. Results of MLC test*, c.p.m. x 10^{-3}

Responding cells	Stimulating cells						
	B6	B10	S-27	S-31	M560	E2G2	B10.A
B6	0	0	6.9	9.3	8.3	61.2	23.4
B6-S-27	10.5	11.8	0	n.d.	13.1	47.4	23.3
S-31	13.0	13.2	n.d.	0	10.7	25.6	13.9
M560	n.d	27.0	n.d.	n.d.	0	58.3	21.7

* $5x10^5$ responding cells were mixed with $5x10^5$ stimulating spleen cells (or $2x10^4$ tumor cells) in a well of a 96-well plate, pulsed with [3-H]TdR on day 5 and harvested 16 hr later. Spleen cells from mice pre-immunized with parental tissue injections were used as responding cells. Irradiated spleen cells (3200 r from a gamma source) or E2G2 tumor cells (10000 r) were used as stimulating cells; n.d., not done.

Table 5. Metastases of E2M clone and immunity to the tumor in C57BL/6J (B6) mice and mutants*

Mouse strain	Total mice	Died of metastases	Survival time in days, mice with metastases	No of mice surviving without metastases	No of immune mice**
B6	18	9	18-25	9	3
bm3	5	0		5	5
B6-S-27	11	2	45-50	9	6
S-31	9	8	12-18	1	1

* Mice were injected intradermally with $1x10^5$ viable tumor cells and the growing tumors were removed surgically on day 8.

** Surviving mice were challenged with intraperitoneal injection of $1x10^5$ viable tumor cells on day 30 after the tumor excision. Ten to fifteen non-immunized mice of each strain were used as a control and they all died of tumor.

Our test results show that the S-27 and S-31 mutants have very weak antigenic difference(s) with non-mutant mice, yet they exert pronounced influence on the survival of mice injected with parental strain lymphoma cells, EL-4, and on the frequency of metastases of this tumor in mice. NK cell activity or natural resistance of mice to the tumor has not been altered by these mutations. Such a combination of features makes these mutants different from any existing H-2 mutant or any other mouse mutant known to us. Experiments are in progress to determine if a tissue-specific antigen(s) detectable in B6 and B10 mice by MLC but not by skin grafting and lacking in the S-27 and S-31 mutants is responsible for the resistance to the tumor in mutant animals. Preliminary data indicate that injections of normal lymphoid tissues of B6 origin do not protect mutants against the subsequent tumor transplantation. This result suggests that the mutant gene product(s) may play a regulatory role in the immune response of mice to the TSA rather than being a target for an immune attack against the tumor. The mutants will be used as a model for studies of genes that govern the host response to the growing syngeneic tumor and its metastases.

Acknowledgments

We thank Dr. Edward A. Boyse for advise. This work was supported by NIH grant GM28017 to I.K.E.

REFERENCES

Egorov IK, Egorov OS (1988) Detection of new MHC mutations in mice by skin grafting, tumor transplantation and monoclonal antibodies: a comparison. Genetics 118:287-298

Hellstrom KE, Hellstrom I (1989) Oncogen-associated tumor antigens as targets for immunotherapy. FASEB J 3:1715-1722

Hewitt HB, Blake ER, Walder AS (1976) A critique of the evidence for active host defence against cancer, based on personal studies of 27 murine tumors of spontaneous origin. Br J Cancer 33:241-257

Klein G (1966) Tumor antigens. Annu Rev Microbiol 20:233-252

Prehn R, Main D (1957) Immunity to methylcholanthrene-induced sarcomas. J Natl Cancer Inst 18:768-798

Transgene Induced MHC Mismatch Results in Rapid Natural Killer Cell Dependent Elimination of Lymphoma Cells

P. HÖGLUND, H. G. LJUNGGREN, C. ÖHLÉN, and K. KÄRRE[1]

[1] Department of Tumor Biology, Karolinska Institutet, Box 60400, 10401 Stockholm, Sweden

INTRODUCTION

We have previously reported that selection of tumor cells against loss of H-2 products renders the cells more sensitive to natural killer (NK) cell mediated lysis (Ljunggren 1985; Kärre 1986). One of the explanations to these findings has been that NK cells scan the surface of target cells for presence of self H-2 markers. Failure to detect self H-2 would result in a completed lytic cycle initiated by a putative NK receptor. However, if identity in H-2 expression between effector and target is achieved, the result would be inhibition of lysis and target cell survival. Several investigators have addressed this question using NK sensitive / H-2 negative variant tumor cells. When H-2 expression was restored by transfection, a decrease in NK sensitivity was observed (Storkus 1989; Shimizu 1989).

In order to further test this theory, we have used the H-2 transgenic strain D8 to study the role of a novel class I gene in the host in natural resistance against transplanted lymphomas (Höglund 1988). The D8 strain was produced by microinjection of the D^d gene from Balb/c into C57Bl/6 (B6) embryos (Bieberich 1987). The H-2 phenotypes of the D8, as well as the other strains and tumors used in this paper, are shown in table 1.

Table 1. H-2 phenotypes of mice and tumors used in this paper

CLASS I LOCUS		K*	D	tg**
MICE	C57Bl/6 (B6)	b/b	b/b	-
	D8	b/b	b/b	d
TUMORS***	RMA	b/b	b/b	-
	RMA-S	-	-	-

* Expression of MHC class I genes.
** Expression of transgene.
*** RMA is a subline of the Raucher virus induced T-cell lymphoma RBL-5, and RMA-S is a H-2 deficient variant of RMA (Ljunggren 1985).

RESULTS AND DISCUSSION

Table 2 shows percent tumor takes in B6 and D8 strains after subcutaneous (s.c.) inoculation of 10^4 RMA and RMA-S cells. The most important conclusions from table 2 is that RMA was rejected in the D8 strain while it was accepted in the B6 strain. This

I.K. Egorov/C.S. David (Eds.)
Transgenic Mice and Mutants in MHC Research

rejection was mediated by NK-cells and cosegregated with D^d expression in backcross analysis (Höglund 1988). The fact that RMA-S was eaqually well rejected in both strains further confirms the importance of missing alleles at the target level.

Table 2. Percent tumor takes after s.c. inoculation of 10^4 cells

STRAIN		B6	D8
TUMOR	RMA	100*	30
	RMA-S	5	0

* Mean value of several independent experiments with no less than 15 mice in any group.

We have now extended these studies by investigating the kinetics of the elimination of RMA in the transgenic D8 mice (Fig. 1). Tumor cells were metabolically labelled with ^{125}IUdR and injected intravenously. At different timepoints after inoculation, the mice were killed and % remaining radioactivity in different organs was determined.

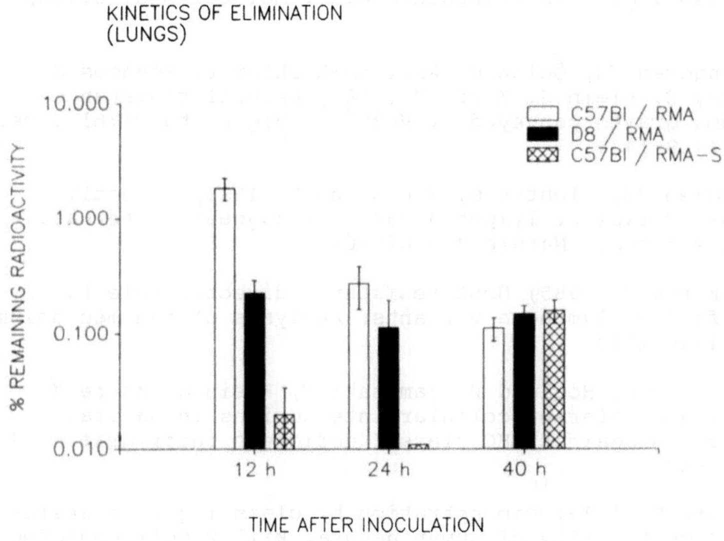

Fig. 1. Kinetics of rapid elimination of radiolabelled tumor cells after i.v. inoculation. The largest differences were always seen in the lungs but were also present in the liver and in the spleen although to a lower extent

Figure 1 shows that when D8 was used as a recipient for RMA inoculation, there was a 10 fold difference in elimination compared with RMA inoculated into B6. In addition, RMA-S was much more efficiently eliminated in the syngeneic B6 strain than the wild type RMA (Ljunggren 1985, 1988). Note that the % remaining radioactivity is expressed on a log scale, which means a 100 fold difference between these two tumors. From this experiment it could be concluded that elimination was present already 12 hours after inoculation of the tumor cells and with the largest differences seen after 24 hours. Furthermore, the differences in both comparisons could be abrogated by anti NK1.1 treatment of the host prior to inoculation (data not shown).

We believe that the data on s.c. growth in table 2 and the rapid elimination data in figure 1 reflect two different aspects of one single mechanism. We speculate that the elimination of RMA-S in B6 and the elimination of RMA in D8 is triggered by the same phenotypic incompatibility between graft and host, i.e. failure of host NK-cells to detect a complete set of self H-2 on the grafted cells. This rejection is rapid and occurs within the first 24 hours after i.v. inoculation of tumor cells. No evidence so far exists in terms of molecular mechanisms of how such a recognition system would work, but one explanation could be that MHC molecules act as downregulators of possible target structures for NK-cells, e.g. Hh-1 antigens.

REFERENCES

Bieberich C, Scangos G, Tanaka K, Jay G (1986) Regulated expression of a murine class I gene in transgenic mice. Mol Cel Biol 6:1339-1342

Höglund P, Ljunggren HG, Öhlén C, Ährlund-Richter L, Scangos G, Bieberich C, Jay G, Klein G, Kärre K (1988) Natural resistance against lymphoma grafts conveyed by H-2D^d transgene to C57Bl mice. J Exp Med 168:1469-1474

Kärre K, Ljunggren HG, Piontek G, Kiessling R (1986) Selective rejection of H-2 deficient lymphoma variants suggests alternative immune defence strategy. Nature 319:675-678

Ljunggren HG, Kärre K (1985) Host resistance directed selectively against H-2 deficient lymphoma variants. Analysis of the mechanism. J Exp Med 162:1745-1749

Ljunggren HG, Öhlén C, Höglund P, Yamasaki T, Klein G, Kärre K (1988) Afferent and efferent cellular interactions in natural resistance directed against MHC class I deficient tumor grafts. J Immunol 140:671-678

Shimizu Y, DeMars R (1989) Demonstration by class I gene transfer that reduced susceptibility of human natural killer cell mediated lysis is inversly correlated with HLA class I antigen expression. Eur J Immunol 19:447-451

Storkus WJ, Alexander J, Payne JA, Dawson JR, Cresswell P (1989) Reversal of natural killing susceptibility in target cells expressing transfected class I HLA genes. Proc Natl Acad Sci USA 86:2361-2364

Highly Immunogenic Transformed Tumor Clones Expressing Allogeneic Class I Histocompatibility Gene Demonstrate a Specific Immunotherapeutic Affect Against the Parental Tumor

K. M. Hui, T. Sim, T. T. Foo, and A. A. Oei[1]

[1] Institute of Molecular and Cell Biology, National University of Singapore, Kent Ridge Crescent, Singapore 0511, Republic of Singapore

Abstract

In this study, the possibility of generating tumor-specific immunity by the introduction of allogeneic class I histocompatibility gene into tumor cells was investigated. Specifically, we have expressed the H-2K^b gene in the non-immunogenic AKR leukemia K36.16. Thirty-eight H-2K^b-transformed K36.16 clones (K^b/K36.16) were isolated and studied individually. Different amounts of the H-2K^b antigens were detected on the surface of these K^b/K36.16 transformed clones. As expected, the expression of the H-2K^b antigens on the K36.16 tumor cells led to the rejection of the transformed clones by semi-syngeneic AKR mice. It was also noted that AKR mice which had been immunized with the K^b/K36.16 clones were able to survive a subsequent challenge of the parental K36.16 tumor cells. More importantly, from the clinical point of view, some of these K^b/K36.16 clones could induce a specific immune response against the growth of the K36.16 cells in tumor-bearing AKR mice. Such immunotherapeutic observation reinforces the feasibility of using gene-transfer as a molecular approach to abrogate tumor growth.

Introduction

The MHC class I antigens are vital for the activity of the CTL to lyse tumor cells (Doherty, et. al., 1984). Therefore, the CTL, at least in the *in vitro* situation, requires a receptor to recognize a complex between the viral/tumor antigenic peptide and the class I MHC molecules for effective lytic function. The class I histocompatibility antigens, however, could also be recognized as distinct, non-self antigens when presented to allogeneic cells (Sherman and Maleckar, 1988). This recognition of non-self class I antigenic differences could trigger a strong graft rejection *in vivo* and a strong primary cytotoxic response *in vitro*.

I.K. Egorov/C.S. David (Eds.)
Transgenic Mice and Mutants in MHC Research

Tumor-associated transplantation antigens (TATA) are usually non-immunogenic or at best, weakly immunogenic in syngeneic host. Many investigators have therefore tried to increase the immunogenicity of tumor cells by DNA-mediated gene transfer experiments (Hui, 1989). However, all these approaches failed to induce an immune response against the poorly immunogenic parental tumor cells growing in the host.

It has been suggested that cellular cooperation might augment tumor resistance (Mitchison, 1970). According to Mitchison, if helper T cells are generated by immunizing hosts with antigens used for the modification of tumor cells, subsequent immunization with "modified" tumor cells could enhance the anti-TATA immune response. Conceivably, immune response to TATA could be improved by coupling additional antigens to the tumor cell surface. This concept, together with the fact that alloreactivity induces a strong immune reaction, have stimulated our interest in the generation of tumor-specific immunity by the introduction of allogeneic MHC genes into tumor cells.

Results

Generation of the K^b/K36.16 transformed clones

A 9.6 kb EcoRI fragment, containing the entire H-$2K^b$ genes, was isolated from the cosmid H8 (Mellor, et. al., 1982) and subcloned into the plasmid pAT 153. This DNA construct, along with the plasmid pTCF (Hui, et. al., 1984), were used to transform the K36.16 tumor cells. Thirty-eight G418-resistant colonies were isolated and propagated in G418-containing medium for further studies. The presence of the H-$2K^b$ antigens on the cell surface of these stably transformed clones was detected by radiobinding assays (Table I). These clones expressed various levels of the H-$2K^b$ antigens and were grouped according to the level of H-$2D^k$ antigens present on their cell surface (Table I).

S1 nuclease protection analyses (Hui, et. al., 1984) also showed that only the K^b/K36.16 transformed clones such as K^b-S2, K^b-S6, and K^b-S20 expressed H-$2K^b$ mRNA, when compared to the K36.16 and AKR negative and C57BL/10 positive controls (Fig. 1). Multiple protected minor bands longer than 230nt were due to incomplete S1 digestion of RNA-DNA hybrids formed between the DNA probe and mRNA molecules other than H-$2K^b$.

Table I. Summary of the results obtained from radio-binding Assays of the K^b/K36.16 clones (Net ^{125}I cpm)[1]

$H\text{-}2D^k$(>2000 cpm)	$H\text{-}2D^k$(1500 - 1000 cpm)	$H\text{-}2D^k$(800 - 700 cpm)	$H\text{-}2D^k$~500 cpm	$H\text{-}2D^k$(200 - 100 cpm)
K^b-S26 (5717)	K^b-S25 (270)	K^b-S6 (3112)	K^b-S21 (2453)	K^b-S24 (3510)
K^b-S33 (7574)	K^b-S2 (328)	K^b-S13 (13528)	K^b-S10 (4832)	K^b-S32 (5559)
	K^b-S3 (992)	K^b-S27 (15170)	K^b-S36 (14217)	K^b-S9 (6226)
	K^b-S4 (2207)		K^b-S22 (15084)	K^b-S34 (7783)
	K^b-S7 (12024)		K^b-S35 (19758)	K^b-S31 (8426)
	K^b-S14 (15284)		K^b-S5 (24408)	K^b-S12 (9083)
			K^b-S38 (24709)	K^b-S11 (9694)
				K^b-S20 (9792)
				K^b-S39 (11217)
				K^b-S30 (11393)
				K^b-S23 (13404)
				K^b-S8 (13972)
				K^b-S18 (14196)
				K^b-S17 (14315)
Controls:-				K^b-S28 (14425)
				K^b-S15 (14851)
K36: $H\text{-}2D^k$: 1007 cpm				K^b-S19 (16113)
EL4: $H\text{-}2K^b$: 5120 cpm				K^b-S16 (16646)
				K^b-S37 (17879)
				K^b-S29 (18633)

[1] 1×10^6 cells were used for each of the radiobinding assays and the tests were done in duplicates.

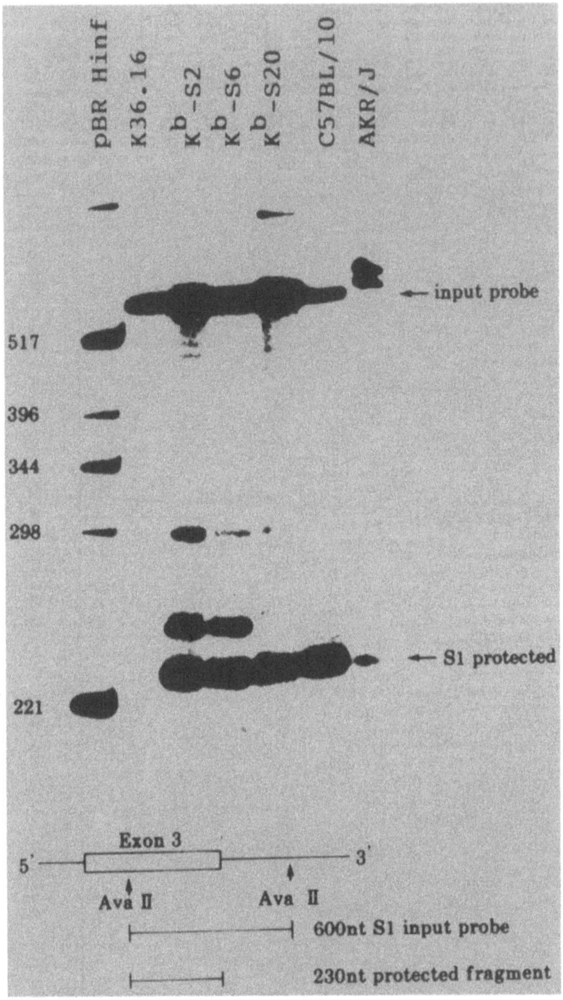

The presence of the H-$2K^b$ antigens was further confirmed by two-dimensional SDS gel electrophoresis of immunoprecipitates obtained using anti-H-$2K^b$ monoclonal antibody and protein-A as secondary antibody (Jones, 1977). Characteristics 45 kd spots (as indicated by arrows, Fig. 2) for the H-$2K^b$ antigens were present only when lysates of normal C57BL/6 spleen cells (Fig. 2A) and the K^b/K36.16 transformed clones (for example K^b-S38) were used (Fig. 2C). Cell lysates prepared from K36.16 tumor cells did not produce these spots (Fig. 2B).

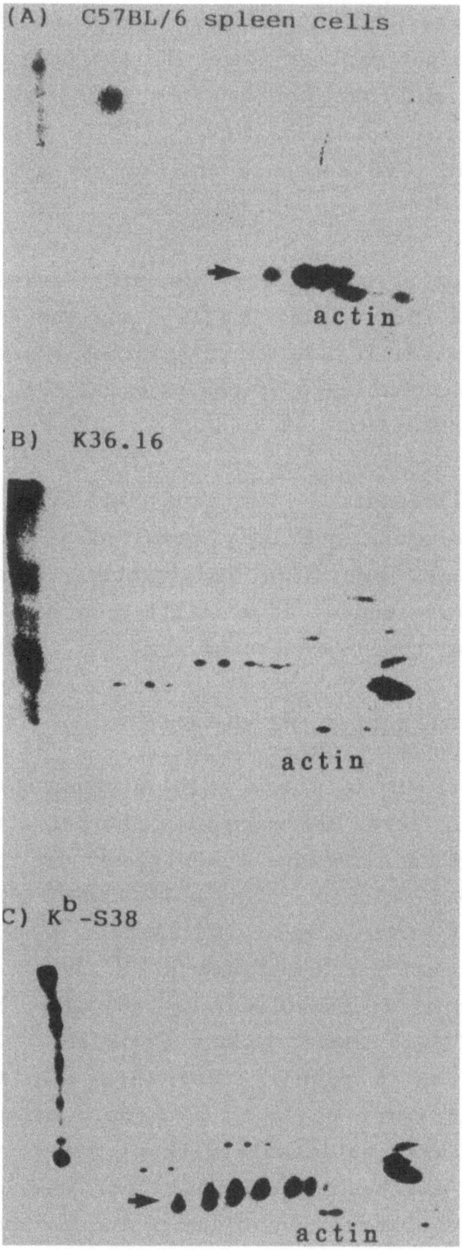

Fig.2

Rejection of the K^b/K36.16 clones by AKR mice

As expected, all the K^b/K36.16 clones which were positive for the H-2K^b antigens were unable to induce tumors in AKR mice (Table II). One of the transformed clones, K^b-S25, did give tumors in AKR mice (Table II). Clone K^b-S25 did not express the H-2K^b antigens on its cell surface (Table I) and did not contain the H-2K^b gene sequence when analyzed by Southern blot (not shown). This clone probably contains only the plasmid pTCF sequence and therefore was not being further studied.

The K^b/K36.16 clones were also able to induce immunity against the original, unmodified, parental K36.16 tumor cells. All the AKR mice that had been immunized with the K^b/K36.16 transformed clones were able to withstand a subsequent challenge of the parental K36.16 tumor cells (Table II).

The introduction of the allogeneic H-2K^b gene into the k haplotype K36.16 tumor cells therefore not only resulted in the rejection of the K^b/K36.16 clones, but more importantly, these transformed clones were able to induce a specific protective tumor-immunity against the original K36.16 tumor cells.

The K^b/K36.16 clones demonstrated an active and specific immunotherapeutic affect on the parental K36.16 tumor cells.

To evaluate the potential of using the K^b/K36.16 clones as immunotherapeutic agents, we had injected AKR mice with the parental K36.16 tumor cells and then tried to eliminate tumor growth by the injection of the different K^b/K36.16 clones. Clones K^b-S3, K^b-S11, K^b-S13, K^b-S21, K^b-S24, K^b-S27, K^b-S29, K^b-S33, K^b-S36 and K^b-S38 which expressed different levels of the H-2K^b and H-2D^k antigens were chosen for this study. For most of the clones tested, with the exception of K^b-S27, when $1x10^4$ K36.16 tumor cells were injected 2d before the injection of the K^b/K36.16 clones, they were able to suppress the growth of the K36.16 tumor cells to a certain extent (Fig. 3). This effect was especially noticeable with clones K^b-S3 and K^b-S38. Morever, mice that had been cured would not develop further tumor even after prolonged observation of up to two months. In addition, the observed immunotherapeutic affect is specific for the K36.16 tumor cells as another AKR leukemia, 369, grew readily in AKR mice even with the K^b-S3 and K^b-S38 cells.

Table II. Induction of tumors in AKR mice by the K^b/K36.16 clones.

Cell lines	# of mice with tumors AKR(1)	2°challenge(2)
K36.16	10/10	ND
pTCF-K36.16	10/10	ND
K^b-S2	0/10	0/10
K^b-S3	0/10	0/10
K^b-S4	0/10	0/10
K^b-S5	0/10	0/10
K^b-S6	0/10	0/10
K^b-S7	0/10	0/10
K^b-S9	0/10	0/10
K^b-S10	0/10	0/10
K^b-S11	0/10	0/10
K^b-S12	0/10	0/10
K^b-S13	0/10	0/10
K^b-S14	0/10	0/10
K^b-S15	0/10	0/10
K^b-S20	0/10	0/10
K^b-S21	0/10	0/10
K^b-S24	0/10	0/10
K^b-S25	10/10	ND
K^b-S27	0/10	0/10
K^b-S29	0/10	0/10
K^b-S30	0/10	0/10
K^b-S31	0/10	0/10
K^b-S32	0/10	0/10
K^b-S33	0/10	0/10
K^b-S34	0/10	0/10
K^b-S35	0/10	0/10
K^b-S38	0/10	0/10

(1) 5×10^5 cells were injected sc per mouse.

(2) Mice which did not have a tumor after one month were further challenged with 5×10^5 K36.16 tumor cells. The number of mice with tumor was recorded after another one month.

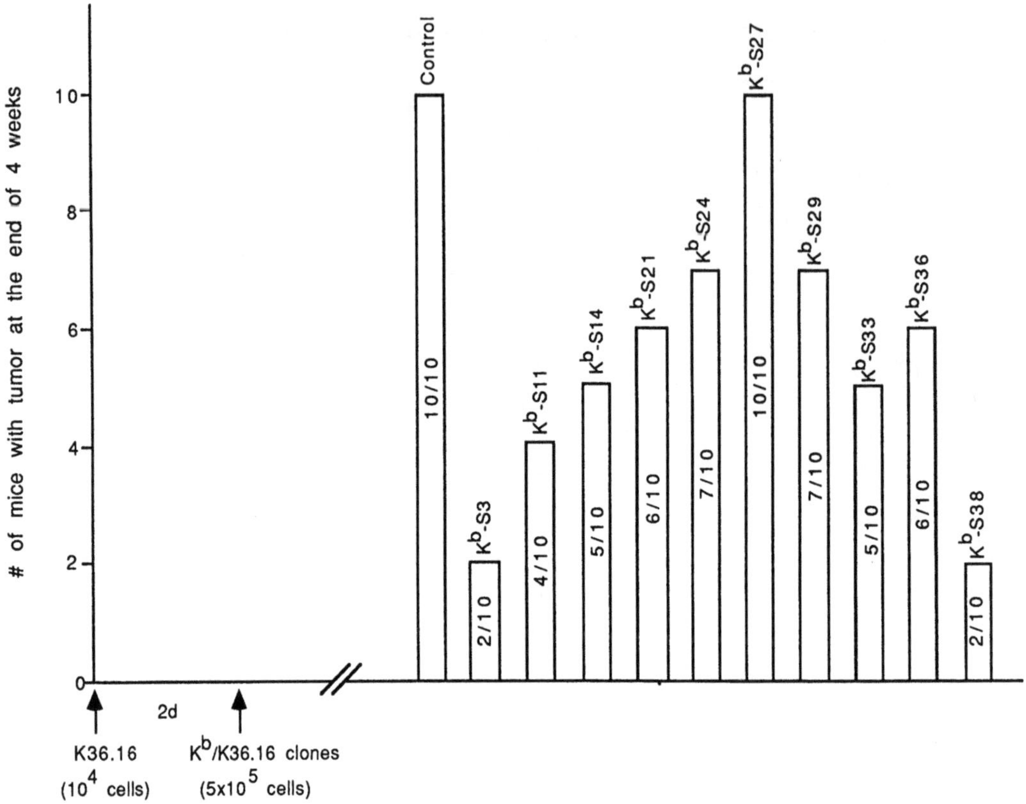

FIGURE 3. K^b-K36.16 TRANSFORMED CLONES COULD SUPPRESS THE GROWTH OF K36.16 TUMOR CELLS.

The K^b/K36.16 transformed clones failed to induce tumor-specific CTL.

To better understand the mechanisms responsible for the rejection of tumor cells *in vivo*, K^b/K36.16 clones which exhibit strong immunogenicity *in vivo* were used to immunize AKR mice and the effector T cells harvested for CTL assays. Although the original K36.16 tumor cells were readily rejected *in vivo*, no CTL activity against the parental K36.16 tumor cells could be demonstrated under several different protocols (not shown). However, the K^b/K36.16 clones and an H-2K^b-positive lymphoma, EL4, were killed very efficiently. Typical results obtained, for example, with clone K^b-S38 and priming *in vivo* for two weeks were shown in Fig. 4. Hence, the *in vitro* CTL activity detected was mainly alloreactivity against the H-2K^b antigens.

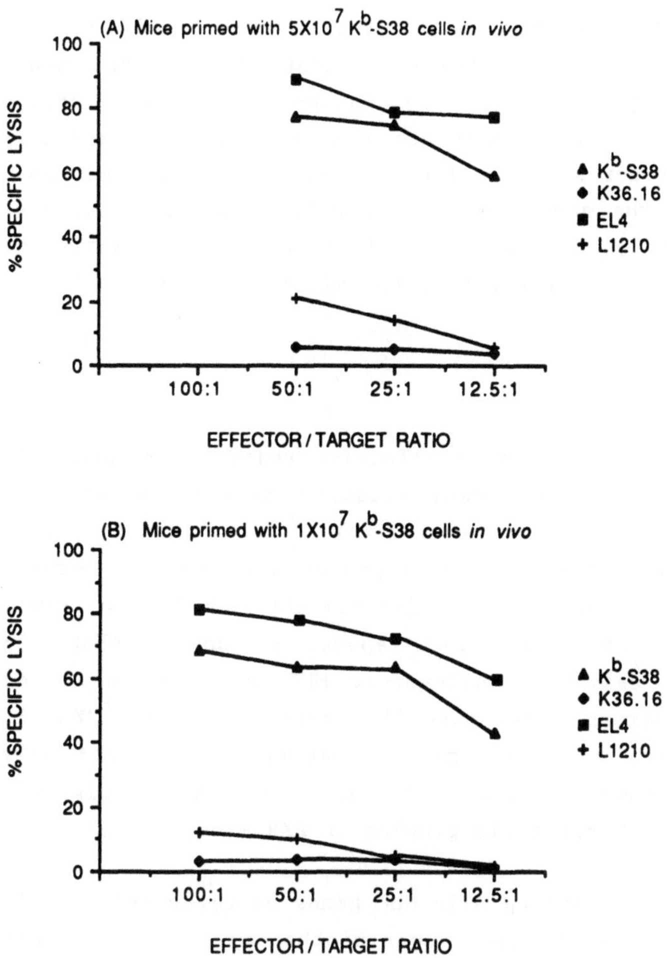

FIGURE 4. GENERATION OF AKR ANTI-K^b-S38 EFFECTOR T CELLS

Presently, the precise molecular and cellular explanation for tumor-rejection is unknown. Some possible mechanisms are : (A) Associative recognition. The TATA become physically associated with the H-$2K^b$ alloantigens after the transformation. As a consequence of the recognition of the H-$2K^b$ antigens by the host, the TATA can be recognised. (B) Cellular co-operation. The recognition of the H-$2K^b$ alloantigens by the host produces helper factors which are necessary for the activation of helper T cells to generate CTL which could subsequently recognize the TATA. (C) Production of tumor-specific cytotoxic lymphokines as a result of the recognition of the immunogenic transformed tumor clones. The first two hypotheses suggest the presence of tumor-specific cytotoxic cells.

In the experiments which we have performed, we cannot rule out the possibility of masking the activity of a less vigorous and minor tumor-specific cytolytic cell population by the strong alloreactivity against the transformed clones. We are currently trying to establish tumor-specific T cell clones from mice that have been immunized with the immunogenic K^b/K36.16 transformed clones. This approach, hopefully, will enable us to detect independently the tumor-specific cytolytic activity from the alloreactivity.

Discussion

Most cancers do not produce an effective rejection response in syngeneic tumor-bearing hosts. Many attempts have therefore been taken to try to enhance the antigenicity of malignant cells. However, no satisfactory experimental approach has been put forward that can generate tumor-specific immunity for the reduction of tumor burden in syngeneic host. In this report, we showed that the introduction and expression of allogeneic MHC molecules on tumor cells could induce strong tumor-specific immunity. The K^b/K36.16 transformed clones were able to induce an immune response not only to the modified tumor cells but also to the original, non-immunogenic K36.16 tumor cells growing in AKR mice.

This approach of expressing allo-MHC genes on tumor cells could potentially be used for effective cancer biotherapy. As often when cancer chemotherapy and radiation doses are increased to eliminate residual tumor cells in patients, a dose-limiting toxicity is realized before eradication of tumor cells is complete. The combination of surgical removal of tumor mass and the generation of immunotherapy by the de novo expression of introduced allo-MHC gene could be an attractive alternative to eliminate residual tumor cells and metastases.

Acknowledgements

This work was supported by the Institute of Molecular and Cell Biology, National University of Singapore.

References

Doherty, P., Knowles, B. and Wettstein, P. (1984) Immunological surveillance of tumors in the context of major histocompatibility complex restriction of T cell function. Adv. Cancer Res. 42: 1.

Hui, K. and Kim, B. (1984) Tumor-specific immunity induced by somatic hybrids. IV. Relationship between immunogenicity and expression of surface tumor-associated antigens. Cellular Immunol. 87: 591.

Hui, K., Grosveld, F. and Festenstein, H. (1984) Rejection of transplantable AKR leukemia cells following MHC DNA-mediated cell transformation. Nature 311: 750.

Hui, K.M. (1989) Re-expression of major histocompatibility complex (MHC) class I molecules on malignant tumor cells and its effect on host-tumor interaction. Bioassays 10. (In press)

Jones, P.P. (1977) Analysis of H-2 and Ia molecules by two-dimensional gel electrophoresis. J. Exp. Med. 146: 1261.

Mellor, A., Golden, L., Weiss, E., Bullman, H., Hurst, J., Simpson, E., James, R., Townsend, A., Taylor, P., Schmidt, W., Ferluga, J., Leben, L., Santamaria, M., Atfield, G., Festenstein, H., and Flavell, R. (1982) Expression of murine $H\text{-}2K^b$ histocompatibility antigen in cells transformed with cloned H-2 genes. Nature 298: 529.

Mitchison, N.A. Immunologic approach to cancer. (1970) Transplant. Proc. 2: 92.

Sherman, L. and Maleckar, J. (1988) Genetic and environmental regulation of the cytolytic T lymphocyte receptor repertoire specific for alloantigens. Immunol. Rev. 101: 115.

Graft Specific MHC Class II Gene Expression During Allograft Response

D. S. Kittur, R. Xu, J. F. Burdick, and W. Adler[1]

[1]Department of Surgery, The Johns Hopkins University School of Medicine and Gerontology Research Center, NIA, FSKMC, 4940 Eastern Avenue, Baltimore, MD 21224, USA

INTRODUCTION

Although MHC class II gene regulation has been studied intensively in vitro, the in vivo regulation of this inducible gene has not. Allografts provide a model to study the in vivo induction of these genes since many studies have shown that allostimuli lead to induction of graft specific MHC class II antigens in the graft tissue. A problem with studying expression of graft specific class II genes is the presence of host derived lymphocytes in the graft. Since these lymphocytes also actively transcribe class II genes, the resultant class II expression in allografts is a combination of graft and host specific class II expression. In this report we demonstrate how this difficulty can be overcome and how the allograft system can be used to study the mechanism of class II gene induction in vivo.

METHODS

Oligonucleotide (oligo) probes were developed that hybridized specifically to I-Aß chain genes of $H\text{-}2^d$ and $H\text{-}2^k$ haplotype mice. Vascularized cardiac grafts were transplanted from B10.Br ($H\text{-}2^k$) to B10.D2 ($H\text{-}2^d$) recipients. Mean survival time for hearts in this combination is 8 days. The allografts were excised sequentially from 2nd to 8th post transplant days. mRNA extracted from these hearts were Northern blotted to nylon filters and probed with $H\text{-}2^k$ specific oligo probe. After autoradiography, the filters were stripped and rehybridized with gamma IFN probe (Courtesy of Dr. P. Gray).

RESULTS

All excised hearts revealed histopathology consistent with varying degrees of rejection. mRNA for $I\text{-}A\beta^k$ was not detected in normal hearts but could be detected in transplanted hearts (Fig. 1) starting at day 3 and increasing by day 4. A high level of $I\text{-}A\beta^k$ mRNA was observed in these hearts until the 8th day after transplantation. Cell surface expression of $I\text{-}A^k$ in these hearts appeared to follow $I\text{-}A\beta^k$ mRNA transcription. When the same filters were rehybridized with gamma IFN cDNA probe, hearts from 2nd to 4th days did not have detectable gamma IFN gene transcripts, while the allografts on days 5 and 6 demonstrated gamma IFN specific mRNA (Fig. 2).

I.K. Egorov/C.S. David (Eds.)
Transgenic Mice and Mutants in MHC Research

DISCUSSION

In vitro regulation of the class II gene is mainly under transcriptional control but mRNA stability is also involved in regulation of these genes. Our data suggests that I-Aß gene expression in allografts is regulated at two levels. The marked early rise in mRNA levels is suggestive of increase in transcription of the I-Aß gene; the sustained rise in mRNA level over 3-4 day period may indicate that post-transcriptional mechanisms such as mRNA stability may play a role in expression of this gene in allografts.

An alternate explanation for the sustained rise in I-Aß mRNA levels is suggested by the data on IFN gamma gene expression in these allografts. As noted in the results section, IFN gamma mRNA are detected on 5th and 6th days after allografting i.e., after the I-Aß gene expression has reached its peak. This late expression of IFN gamma gene could indicate that: a) IFN gamma secretion at this time may further stimulate transcription of I-Aß gene or b) IFN gamma may stabilize the previously transcribed I-Aß mRNA and thus be a factor in posttranscriptional regulation of the I-Aß gene in allografts.

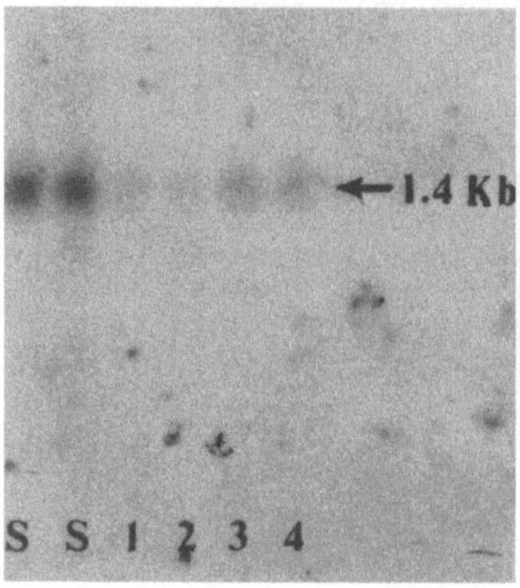

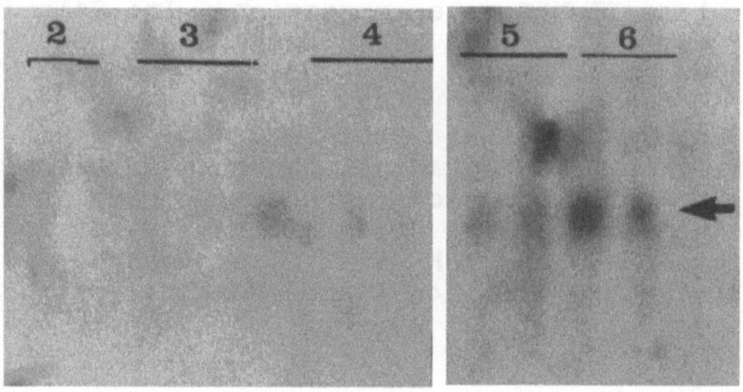

Fig. 1. Graft specific I-Aß^k expression in cardiac allografts. 5μgms of mRNA from cardiac allografts excised at day 5 (lanes 1,2) and day 6 (lanes 3,4) probed with I-Aß^k specific oligo. Lanes 5,S contain 5μgms of B10.Br spleen mRNA.

Fig. 2. Gamma IFN mRNA expression in cardiac allografts. Northern blots shown in Fig. 1 (marked on top of lanes as 5 and 6 indicating number of days after transplantation) and similar blots of mRNA from post transplant days 2,3 and 4 (marked on top of lanes) probed with gamma IFN probe. Arrow indicates expected position for gamma IFN mRNA migration.

Our observation that IFN gamma mRNA transcripts were detected after surface expression of I-A suggests that IFN gamma secretion is the result of increased surface I-A expression rather than the cause of I-Aß gene transcription. This is contrary to a present hypothesis that class II expression in allografts is induced by IFN gamma. However, it is possible that very low levels of IFN gamma mRNA may be undetectable with the northern blot assay used in these experiments. Nonetheless, these experiments reveal that allostimulation results in increased IFN gamma expression which may result in a cascade of further MHC antigen induction and subsequent rejection of the grafted tissue.

In summary, graft specific MHC class II gene expression in vivo can be studied in an allograft model. It may be possible to apply insights derived from these studies to alter immunologically relevant phenomenon and develop new therapeutic strategies for prevention graft rejection.

REFERENCES

Burdick JF, Clow LW (1986) Rejection of murine cardiac allografts. Transplantation 42:67-72.

Collins T, Pober JS, Strominger JL (1987) Physiologic regulation of class II MHC gene expression. In: Solheim BG, Moller E, Ferrone S (eds) HLA class II antigens. Springer-Verlag, Berlin Heidelberg New York, p 14.

Gray PW, GoeddelDV (1983) Cloning and expression of murine immune interferon cDNA. Proc Natl Acad Sci USA 80:5842-5846.

Halloran PF, Wadgymar A, Autenried P (1986) Inhibition of MHC product induction may contribute to the immunosuppressive action of ciclosporin. Prog Allergy 38:258-268.

Kittur DS, Huang SH, Hill G, French AW, Adler W (1987) MHC class II induction in allografts - RNA analysis. Surgical Forum 38:339-341.

Influence of MHC and Non-MHC Genes on Tumorigenesis and the Use of the Recombinant Congenic Strains as a Novel Tool for the Genetic Analysis of Tumor Susceptibility

P. DEMANT[1], L. C. J. M. OOMEN[1], C. J. A. MOEN[1], M. A. VAN DER VALK[1], A. A. M. HART[2], and L. F. M. VAN ZUTPHEN[3]

[1] Departments of Molecular Genetics and [2] Clinical Oncology, The Netherlands Cancer Institute (Antoni van Leeuwenhoek Huis), Plesmanlaan 121, 1066 XC Amsterdam, The Netherlands
[3] Department of Laboratory Animal Science, University of Utrecht, Utrecht, The Netherlands

One of the main developments in the contemporary biology is the rapid progress in the structural analysis of the mammalian genome, concentrated mainly on two species - the human and the mouse. A specific potential of the analysis of the mouse genome resides in the possibility of combining the structural analysis with the study of the genetic determination of various functional traits. This contribution of mouse studies is especially valuable in view of the limited possibilities of direct biological experiments in humans.

The identification of mutant genes either by using as a starting point the extant phenotypic differences between inbred strains, or by producing *de novo* mutants in an inbred strain, has been a powerful approach to the analysis of the genetic determination of many biological phenomena. These methods were applied intensively in the analysis of the Major Histocompatibility Complex (MHC) in the mouse (H-2) and in other species. The analysis of the MHC has been very attractive because of its significance for tissue transplantation, basic immunology, and susceptibility to a variety of diseases. Another, probably not less important factor has been its accessibility for analysis, greatly facilitated by the production of the mouse H-2 congenic strains (Snell, 1958) and in human populations characterization of HLA haplotypes.

These methodological and conceptual advantages of the MHC do not exist with most other genes. Therefore the genetic control of a very large number of biologic phenomena remains only partly elucidated. Most of them are influenced by several non-linked genes which have remained largely unknown. A number of these traits are influenced by both the MHC as well as by several unknown non-MHC genes. Here we shall discuss, as a paradigm for genetics of such biological traits, the relative role of the MHC and non-MHC genes in the susceptibility to lung tumors in the mouse. Subsequently, we shall also describe a novel genetic tool - the Recombinant Congenic Strains (RCS) (Demant and Hart, 1986) - designed to identify those non-MHC genes which control tumor and disease susceptibility, and which eluded the analysis by other genetic methods.

The link between heredity and the development of tumors in mice was first observed with lung tumors (Tyzzer, 1907) and mammary tumors (Lathrop and Loeb, 1913-1914). In spite of considerable efforts and progress made since then, the most important genes which are responsible for the lung and mammary tumor susceptibility remain still mostly unknown. Similarly, genes responsible for susceptibility to tumorigenesis in colon, intestine, liver, and a number of other tissues are largely or completely unknown (Demant et al., 1989). Two approaches to this problem have been attempted. The first is to study whether the already known genes and mutations affect tumor incidence. For example, the classical paper of Lilly et al. (1964), describing the influence of the *H-2* complex on susceptibility to the virally induced leukemia, started a flood of studies investigating the role of *H-2* in tumorigenesis. Although such analysis is of unquestioned importance, the disadvantage of this approach is that one asks the question "what role plays this particular gene in tumorigenesis ?" instead of asking "which genes have the largest effects on tumorigenesis ?" The second approach, although often attempted, still awaits solution.

I.K. Egorov/C.S. David (Eds.)
Transgenic Mice and Mutants in MHC Research

H-2 AND *NON-H-2* GENES AND LUNG TUMOR SUSCEPTIBILITY

That the susceptibility to lung tumor development in the mouse, spontaneous or chemically induced, is genetically controlled, has been recognized rather early, and it turned out that several hitherto unidentified genes are responsible (for review see Demant et al., 1989). In addition, the investigation on the possible influence of some of the **known** genes on susceptibility to lung tumorigenesis, has revealed a role for the mutations *obese*, *dwarf*, *lethal yellow*, the *H-2* complex, and the *K-ras-2* gene. It is not clear to what extent of the difference in susceptibility between inbred strains can be attributed to these five genes, and what is the importance of still unknown genes.

A specific feature of lung tumorigenesis in the mouse is the development of several types of lung tumors. Two major lung tumor types, alveolar and papillary, can be found. These two tumor types have been reported to differ in their biological behaviour (for review see Kauffman et al., 1979). The papillary tumors appear to be more malignant than the alveolar tumors. The morphological characteristics of tumor cells, as revealed ultrastructurally, also differ between the tumor types. Despite these important differences in behaviour and cellular characteristics, both tumor types are believed to originate from alveolar type II cells (Rehm et al., 1988). The alveolar and papillary lung tumors occur in variable proportions in different inbred strains or H-2 congenic strains (Oomen et al., 1983), and hence their development is under different genetic control.

Recently we compared the occurrence of various lung tumor types in N-ethyl-N-nitrosourea-(ENU-) treated mice of the strains O20 and B10.O20/Dem (both $H\text{-}2^{pz}$). The strain B10.O20 produces a moderate number of tumors, all of them of the papillary type. The O20 strain produces not only many more tumors, but also a greater variety of tumor types - papillary, malignant papillary, alveolar, and alveolar malignant (Fig. 1). As all these tumors arise from the same target cell, the alveolar type II cells, obviously the non-MHC genes which differ between these two strains affect the differentiation and progression pathways leading to the resulting different tumor phenotypes. It is of considerable importance to identify the responsible genes, to establish whether the same or different genes are responsible for the quantitative and qualitative features of tumorigenesis, to analyze the mechanisms of their action, and to search for their homologous genes in human. The strain B10.A, which has the same genetic background as

Histological type of lung tumors
in two RCS parent strains

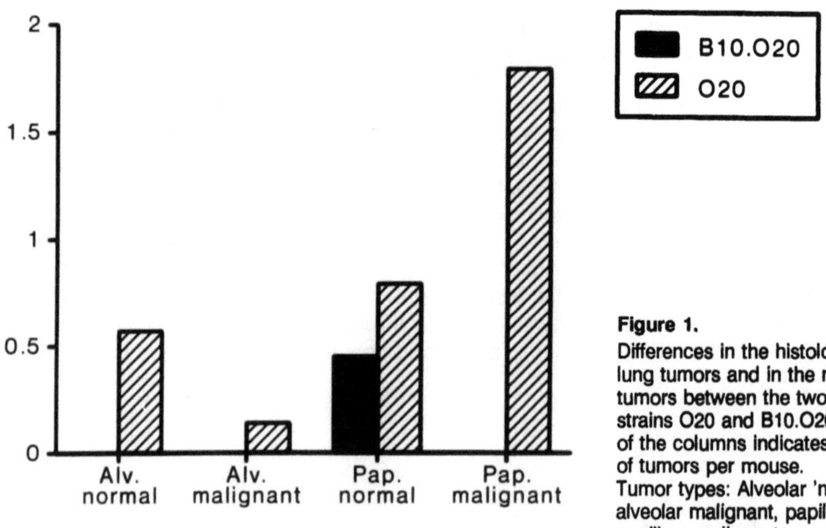

Figure 1.
Differences in the histological type of lung tumors and in the number of lung tumors between the two RCS parental strains O20 and B10.O20. The height of the columns indicates the number of tumors per mouse.
Tumor types: Alveolar 'normal', alveolar malignant, papillary 'normal', papillary malignant.

B10.O20 but carries the *H-2*^{a} haplotype, develops both alveolar and papillary tumors after prenatal ENU-treatment. The recently developed strain O20.A/Dem, which carries the *H-2*^{a} haplotype on the O20 genetic background, is thus expected to produce even more tumors of various types than the strain O20, and to be a suitable tool for analysis of the interaction of MHC and non-MHC genes in determination of the tumor phenotype.

How do the H-2 and the non-MHC genes affect the development of lung tumors ? Several studies (reviewed by Demant et al., 1989) show that number, proliferation rate, and differentiation stage of the target cells may be the important factors in the geneticallly determined susceptibility to carcinogen-induced lung tumorigenesis. Together with the results obtained with tumors induced in transplanted lung tissue and in chimeric mice (see Shapiro and Kirschbaum, 1951; Heston and Dunn, 1951; Mintz et al., 1971), these findings indicate that the genes involved in lung tumorigenesis indeed act primarily at the target cell level.

We addressed the question how the *H-2* complex influences the development of lung tumors (Oomen et al., 1989). In all studies in which appropriate *H-2* recombinant strains were used (Faraldo et al., 1979; Oomen et al., 1983; Miyashita and Moriwaki, 1987; Oomen et al., 1988; Miyashita et al., 1989) the results indicate involvement of more than one gene in the *H-2* complex. We showed that the two lung tumor types (alveolar and papillary) are differently influenced by *H-2* genes in mice from *H-2* congenic strains on the C57BL/10 background, treated either prenatally (Oomen et al., 1983) or postnatally (Oomen et al., 1988) with the carcinogen N-ethyl-N-nitrosourea (ENU). The genes of the *H-2* complex also affect the proportion of alveolar and papillary tumors. After prenatal treatment, the incidence and number of alveolar tumors were influenced by the *H-2* haplotype. For papillary tumors the mean size, but not the incidence or number, was haplotype-related and this *H-2* effect on size of papillary tumors may be due to an *H-2*-associated decrease in growth rate of papillary tumors which probably sets in after two months of age (Oomen et al., 1983).

Recently we obtained evidence that the influence of the *H-2* complex on lung tumor susceptibility may, to a considerable extent, be related to modulation by *H-2* of the glucocorticoid hormone effects on target cells. Glucocorticoid hormones are important regulators of prenatal development and postnatal functioning of lung epithelium. We found that *H-2* influences the enhancing effect of the glucocorticoid hormone on lung differentiation. The stimulatory effect of a prenatal glucocorticoid treatment on the development of alveolar space in fetal lung is significantly affected by the *H-2* haplotype: the increase in alveolar space is three times higher in the C57BL/10 strain (B10, *H-2*^{b}) than in the congenic strain B10.A (*H-2*^{a}). We also found that, when carcinogen and glucocorticoid hormone are administered transplacentally to mouse fetuses, this hormone treatment influences the coincidental ENU induction of lung tumors (Oomen et al., 1989). The effect of glucocorticoid treatment is lung tumor type-specific; it affects the papillary tumors but not the alveolar tumors. The number (multiplicity) of papillary tumors is significantly affected by the hormone treatment, and the effect of treatment is influenced by *H-2* haplotype: in the strain B10 the mean number of the papillary tumors is increased, whereas a decrease occurs in mice of the B10.A strain (Oomen et al., 1989). Taken together these findings suggest that the *H-2* complex affects one or more steps in lung organogenesis and tumorigenesis through influence on hormonal regulation of cell differentiation.

Our experiments indicate that the influence of the MHC genes on the lung tumor development may involve previously unsuspected mechanisms. As some of the susceptibility genes map to individual *H-2* regions, mainly the *IE-S* segment (for references see Demant et al., 1989), it is thus now possible to study the role of the individual genes in this segment, using e.g. the transgenic mice expressing the exogeneous class II genes (Le Meur et al., 1985). The role of other genes located in the center of the *H-2* gene complex (*TNF*α, *TNF*β, *Neu*-1, and the two novel S-region genes decribed by Tsuge et al., 1987 and Lévi-Strauss et al., 1988) needs to be studied as well.

Although the influence of the *H-2* complex on the induction of lung tumors has been extensively studied, the role of the *H-2* is relatively minor, compared to the large differences caused by non-*H-2*-genes. For instance, the highly susceptible strain A/J (*H-2*^{a}) and the highly resistant strain C57BL/6J (*H-2*^{b}) differ approximately 30-40 fold in the number of tumors per mouse induced by a single injection of 1 mg urethane/g body weight. This difference is caused by multiple genes (Malkinson et al., 1985). However, when the H-2 haplotypes of these strains are compared on the same genetic background, they cause about a 2-4 fold difference (Miyashita and Moriwaki, 1987). Similarly, the *K-ras-2* allele is responsible for about a 3-fold increase in the number of urethan-induced tumors, both in the Recombinant Inbred Strains (RIS) and in F_2 hybrids (Ryan et al., 1987). The identification of the other genes contributing to the difference between A/J and C57BL/6J needs to be completed in order to analyze their specific effects on tumor susceptibility.

SEARCH FOR TUMOR SUSCEPTIBILITY GENES AND THE USE OF RCS

In most instances of strain differences in tumor susceptibility several genes are involved. Therefore definition and mapping of the tumor susceptibility genes turned out to be very difficult, or impossible, and these genes remain largely unidentified.

Several approaches have been used in the genetics of tumor susceptibility. Various methods for statistical analysis of such multigenic quantitative differences in segregating populations were developed (reviewed by Roderick and Schlager, 1966; Falconer, 1963). A quantitative phenotype, however, cannot be established reliably in a single mouse, and it is also difficult to characterize such a population for a sufficient number of genetic markers spread over the whole genome. Bailey (1965; 1971) recognized the need for a better analytical genetic tool and devised the Recombinant inbred strains (RIS). A series of RIS is produced, using a number of pairs of F_2 mice from a cross between two inbred strains (Fig. 2A). Each RIS received approximately half of its genes from each parental inbred strain. Consequently, the strain distribution patterns in the RIS of a newly studied gene can be compared with that of all previously typed genes, greatly facilitating the detection of linkage.

Although gene mapping with the help of RIS has been generally very fruitful, unfortunately the RIS did not allow the definition of important non-MHC tumor susceptibility genes. In most studies the tumor incidences of the different RIS form a continuous range of, rather than well-defined classes (Demant and Hart, 1986). For this reason the Recombinant Inbred Strains did not allow the identification of the genes responsible for lung tumorigenesis, even in a very extensive experiment (Malkinson et al., 1985). This is primarily due to additive and non-additive interactions between the several genes controlling tumor susceptibility, which obscure the correlation between the phenotypes of RIS and their genotypes. RIS with similar phenotypes may have different genotypes. A theoretical example of the analysis of tumor susceptibility controlled by three non-linked loci by a series of 16 RIS is given in Table 1A. Genetically quite different RIS strains (1, 2 versus 11, 12, and 5, 6, versus 15, 16) are phenotypically similar. The

Fig 2A + 2B

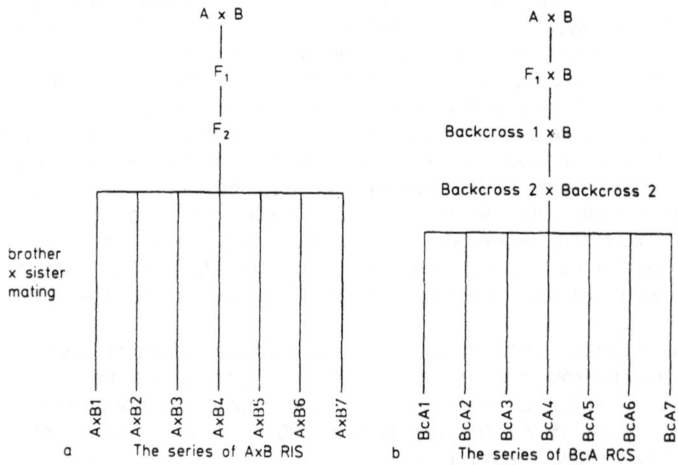

Table 1. TUMOR SUSCEPTIBILITY GENES ANALYZED WITH RECOMBINANT INBRED STRAINS AND RECOMBINANT CONGENIC STRAINS - A HYPOTHETICAL MODEL WITH THREE NON-LINKED LOCI WITH ADDITIVE EFFECTS

PARENTAL STRAINS: A = tumor incidence 90%

B = tumor incidence 10%

Effects of the individual genes on tumor incidence:

LOCUS 1:	ALLELE A 50%	ALLELE B 0%
LOCUS 2:	ALLELE A 0%	ALLELE B -70%
LOCUS 3:	ALLELE A 40%	ALLELE B 80%

A. RECOMBINANT INBRED STRAINS:

	STRAINS															
	1	2	3	4	5	6	7	8	9	10	11	12	13	14	15	16
LOCUS 1	A	A	A	A	A	A	B	B	A	A	B	B	B	B	B	B
LOCUS 2	A	A	A	A	B	B	A	A	B	B	A	A	B	B	B	B
LOCUS 3	A	A	B	B	A	A	A	A	B	B	B	B	A	A	B	B
* INCIDENCE %	90	90	100	100	20	20	40	40	60	60	80	80	0	0	10	10

B. RECOMBINANT CONGENIC STRAINS:

	STRAINS															
	1	2	3	4	5	6	7	8	9	10	11	12	13	14	15	16
LOCUS 1	B	A	B	B	B	B	B	B	B	B	B	B	B	B	B	B
LOCUS 2	B	B	A	B	B	B	B	B	B	B	B	B	A	B	B	B
LOCUS 3	B	B	B	B	A	B	B	B	B	B	B	B	A	B	B	B
* INCIDENCE %	10	60	80	10	0	10	10	10	10	10	10	10	40	10	10	10

* The tumor incidence in each strain is the sum of the effects of individual alleles at the three relevant loci.

linkage indicated by the RIS may be spurious and misleading. For example, the locus number 4 in Table 1 is not involved at all in tumor susceptibility, but exhibits the best correlation with it. As these problems are recurrent when analysing susceptibility to tumors or to disease (for references see Demant and Hart, 1986), we have undertaken a novel approach, which ought to allow a better resolution and subsequent identification of the multiple genes involved in these phenomena.

A new analytic system, the Recombinant Congenic Strains (RCS), has been developed, which is able to transform the multigenic effect into a set of single gene effects, which could be subsequently mapped and analyzed separately (Demant and Hart, 1986). The RCS, similarly as the RIS, are produced from a cross between two inbred strains. While each RIS received genes from each of the two parental strains in equal proportions, in the RCS the genes of one parental inbred strain (the 'donor' strain) are randomly 'diluted' in small portions in the genetic background of the second parental inbred strain (the 'background' strain). This is done by backcrossing repeatedly the 'donor' strain to the 'background' strain, and subsequent brother-sister mating (Fig. 2B). Each RCS contains a relatively limited set of genes (appr. 12.5%) originating from the donor strain. Each of the several genes affecting tumor susceptibility in the 'donor' strain will likely be transferred into a different RCS. Those RCS which received such genes from the 'donor' strain will differ from the 'background' strain in tumor susceptibility, while the majority of the RCS will be phenotypically identical to the 'background' strain. The segregation of the tumor susceptibility genes in RIS and RCS and the resulting phenotypes, using a theoretical situation with 3 unlinked loci, are given in Table 1B. As the RCS become typed for a large number of loci, the strain distribution of each new trait/gene can be compared with that of the previously typed genes to obtain indication of linkage. This indication will require a confirmation in segregation test.

Tumorigenesis differs between inbred strains not only quantitatively, but also qualitatively. Tumors of the same origin may differ between strains in the prevailing type of their differentiation or stage of progression. This 'qualitative' aspect of tumor susceptibility must be kept in mind, although for practical purposes the genetics of tumor susceptibility must be treated as a quantitative trait. Actually, as illustrated by the difference in the phenotypes of the lung tumors found in the strains O20/A and B10.O20/Dem the study of the genes determining 'qualitative' aspects of tumorigenesis is of considerable interest.

Presently, three series of RCS are being produced in our laboratory: C3HcB10/Dem, O20cB10.O20/Dem, and BALB/ccSTS/Dem, with the background and donor strains C3H/Sn and C57BL/10Sn; O20/A and B10.O20/Dem, and BALB/cHe and STS/A, respectively. The observed segregation of the genes of the two parental strains among the RCS does not appear to deviate from the theoretical expectations (Table 2).

Our own pilot experiments, as well as the literature, indicate that these strains are suitable to study genetics of susceptibility to tumors of lung and small intestine, colon tumors, liver tumors, virally and hormonally induced mammary tumors, and virally induced leukemias (Table 3). However, they can be used also to study genetics of various aspects of the immune response, the non-MHC genes responsible for the graft-versus-host reaction, and other functional traits.

ACKNOWLEDGEMENT

We thank Mrs. M. Butzelaar, Mrs. L. Rijswijk, Mrs. A. de Moes-Bezemer, Mrs. E. Delzenne-Goette, and Mr. J. de Moes for technical assistance, and Mrs. M. Sonne for preparation of the manuscript.

Table 2. SEGREGATION OF SIX LOCI IN
CcS/Dem (BALB/c-STS) RECOMBINANT CONGENIC STRAINS[+]

STRAIN	ALLELE					
	Pep-3 Chr.1	Pgm-1 Chr.5	rd Chr.5	Gpi-1 Chr.7	Es-3 Chr.11	Gdc-1 Chr.15
A. PARENTAL						
BALB/cHeA	C	C	C	C	C	C
STS/A	S	S	S	S	S	S
B. RCS						
CcS-A	C	C	C	C	C	CS
CcS-B	C	C	C	CS	C	C
CcS-D	C	C	C	C	C	C
CcS-E	C	C	C	C	S*	C
CcS-F	C	S	C	C	C	C
CcS-G	S*	S*	C	S*	C	C
CcS-H	S*	C	C	S*	C*	C
CcS-I	CS	CS	C	CS	S	C
CcS-J	C	C	C	CS	C	C
CcS-K	C	C	S	C	C	S
CcS-L	S	C	C	S	C	C
CcS-M	C	C*	C	C*	C	C
CcS-N	C	C	S	C	C	S
CcS-P	C	C*	C	C	C	C
CcS-Q	C	C	C	C	C*	C
CcS-R	C	C	C	C	C	C
CcS-S	C	C	C	C	C*	C
CcS-T	C	CS	C	C	C*	C
CcS-U	S*	C	C	C	C	C
CcS-W	C	C*	S	CS	C	C

[+] Tested at 5-7th generations of inbreeding.

* Predominant allele (> 80%).

Table 3. TUMOR SUSCEPTIBILITY CHARACTERISTICS OF THE INBRED STRAINS USED TO PRODUCE THE RCS

	HcB/Dem series		OcB/Dem series		CcS/Dem series	
	C3H/HeASn	C57BL/10ScSnA	O20/A	B10.O20/Dem	BALB/cHeA	STS/A
lung tumors	N.D.	Resistant	Susceptible	Resistant	Susceptible	Susceptible
intestinal tumors	N.D.	Susceptible	Resistant	Susceptible	Resistant	Resistant
mammary tumors	Susceptible	Resistant	Susceptible	Resistant	Susceptible	Resistant
colon tumors	N.D.	N.D.	N.D.	N.D.	Resistant	Susceptible
Gross MuLV leukemia	Susceptible	Resistant	N.D.	N.D.	N.D.	N.D.
liver tumors	Susceptible	Resistant	N.D.	N.D.	N.D.	N.D.
plasmocytoma	N.D.	N.D.	N.D.	N.D.	Susceptible	Resistant
exocr.pancreatic nodules	N.D.	N.D.	Resistant	N.D.	Susceptible	N.D.

The data stem from our preliminary screening (M.A. van der Valk, L.C.J.M. Oomen et al., in preparation) with the exception of colon tumors (Fleiszer et al., 1987) and plasmocytoma (Hilgers and Arends, 1985). The lung, intestinal, liver, and mammary tumors were induced by N-ethyl-N-nitrosourea treatment, accompanied, in the case of mammary tumors, with hormonal stimulation. The mammary tumors in the C3H/C57BL/10ScSn combination were induced by C3H-MTV (Mühlbock and Dux, 1981). The data on MuLV-induced leukemia are from literature (for references see Lilly and Mayer, 1980), the preneoplastic pancreatic nodules of exocrine cells in mice treated with ENU and hypophyseal isografts were observed by Dr. M.A. van der Valk at The Netherlands Cancer Institute (unpublished).

REFERENCES

Bailey DW (1965) A search for genetic background influences on survival time of skin grafts from mice bearing Y-linked histoincompatibility. Transpl 3: 531-534

Bailey DW (1971) Recombinant inbred strains: An aid to finding identity, linkage, and function of histocompatibility and other genes. Transpl 11: 325-327

Demant P, Hart AAM (1986) Recombinant congenic strains: A new tool for analyzing genetic traits determined by more than one gene. Immunogenetics 24: 416-422

Demant P, Oomen LCJM, Oudshoorn-Snoek M (1989) Genetics of tumor susceptibility in the mouse: Major histocompatibility complex and non-MHC genes. Advances in Cancer Research 53: 117-179

Falconer DS (1963) Quantitative inheritance. In: Burdette WJ (ed) Methodology in Mammalian Genetics. Holden-Day, San Francisco, pp 193-216

Faraldo MJ, Dux A, Mühlbock O, Hart G (1979) Histocompatibility genes (the H-2 complex) and susceptibility to spontaneous lung tumours in mice. Immunogenetics 9: 383-404

Fleiszer D, Hilgers J, Skamene E (1987) Multigenic control of colon carcinogenesis in mice treated with 1,2-dimethyl-hydrazine. Curr Topics Microbiol Immunol 137: 243-249

Heston WE, Dunn TB (1951) Tumor development in susceptible strain A and resistant strain L lung transplants in LAF1 hosts. J Natl Cancer Inst 11: 1057-1071

Hilgers J, Arends J (1985) A series of recombinant inbred strains between the BALB/cHeA and STS/A mouse strains. Curr Topics Microbiol Immunol 122: 31-37

Kauffman SL, Alexander L, Sass L (1979) Histologic and ultrastructural features of the Clara cell adenoma of the mouse lung. Lab Invest 40: 708-716

Lathrop AEC, Loeb L (1913-1914) The incidence of cancer in various strains of mice. Proc. Soc. Exp. Biol. Med. 11: 34-38

Le Meur M, Gerlinger P, Benoist C, Mathis D (1985) Correcting an immune-response deficiency by creating E_α gene transgenic mice. Nature 316: 38-42

Lévi-Strauss M, Carroll MC, Steinmetz M, Meo T (1988) A previously undetected MHC gene with an unusual periodic structure. Science 240: 201-202

Lilly F, Boyse EA, Old LJ (1964) Genetic basis of susceptibility to viral leukemogenesis. Lancet *ii*: 1207-1209

Malkinson AM, Nesbitt MN, Skamene E (1985) Susceptibility to urethan-induced pulmonary adenomas between A/J and C57BL/6J mice: Use of AxB and BxA recombinant inbred lines indicating a three-locus genetic model. J Natl Cancer Inst 75: 971-974

Mintz B, Custer RP, Donnelly AJ (1971) Genetic diseases and developmental defects analyzed in allophenic mice. Int Rev Exp Pathol 10: 143-179

Miyashita N, Moriwaki K (1987) H-2-controlled genetic susceptibility to pulmonary adenomas induced by urethane and 4-nitroquinoline 1-oxide in A/Wy congenic strains. Jpn J Cancer Res (Gann) 78: 494-498

Miyashita N, Moriwaki K, Migita S (1989) The H-2 class II genes and the susceptibility to the development of pulmonary adenoma in mice. Immunogenetics 29: 14-18

Oomen LCJM, Demant P, Hart AAM, Emmelot P (1983) Multiple genes in the H-2 complex affect differently the number and growth rate of transplacentally induced lung tumours in mice. Int J Cancer 31: 447-454

Oomen LCJM, van der Valk MA, Hart AAM, Demant P, Emmelot P (1988) Influence of mouse major histocompatibility complex (H-2) on N-ethyl-N-nitrosourea-induced tumor formation in various organs. Cancer Res 48: 6634-6641

Oomen LCJM, van der Valk MA, Hart AAM, Demant P (1989) Glucocorticoid hormone effect on transplacental carcinogenesis and lung differentiation: Influence of histocompatibility-2 (H-2) complex. J Natl Cancer Inst 81: 512-517

Rehm S, Ward JM, ten Have-Opbroek AAW, Anderson LM, Singh G, Katyal SL, Rice JM (1988) Mouse papillary lung tumors transplacentally induced by N-Nitrosoethylurea: Evidence for alveolar type II cell origin by comparative light microscopic, ultrastructural, and immunohistochemical studies. Cancer Res 48: 148-160

Roderick TH, Schlager G (1966) Multiple factor inheritance. In: Green EL (ed) Biology of the Laboratory Mouse. McGraw-Hill, New York, pp 151-164

Ryan J, Barker PE, Nesbitt MN, Ruddle FH (1987) KRAS2 as a genetic marker for lung tumor susceptibility in inbred mice. J Natl Cancer Inst 79: 1351-1357
Shapiro JR, Kirschbaum A (1951) Intrinsic tissue response to induction of pulmonary tumors. Cancer Res 11: 644-647
Snell GD (1958) Histocompatibility genes of the mouse. II. Productions and analysis of isogenic resistant lines. J Natl Cancer Inst 21: 843-877
Tsuge I, Shen FW, Steinmetz M, Boyse EA (1987) A gene in the H-2S-H-2D interval of the major histocompatibility complex which is transcribed in B cells and macrophages. Immunogenetics 26: 378-380
Tyzzer EE (1907) A study of heredity in relation to the development of tumors in mice. J Med Res 17: 199-211

Subject Index

Aα gene 133–141
  promoter 134
  regulation of expression 133–141
Aβ gene 143–152, 296
abscesses 263, 264, 272
accessory molecule 131, 188, 250
  see also CD4, CD8, Lyt-2, LFA.1
additive effects 303
adenocarcinoma 95
  kidney 95, 96
  lung 95, 96
  mammary gland 96
  seminal vesicle 96
affinity 43
α_1 domain 77
  mutations 67
α_2 domain 77
  mutations 67
α-helix, *see* crystal structure model
allo-antigens, *see also* CTL
  reactivity, models for 42, 57
  recognition 33, 74
allografts 226
alloreactivity 286, 292, 299
Alu repeat 156
alveolar 300, 301
  malignant 300
amino acid sequence/substitutions
  Aβ molecule 37, 239
  allelic variation 42, 43
  functionally silent 56–58
  H-2 mutants 7, 15, 16, 26, 28, 34, 35, 44, 48–50, 56–58, 214–216
  HLA variants/mutants 69, 70, 77–87, 254, 260
A molecules, *see* class II molecules
ankylosing spondylitis 259, 260
antigen
  binding/recognition cleft/site 14–16, 31, 50, 56, 57, 72, 254
  presentation 56, 114, 235, 251–252, 260
  presenting cell (APC) 39, 151, 219, 247–255
  recognition, *see* recognition of antigens
antigenicity 68
anti NK1.1 treatment 284
arthritis 259–274
  acute 261
  chronic 261
  experimental model of 261
  inflammatory 261
  peptoglycan-induced 271
  post infectious 259
  post infectious reactive 275
  reactive 268, 271
  type II collagen-induced 269, 270
arthropathy, reactive 261, 262
  animal model 261, 265
associative recognition 293
autoantibodies 260
autoimmune response 265

B cell
  B-T hybrids 134
  class II negative mutant 135, 155
  clone 36
  signaling 36
β_2-microglobulin 77, 109, 114
  expression 120
  human 30, 120, 185–187, 199–212, 268, 269, 272
  mouse 176, 261, 268–272
β-pleats, *see* crystal structural model
binding constant, Kd 43, 45
blockage of pregnancy 62
blot-hybridization 117–120
box
  CCAAT 103, 126, 134, 138
  TATA 103, 126
  W 135
  X 134–141
  Y 134–141
BSA 147

cAMP induction 138
cancer
  biotherapy 294
  immunotherapy 276, 285, 290, 294
  resistance 277
carcinogen-induced lung tumorigenesis 301

carcinoma (tumor/cells)
  embryonal 125
  prostate/seminal vesicle 95
CD4 59, 221, 248
CD8 42, 59, 114, 169, 183–185, 221, 248
cell surface expression
  $A\beta^k$ 143–152
  $DR\alpha/E\beta$ 251–255
  $H\text{-}2D^P$ 29
  HLA-B7/hβ2m 191–193
  HLA-B27/hβ2m 205–212
  I-E 219–235
  Qa, Tla 109
chemical mutagens 3
CH12.LX 36
chimeric genes, *see* hybrid genes
chloramphenicol acetyltransferase (CAT)
  gene 96, 105
CIS-active sequences/regulatory elements
  114–115, 120, 125, 134–141, 157
class I antigen/molecule
  domains 40–45, 165, 168, 170, 183–185, 239
  genetically engineered/recombinant, *see* hybrid molecules
  membrane expression 39
  see also cell surface expression
  soluble 39–45
class I genes
  expression 102–107, 114, 125–131, 173
  inserted into yeast 18–22
  mutation, *see* H-2 mutation
  regulatory element (CRE), *see* element
  repression of transcription 125
  transcription 125–131
  transfection 18
class II antigen/molecule
  Aβ/Aα pairing 150, 229–245
  allelic restriction 241
  α chain 133, 247
  β chain 133, 247
  cell surface expression 152–193, 219–245
  Eβ/Eα pairing 229–245
  heterodimers, murine/human 250–255
  inactivation mutations 250
  mixed isotype hybrid dimer 230–245
class II gene, *see also* A, E genes
  expression 133–152, 219–227
cleavage site, proteolytic 55, 56
clonal diversity 75
clones, *see* B-cell, T-cell
CML assay 57, 70–72, 74, 82, 90, 167, 171, 175–177, 186–188, 194, 195, 200, 211, 278, 293
collagen, type II 269
compensatory effect 73
complex, TCR:HLA 74
computer algorithm 17
congenic strains of mice 59, 229–235
consensus/response sequence (ICS/IRS)
  CRE 138
  interferon-α/β 119, 121–122
  interferon (ICS) 125
  X box 138
conserved sequences (UPS1, UPS2) 103
coordinate expression 133
cosmid clones, *see also* probes, DNA
  cos B16A (BRw53βIII) 248
  cos H8 286
  cos $H\text{-}2^k$8.1 238
  cos II-5407 (DQ3α/β) 248
  cos SB-2 (DP4α/β) 248
  cos T9C (DRα) 248
  cos 4-1 (DR3β) 248
cross-competition assay 137
crossover, unequal 14
  see also recombination
crystal structure model of HLA-A2
  α-helix 51, 52, 56, 70, 72, 77, 79
  β-pleated region/strand 51, 52, 56, 77
  class I molecule 14, 26, 56–58, 73–75, 77, 79
  class II molecule 254
  groove 72
  loop structure 51
  peptide binding pocket 72, 75, 77, 165
  side chains 72, 79
CTL (cytotoxic T lymphocyte) 3, 34, 57, 114, 260, 285, 292–294
  alloreactive 44, 57, 59–61, 67–75, 82–87, 165, 206, 255
  blocking/inhibition by mAbs 193–195, 199–203
  class I restricted 53, 274
  cold target inhibition 201
  effector cell 28
  peptide inhibition of 44
  peptide recognition 81–87
  precursor frequency 165, 167, 180, 201
  reaction patterns of clones 69, 71, 72, 81, 83, 186–188, 194, 195
  recognition 70, 89
  response 35, 268
  see also T cell
  virus specific 29, 184, 211
  xenoreactive 78, 82–87, 165–213
CTL-defined (CD) genes/antigens 60–61
cycloheximide 127
cytotoxicity assay, *see* CTL

DAUDI cells 175
densitometry 111
diarrhea 272
differentiation 37

disease
association 259
models 273
susceptibility 237, 259–306
DNA
clones/probes, *see* probes, DNA
mediated gene transfer, *see* transfection assay
palindromic sequences 18
sequencing 18, 26, 48–49, 52, 68, 128, 134, 139
DNaseI hypersensitive sites
Aα gene 134–136
domains
class I, II, *see* class I, II
dot blot analysis 116
D^p, *see* H-2D^p
DQ, *see* HLA-DQ

Eα gene
regulation of expression 136
Eβ gene
cell surface expression 229–235
regulation of expression 136
EBV cell lines 175, 193, 194
effector cell, *see* CTL
electrophoresis, 2D-SDS 288
element, DNA sequence 92, 96, 102, 125–131
BRE 138
cAMP responsive 103, 125–131, 138
class I regulatory (CRE) 125
control 101
CRE/ATF 139
enhancer 99, 100, 103, 111–113, 121, 122, 126, 136
estrogen response 129
interferon response 103–107
negative regulatory (NRE) 100, 103–107
positive regulatory 103–107, 126
regulatory 100, 109–114, 209
silencer 103
EL-4 cells
metastases 281
transfectants 179–185
HLA-A2.1, -A2.2
HLA-B7
HLA-B7/H-2L^d chimeric molecule
H-2L^d
H-2L^d/HLA-B7 chimeric molecule
ELISA assay 148
embryogenesis 125
ENU 300, 301
envelope proteins 260
epitope 3, 265
allospecific CTL 74, 165–203
CTL-defined 77–87, 89–90, 167–170
K^b serological 50, 52
L^d serological 50, 52
serological 49, 54, 77–87
T cell 67, 68
xenospecific CTL 165–203
ethnic group 259
exon
class I genes 14, 15, 18, 40, 49–52, 68, 111, 117, 168
shuffling 39, 40, 50–52, 77–179
expression of transgene, *see* transgene
expression vector 92
class I 49, 52–55

FACS analysis 27, 57, 68, 145, 146, 174, 188, 192, 206, 210, 214–215, 220, 222, 231–232, 238, 240–242, 249, 251, 268
factor
binding site 100, 127, 134
cognate binding 107
constitutive 125–131
DNA binding 134, 139
interferon-inducible 125
nuclear 135–141
trans acting 106, 107, 125–131
family, multigene 14, 102, 109
fibrosarcome
Bc20 (H-2^b) 166
3′ flanking sequences
class I genes 114, 121
5′ flanking sequences
class I genes 103, 110–116
deletions 104
flow cytometry, *see* FACS
F9 cells 127, 128
footpad swelling assay 29
functions of MHC molecules 3, 33, 38, 39, 43, 45, 48, 53, 102, 114, 122, 143, 161, 210–212, 249, 251

Gel
mobility shift assay 127, 128, 130, 135
retardation assay 137
gene conversion 8, 14–22
in yeast 18, 19
models of 20–22
gene dosage 143
gene expression, *see* cell surface expression, regulation of expression, tissue-specific-expression
gene order 61
genetic structure, minor H loci 59
genetics of tumor susceptibility 302
genomic clone 110, 143, 151, 205, 238
see also probe
glucocorticoid hormone 301
glycoprotein

cell surface 33, 47, 50, 54, 114, 133
graft rejection 3, 4, 17, 26, 59, 110–112, 173, 202, 214, 239, 242, 245, 276, 277, 285

haplotype-matched/mismatched combinations 237–245
H-2 mutation/mutant frequency 3–10, 15
in vitro 26–31, 33–35, 47–55
in vivo 3–10, 14–15, 19–22, 26, 62–64, 89
rates, *see* frequency
spontaneous 4, 5
H-2D gene
mutants 3–6, 8, 9
protein 39, 42
$H\text{-}2^{d}$ mutants 5–6
$H\text{-}2D^{b}$ gene product
expression 174
$H\text{-}2D^{P}$ mutants
in vitro 26–31, 33
$H\text{-}2K^{b}$ mutants 5–10, 40
"bg" group 7
$H\text{-}2K^{bm1}$ mutant 62–64, 214–216
reverse mutation to K^{b} 9–10
$H\text{-}2^{bm3}$ mutant 56–58, 277–287
$H\text{-}2^{bm5}$ mutant 62–64
$H\text{-}2^{bm7}$ mutant 62–64
$H\text{-}2K^{bm8}$ mutant 62–64
variants of 48–50, 56–58
$H\text{-}2K^{bm10}$ mutant
gene expression 43
sequence 44
$H\text{-}2^{bm12}$ mutant 5, 62, 237
$H\text{-}2^{bm23}$ mutant 277
$H\text{-}2^{bm26}$ mutant 279
$H\text{-}2K^{km2}$ mutant 16, 18
H-2 negative variant tumor cells 282
H-3 locus 59–61
H-4 locus 59–61
hematopoeisis, extramedullary 141, 152
heterocyclic effect 34, 73, 74
(H,G)-A--L synthetic antigen 143, 147–148
Hh-1 antigens 284
hind limbs 262, 263
histocompatibility
antigens/loci, major 39
antigens/loci, minor 59–61, 279, 285
gene mutation in vivo 3, 276
gene mutation rates 3–13
human minor 136
H-2 279
non-H-2 3
HLA-A2 antigen/gene 70
A2.1, A2.2, A2.3 77, 179–188
carbohydrate moiety 186–187
crystal structure of, *see* crystal structure model
mutants 78–87, 186
see also transgenic mice
peptides 80
response to (in TGM) 167
HLA-B1 Antigen/Gen
expression 114–122, 191–193
see also transgenic mice
HLA-B27 antigen/gene 67–75
B*2701-05 genes 67–74, 170, 193–212
disease association and arthritis 205, 212, 259–275
expression 173, 208–210
mutants 67–75, 274
see also transgenic mice
peptides 260, 274
subtypes 67–75, 260–274
HLA-B51 antigen/gene 89–90, 206
HLA-Bw52 antigen/gene 89–90
HLA-Cw3 antigen/gene 162, 165–171, 191–196
response to (in TGM) 191–196
HLA-DQa gene
expression 155–157
hormone receptor superfamily
DNA binding region 128, 129
nuclear 128, 129
steroid/thyroid 128
HPLC 42
hybrid gene/molecule
class I 14, 18, 19, 39–45, 47–55, 89–90
class II 36–38
class I/class II 53–55
HLA/H-2 165, 168, 179–188
H-2/Q10 40
hybridoma
alloreactive T cell 39, 42–45
hygromycin-resistance gene 156
hyper-polymorphic residues 14
hypersensitivity, delayed type 28

$I\text{-}A^{b}$ mutation, *see* $H\text{-}2^{bm12}$
Ia chains 36
see also class II antigen, $A\alpha$, $A\beta$, $E\alpha$, $E\beta$
I-A gene/antigen, *see also* $A\alpha$, $A\beta$
in T cell clonal deletion 229–235
I-E gene/antigen, *see also* $E\alpha$, $E\beta$
in T cell paralysis/clonal deletion 219–235
Ig secretion 36
IL-2 42
IL-4 131, 136, 237
immune response 3, 114, 143, 147–148, 151, 237, 261, 265, 281
genetic control of 3
infection
susceptibility to 149
inflammatory
bowel disease 261
response 269

influenza
  specific CTL 18, 166, 180–181, 184, 185, 206, 273
  virus A/JAP/57 80
interaction
  T cell receptor/MHC molecules 30, 43, 183
  CD8 dependent 42, 183
interferon (IFN)
  α 119, 125
  β 119, 125
  γ 114, 118–122, 125, 133, 135–136, 188, 237
intracellular transport 41
intron 51
inverted repeat 126

joint inflammation 268
juvenile chronic polyarthritis 289

K^b, *see* H-2K^b
kinetics of secretion, H-2L^d/Q10^b molecule 41
Klebsiella 260, 265
  K. pneumoniae 260
  nitrogenase 260, 261
K-ras-2 gene 300, 302

L cells 26
  transfectants 28, 29, 33, 39, 41, 43, 49, 52, 54, 105, 110, 166, 167, 186
leucine-zipper proteins 130, 139–141
leukemia
  AKR 285, 290
  gross MuLV 306
  virally induced 299, 304
LFA.1 molecule 193
library
  phage expression 125
    λ gt11 (murine cDNA) 122, 137
    λ ZapII 147
limiting dilution analysis 68, 167, 180–185, 198–207
  split-well 198–203
loop structure, *see* crystal structure model
loss of H-2 products 282
lymphoblastoid cell lines (human) 68, 78, 179, 185, 207
lymphocytic choriomeningitis virus (LCMV) 29, 30, 33, 35
lymphokines 133, 138, 141, 293
lymphoma 95, 96, 277
  B-cell 93, 156
  Ia-negative mutant 156
  see also EL-4
  T-cell 93, 134
Lyt-2 (Ly-2/3) molecule 165–172, 193

male sterility 161–164
malignant proliferation 98
mammary carcinoma C127 185
marker genes
  expression of 100
mastocytoma tumor/cells 134, 185, 191
mating choices 62
metastasis (tumor) 276, 280, 294
  host resistance to 276–281
MHC
  allelic polymorphism 14, 15, 214
  class I, II, *see* class I, II
  diversity of alleles 14, 62
  evolution 15
  human, *see* HLA
  mouse, *see* H-2
  restriction, *see* restriction
  swine (SLA), *see* PD1, PD7, PD14
microgene conversion, *see* gene conversion
micro-recombination 6
MLC/MLR, *see* proliferative response
Mls locus 229–235
molecular mimicry 260, 261
monoclonal antibodies (mAb), specificity
  β2m 192
  HLA class I 78–80, 118, 173, 174, 183, 192, 210, 260
  HLA class II 248, 249
  H-2 class I 27, 49, 50, 52, 168–169, 183, 192, 210, 215–216
  H-2 class II 37, 145, 146, 230–232, 238
  LFA.1 193, 194
  Ly-2/3 193, 194, 220
  SLA (PD1) 112, 200
  TcR Vβ chain 219, 221
  T3 molecule 222, 223
mortality 262, 272, 273
mosaicism
  germline 115, 144
  somatic tissue 115
motif, *see* element, regulatory sequences
mouse mammary tumor virus (MMTV) 92–101
  LTR 93, 100
  promoter 100
  tissue-specific expression 92, 99
mRNA
  class I 102
  class II 143, 145, 296
  ICSBP 130
  IFN-γ 296
  species 129
mutagenesis
  random by oligonucleotides 26–27, 33
  site directed 26–27, 37, 47, 48, 50–54, 67, 70, 77–78, 82, 126, 214, 274

mutation/mutants
 deletion 8, 20, 39, 104–107, 112
 double 73
 HLA genes, *see* HLA
 H-2 genes, *see* H-2 mutation
 frameshift 39
 insertional, *see* transgenic mouse
 I region, *see* H-2^{bm12}
 reverse 9–10
 spontaneous 3–13, 276
 substitution 8
 site directed, *see* mutagenesis
myeloma tumor/cells 134

natural resistance 278–279, 281–283
negative regulation 96
NK (natural killer) cells 278, 281–284
 mediated lysis 282
Northern blot analysis 104, 209, 297
nuclear proteins 111

odortypes/odor phenotypes 62, 63
olfactory recognition 62
oligonucleotide 26, 47, 51, 53, 67, 68, 127, 136–139, 242, 297
oncogene 92, 93
 c-erb 128
oncogenic potential 92

pancreatic
 acinar cells expressing I-E 221
 B cells expressing I-E 221
P815 cells 199, 206
PD1 (swine MHC) antigen/gene
 expression 103–107
 regulatory elements 106
 transgene 103
PD7 gene 109–113
PD14 gene 109–113
peptide (antigenic)
 binding 56, 81, 237
 binding site 30, 56, 74–75
 endogenous 75
 H-$2K^b$ 43, 45
 immunogenic 30
 interaction with MHC 80, 81
 presentation of 34
 procession 198
 recognition of 80–87
 self 30, 43, 87, 165
plaque assay 36, 37
plasmid
 I-A genes 36
 pFR400 (SV-40 early promoter) 161
 pSV2neo 36
 pTZ19R 214
 puc18 238
plasmocytoma 306
polyarthritis
 acute 270
 T-cell dependent erosive 270
polymerase chain reaction (PCR) 47–49, 54
polymorphism 14–16, 60, 68
probes/clones, DNA
 $A\alpha^k$ from cos H-2^k8.1 238
 $A\beta^{k}$*MB 238
 class I 18, 117, 206, 209
 class II 143, 238
 CRE/ICS-binding proteins 127
 DP7.4, PD7.7 109–110
 Hf220 136
 HLA-B7 clone pJY150 117
 HLA-B27 261
 HLA-B27 clone cd2.6 205
 HLA-B specific 206, 209
 H-2RIIBP 128
 H-$2K^b$#152 variant 214
 human actin cDNA 206
 interferon-γ 296
 pIAβ^k-1 143
 PD1-specific 103
 XAα(Aα gene X box) 137, 138
 λmXBP-1 and 2 (murine X-box binding protein) 138
processing system 187
proliferative response 28, 60, 220–223, 239, 242–244, 249, 251–253, 280
promoter 100, 105, 136
 adenovirus early 139
 dq-7 155
 eukaryotic 134
 H-$2L^d$ 126
 tissue-specific 157
protein
 CRE-binding 125, 127
 DNA-binding 127
 ICS-binding 125, 127–131
 region II binding 128
 sequence, *see* amino acid sequence
provirus, endogenous 93, 96
psoriasis 259

Q7 gene 126
Q10 gene 39–44, 126
Qa subregion genes 14, 15, 102, 125

radio-binding assays 286, 287
radiographic sacroiliitis 259
receptor, T cell (TCR) 14, 15, 30, 35, 39, 56, 72–74, 77, 81, 133, 152, 179, 183–185, 247
 α and β gene rearrangement 194, 196
 allospecific 40, 43–45
 antigen specific 40

crosslinking experiments 224
MHC-restricted 90
see also repertoire
T3/antigen receptor complex 223
Vβ17a gene 219–226
Vβ5 gene 220
Vβ8 gene 220
Vβ6 deletion 229–235
Vβ11 deletion 233–234
recognition of peptide 81
recombinant congenic strains 299, 303–305
recombinant inbred strains 302–304
recombinants, intra-H-2 230
recombination 18, 20–22
between H loci 60, 61
hotspot 21
in vitro 26
meiotic 18
reciprocal 14
regulation of expression
class I genes 103–107, 125–131, 296
class II genes 133–157
post-transcriptional 119, 122, 136, 186, 298
tissue specific 96
transcriptional 102, 119, 122, 133
regulatory elements, *see* element
regulatory sequences/motifs
class I genes 125, 130
class II genes 133–141
Reiter's syndrome 259–261
repertoire
allogeneic CTL 89
B cell 179
selection of 248–255
TCR 87, 191, 247
T cell 152, 166, 198, 211, 247, 250–255
reporter gene 105, 156–157
restriction map
HLA-B7 115
HLA-B27 (cosmid cd2.6) 207
H-2D^k, H-2K^k 19
K^b/L^d 49, 53
PD1 106
PD7.4 110
PD7.7 110
retroviral R1, R2 156
restriction, MHC
class I restricted 30
class II restricted 59, 143, 237, 255
HLA restricted 78, 80, 173, 184, 274
MHC restricted 3, 33, 34, 36, 59, 165–203
retrovirus, endogenous 92
rhabdomyosarcoma
cell line RD 185, 187
rheumatic disease 274
RNA
H-2 120
HLA-B7 116, 117, 120
PD1 103
probes 18
RNase T1 protection assay 94, 95

Salmonella 259
segregation analysis 60
selective forces 15–17, 61
sensory discrimination/distinction 62, 64
sequence, *see* amino acid or DNA sequence
Shigella 259, 265
antibody level/titres 272
S. flexneri 260, 261, 271, 272
S. sonnei 260
signal transmission 37, 38
simian CV101 cell line 187
site directed mutagenesis, *see* mutagenesis
skin grafting 3, 4, 26, 110, 174, 216, 239, 261, 277
slot-blot analysis 118, 121
somatic cell hybrids 134, 187
S1 nuclease analysis 111, 286
Southern blot analysis 116, 196, 208
specific pathogen-free animals 149
spermatogenesis 162–164
splicing 47–49, 54, 138
spondylitis 259
ankylosing 268
spondyloarthropathies 259
stereochemical requirements 74
streptococcal cell wall peptidoglycan 269–271
structure/function relationships
class I MHC molecular 77
class II (Ia) molecules 38, 237
H-2 molecules 54, 214–216
HLA molecules 274
substitutions
charge 86
conservative 86
neutral 86
see amino acid sequence
see DNA sequencing
SV40
early region genes 92, 93, 161
large T antigen 92, 93
promoter 156
small t antigen 92, 93
transforming ability 94
5′ LTR 155–156

T cell
activation 219
autoreactive 247
B-T hybrids 134
clonal anergy 219–226

clonal deletion 219, 221–226, 229, 247, 252, 255, 265
clones 29, 45, 59–61, 68–71, 80, 187, 188, 193
cytotoxic, *see* CTL
helper 59–61, 247
Ia-positive 156
paralysis 219
receptor, *see* receptor, T cell
recognition by 72, 75, 179, 247–255
subsets 59
technology
recombinant DNA 173
transgenic 173
testosterone 162
tetrad, recombinant 19
TH-defined (HD) genes/antigens 60–61
three-dimensional model, *see* crystal structure model
thymoma 134
tissue-specific antigens 186
tissue-specific expression
class I antigens 39–45, 103–107, 114, 208
class I, non-classical 109–113
Tla subregion 14, 102, 125
tolerance, immunological
generation of 44, 185, 237
mechanism 39, 44–45, 219–235, 253
see also T-cell
self 44, 247, 253
transcomplementation 244
transcription 92–101, 110–113
activation domain 129
activating factor 139
initiation 105, 109, 134
regulation 102, 125, 136, 208
suppression 100
unit dg-7/hph 155
transfectant, *see* transfected cell
transfected
cell 28, 29, 37, 42, 50, 57, 70–72, 80–81, 89–90, 102, 156, 166, 286, 288, 292, 294
gene 34
gene product expression 27, 29, 38
transfection assay 26, 40, 68, 134, 155, 240, 286
transgene
CD1 93
copy number 110, 115, 116, 121, 143, 145, 146–152, 207–209, 214, 241, 244, 249
expression 93–101, 144, 172, 207–209, 214–216, 265
ET450 93, 99
H-2D^d 44
H-2K^k 45
HLA-A2.1 79, 179
HLA-B27/L3 170
HLA-B27.2 162, 173, 199
HLA-B27.5 162
HLA-Cw3 162, 165–171
Hu-β2m 162, 173, 199
insertion/integration site 117–119, 122, 144, 206, 249
LST 95, 99
LTag 92, 93, 95, 99
LTM 95, 99
PD1 103–107, 162, 196–198
position effect 117, 119, 122, 206, 244, 249
p1BCAT 93, 99
toxicity 151, 152, 244
SLA-PD1, *see* PD1
transgenic mouse 92–101
Aβ^k 144
Aα^k 147, 233, 239
Aβ^k, Aα^k 147–148, 233
Aβ^k 239
BG.Eα^k 232
B27dTGM 199
class I 39, 44, 103–107
class II 143–152, 219
C57BL/10 DQ3.2α/β 248
El-I-E 219
H-2 (D8) 282
HLA-A2.1 79, 179–188
HLA-B27 162, 198–213, 260–274
H-2^b.B27/a3 TGM 171
HLA-B7/hβ2m 191
insertional mutation 161
INS-I-E 219
LTag 92
reproductive performance 161–164
survival of 149
27TGM 173
MTGM 173
27MTGM 173
-1.4kbEα^d 219
107-1 219
Translation
in vitro 128
transmembrane region/exon 39, 40
transplantation antigens 33, 102, 110–114, 174, 182, 216, 261
trans-regulatory mechanisms 120, 125–131, 134, 157
truncation 121
tumor
associated transplantation antigens (TATA) 286, 293
chemically induced 276
excision 280
immunity 280, 285–286, 290, 294

progression 300
resistance 286
specific antigens 276, 281, 294
virally induced 276
tumorigenesis 93
H-2 role 299
tumor necrosis factor 125
tumors in animals
colon 304, 306
incidence 299
liver 304, 306
lung 299, 300–308
mammary 299, 304, 306
small intestine 304, 306
susceptibility 299, 300, 304
susceptibility genes 302–304
urethane-induced 302

Undulating (un) locus 60
urethane 301

vaccinia viruses 273
variants, *see* mutants
vector
pBR322 67
pUC18 67
pBS 128
pLJ 155
retroviral R1, R2 155, 156
viral
antigens 34
epitopes 29
virus
infected cells 29
specific CTL, *see* CTL

wellharring (we) locus 60

xeno antigens/responses
H-2 restricted (HLA specific) 166–203, 248
see also CTL
xeroderma pigmentosum fibroblast M1 185

Yersinia 259–265
entorocolitica 260, 261, 265, 272, 273
IgG antibodies to 261
IgM antibodies to 261
survival of mice 273

Zn finger domain 128–129

MIX
Papier aus verantwortungsvollen Quellen
Paper from responsible sources
FSC® C105338

If you have any concerns about our products,
you can contact us on
ProductSafety@springernature.com

In case Publisher is established outside the EU,
the EU authorized representative is:
Springer Nature Customer Service Center GmbH
Europaplatz 3, 69115 Heidelberg, Germany

Printed by Libri Plureos GmbH
in Hamburg, Germany